Enzymes in Food and Beverage Processing

Enzymes in Food and Beverage Processing

Robert L. Ory, EDITOR
Southern Regional Research Center

Allen J. St. Angelo, EDITOR
Southern Regional Research Center

A symposium sponsored by the Division of Agricultural and Food Chemistry at the 172nd Meeting of the American Chemical Society, San Francisco, Calif., Aug. 30–31, 1976

ACS SYMPOSIUM SERIES **47**

AMERICAN CHEMICAL SOCIETY
WASHINGTON, D. C. 1977

Library of Congress CIP Data

Enzymes in food and beverage processing

(ACS symposium series; 47 ISSN 0097-6156)
Includes bibliographical references and index.

1. Food industry and trade—Congresses. 2. Enzymes—Industrial applications—Congresses.
I. Ory, Robert L., 1925- II. St. Angelo, Allen J., 1932- . III. American Chemical Society. Division of Agricultural and Food Chemistry. IV. Series: American Chemical Society. ACS symposium series; 47.

TP368.E59 664 77-6645
ISBN 0-8412-0375-X

PRINTED IN THE UNITED STATES OF AMERICA

ACS Symposium Series

FOREWORD

The ACS Symposium Series was founded in 1974 to provide a medium for publishing symposia quickly in book form. The format of the Series parallels that of the continuing Advances in Chemistry Series except that in order to save time the papers are not typeset but are reproduced as they are submitted by the authors in camera-ready form. As a further means of saving time, the papers are not edited or reviewed except by the symposium chairman, who becomes editor of the book. Papers published in the ACS Symposium Series are original contributions not published elsewhere in whole or major part and include reports of research as well as reviews since symposia may embrace both types of presentation.

CONTENTS

PREFACE

Many desirable, or undesirable, changes in flavors and properties of fruits, vegetables, oilseeds, cereals, and uncooked animal products are catalyzed by one or more enzymes. Whether activated intentionally or not, these enzymes affect the ultimate quality of the food or beverage in which they are present. From the dawn of history man has used enzyme systems (albeit unknowingly) for food preservation, for fermentation, and for breadmaking. However, great advances made in enzyme chemistry over the past few decades have enabled food scientists to select specific enzymes to achieve a desired end product without the guesswork inherent in the older systems.

Today winemaking, cheese processing, and breadmaking are no longer just an art perpetuated in unknown microbial cultures, but they are now under strict scientific control. In addition to these older applications of enzyme systems, we now use proteases for meat tenderization and for stabilization of chill haze in brewing, two processes so well established that they are no longer considered unusual applications of enzymes. Even the development of the unique flavor of milk chocolate is believed to have been caused by a lipase that was in the milk added during processing. Today it is hard to recall a time when we did not have milk chocolate.

The number and types of foods and beverages we consume today are continually changing; foods are becoming more refined. They are being processed by heat or by freezing as frozen-raw or precooked convenience food items designed for long storage and easy preparation at home. The ultimate quality of these products will often depend upon the quality of the material prior to processing or on certain processing steps during their preparation. Food scientists are playing an important role in the expanding food industry by improving the flavor, texture, organoleptic, and nutritional quality of virtually all types of beverages, fruits, vegetables, animal and plant protein foods, and even animal feeds. Improvements can be achieved by the use of chemicals, by processing conditions, or by applications of selected enzymes that can directly or indirectly affect the quality of the final product.

The title and subjects for this symposium were selected in early 1974 when we were asked to organize it under the sponsorship of the Division of Agricultural and Food Chemistry. The participants, experts in their respective fields, were chosen to provide a well-balanced program covering the major beverages, plus fruits, vegetables, oilseeds, and both animal and plant protein foods. Noticeably absent are papers reporting enzyme applications in winemaking, meat tenderization, chill-haze removal in brewing, and traditional cheese manufacturing. We felt that these areas have been thoroughly reviewed in other publications and symposia ("Food Related Enzymes," J. R. Whitaker, Ed., *Advances in Chemistry Series* (1974) **136**; 170th ACS National Meeting, Chicago, symposium on "Microbial and Enzymatic Modification of Proteins"). Because of the growing number of enzymes being utilized in other areas of food and beverage processing, we tried to assemble a balanced program to highlight some of these newer applications. The topics include applications of enzymes related to flavor and quality changes in products used in preparing beverages, enzymes that affect flavor and texture of fresh fruits and vegetables, meats, fish, oilseeds, and legumes (now being considered as sources of protein for supplementation of traditional foods), plus some reports of current research on potential future applications of enzymes in immobilized systems, as biological indicators, and in conversion of abattoir wastes—a pollution problem—into food or feed-grade products.

This book, therefore, is a development of the symposium presented at the August, 1976, ACS Meeting in San Francisco. It contains new information on the roles and utilization of endogenous and exogenous enzymes derived from plant, animal, and microbial sources for improving the quality of certain foods. We hope it is useful for food and beverage processors, food chemists, and others engaged in these or related areas of research. We have not reached the limits of our ability to use enzymes to produce new and better quality foods and beverages. Our objective in this book is to provide information that will stimulate others engaged in research on food enzymes to come closer to those limits.

New Orleans, La.
November 1976

ROBERT L. ORY
ALLEN J. ST. ANGELO

1

Enzymes Affecting Flavor and Appearance of Citrus Products

J. H. BRUEMMER, R. A. BAKER, and B. ROE

U. S. Citrus and Subtropical Products Laboratory, Winter Haven, Fla.

Enzymes that affect the quality of citrus products can be classified as: 1) enzymes in the intact fruit, stimulated by handling conditions, 2) enzymes in juice, stimulated by substrates, and 3) commercial enzymes used in processing.

Intact Fruit

Before citrus fruit are processed for juice they are subjected to handling conditions that adversely affect product quality. Although some are hand picked and bagged, others are harvested after they are shaken from the tree. As the fruit fall they may bounce off limbs and branches. They may also bounce when dumped into tractor trailers for transporting to the processing plant. In trailers the fruit at each layer are subject to pressure from weight of fruit above.

Physical stresses increase fruit respiration. Vines et al. (1) reported that oranges dropped 48 in. onto a smooth surface or compressed with 29 lb pressure for 30 sec respired 100 and 70% more CO_2 than controls. The dropped or squeezed fruit did not return to normal respiration within 7 days. Flavor of juice prepared from oranges dropped 36 in. to a hard surface and stored overnight at 4°C was readily distinguished by a trained taste panel as different from juice flavor of fruit not dropped (2).

Fruit respiration is also stimulated by abscission chemicals applied to the tree to loosen fruit for harvest. The chemicals promote ethylene formation in citrus fruit and ethylene stimulates respiration. Flavor of juice prepared from oranges sprayed with various abscission agents was readily distinguished by a trained taste panel as different from juice flavor of unsprayed fruit (3).

The heat and CO_2 generated during respiration are not readily dissipated from the fruit in tractor trailers, which often rest for several days in processing plant yards before they are unloaded. One consequence of these handling conditions is accummulation of fermentation products in the fruit from anaerobic metabolism.

Alcohol:NAD Oxidoreductase E.C.1.1.1.1. During anaerobic metabolism in citrus, the alcohol:NAD oxidoreductase reaction (AOR) replaces the O_2-respiratory chain that reoxidizes reduced NAD, which functions in oxidation reactions of carbohydrate, fat, and protein metabolism. The major substrate for the enzyme is acetaldehyde formed from the reductive decarboxylation of pyruvic acid.

Juice from grapefruit held anaerobically for 40 hr at 40°C was rejected by a trained panel of 11 tasters and described as "fermented" and "overripe" (4). Biochemical changes accompanied the flavor change. The acidity was 20% lower and ethanol content was 15 times higher than in juice from control fruit. In another study, the ethanol content in juice from fruit held for 16 hr at 40°C in CO_2 was 10 times that of juice from fruit held in air (Table I).

Table I
Biochemical Changes in Grapefruit During Anaerobic Metabolism (5).

Atmosphere*	°Brix % sugar	Titratable acidity % citric acid	Malate ion mM	Ethanol mM
Air	7.18	1.46	6.0	2.2
CO_2	6.67	1.34	3.0	23.6

*15 grapefruit in each atmosphere.

Declines in sugar and malate concentrations under CO_2 indicate that glycolysis and malate decarboxylation produced the pyruvate. Grapefruit has a very active "malic enzyme" (E.C. 1.1.1.40) that catalyzes this decarboxylation (5).

Citrus fruit contain other aldehydes that are reactive with AOR. However, they do not compete on the basis of concentration and affinity for the enzyme (8). Two of these, octanal and decanal, are important flavor components of orange juice and are present at 12 and 7 times their flavor thresholds (6). Even so, orange juice tissue contains about 50 times as much acetaldehyde as either octanal or decanal (Table II).

The reactivities with AOR of various aldehydes prepared from orange juice cells relative to the reactivity of acetaldehyde are shown in Table III.

Table II
Flavor Thresholds of Aldehydes and Their Concentrations in Orange Juice.

Aldehyde	Flavor Threshold (6) ppb	Conc. in OJ (7) ppb
Acetaldehyde	-	3000
Octanal	5	60
Decanal	7	50

Table III
Relative Reactivities of Aldehydes with Citrus AOR (8).

Substrate	Conc. mM	Rel. Activity*	Rel. Affinity**
Acetaldehyde	74	100	100
Butanal	37	80	8
Hexanal	28	80	6
Octanal	21	33	5
Decanal	18	6	4

*Specific activity of AOR with acetaldehyde was 15 μmoles NADH oxidized/min/mg protein.
**Substrate affinities calculated from Michaelis constants for the reactions (8).

Butanal and hexanal were almost as reactive as acetaldehyde; octanal and decanal were decidedly less reactive. Competition of the aldehydes for AOR can be estimated from their relative affinities as substrates for the reaction (Table III). These relationships indicate that acetaldehyde would dominate the AOR reaction, and that relatively little of the flavor compounds, octanal or decanal, would be lost in the fruit through this reaction.

Malate:NAD Oxidoreductase (E.C. 1.1.1.37). Inhibition of malate:NAD oxidoreductase (MOR) probably caused the decrease shown in Table I of citric acid during anaerobic metabolism of citrus fruit. By anaerobic treatment of citrus fruit, the ratio of reduced to oxidized NAD^+ increased by 100% from 0.21 to 0.43 (9). MOR activity in extracts of juice cells was completely inhibited by reduced NAD^+ at 5% of the total NAD^+ content (Table IV) (9).

Table IV
Effect of NADH on Malate:NAD Oxidoreductase Activity (9).

NADH μM	$\frac{NADH}{NAD}$ x 100	MOR activity % max
0	0	100
1	0.2	73
5	1	65
10	2	10
25	5	0

Thus, during anaerobic metabolism the increase in reduced NAD^+ relative to the oxidized form probably decreased the rate of oxidation of malate to oxaloacetic acid in the oxidative pathway for the synthesis of citric acid. If citric acid was metabolized faster than it was synthesized, its net concentration would have been reduced. Malate would not have accumulated under these conditions because the very active "malic enzyme" decarboxylates malate to pyruvate.

Enzymes in Juice

Citrus juice is a tissue homogenate and must be heated for the inactivation of its enzymes which are released from their normal restraint in the tissue. Citrus juice tissue contains peroxidase (E.C. 1.11.1.7), diphenol oxidase (E.C. 1.10.3.1), pyruvic decarboxylase (E.C. 4.1.1.1), carboxyesterase (E.C. 3.1.1.1), and pectinesterase (E.C. 3.1.1.11). When released, these enzymes catalyze reactions, involving juice constituents, that could affect flavor and appearance.

Peroxidase. When orange juice was held at 30°C, peroxidase activity declined rapidly but stabilized after 1 hr to about one-third of the original activity (10). The amount of active soluble peroxidase in commercial juices before pasteurization depends upon the extractor finisher force. Hard extraction to obtain high juice yields increased peroxidase activity of the juice; light extraction pressure reduced the amount of soluble peroxidase (10). Juice quality was inversely correlated with juice yield and peroxidase activity.

Ascorbic acid and the phenolic acids, caffeic and *p*-coumaric acids, were effective donors in the peroxidase-H_2O_2 reaction with enzyme preparations from orange juice tissue (10). However, ascorbic and caffeic acids were unreactive when added to orange juice. These compounds did not oxidize to a measurable extent when the juice was held at 30° for 4 hr. Citrus peroxidase is not detrimental to juice flavor if juice is processed within 4 hr of extraction.

Diphenol Oxidase. Extracts of orange juice cells oxidized *o*- and *p*-diphenols and *o*-methoxyphenolics (11). Table V lists a number of simple diphenolics and flavonoid compounds that supported O_2 uptake with the enzyme preparation. All of the flavonoids and many of the simple diphenolics have been reported present in citrus (12). Rates for the *o*- and *p*-diphenolics were about equal, and were higher than for the *o*-methoxyphenolics (11). Rates for the *o*-methoxyphenolic compounds decreased in order as the substituted group changed from acid to amine to α-β-unsaturated acid. Quercetin and hesperidin supported the same rate of O_2 uptake as their aglycones did.

Diphenol oxidase was unstable at acid pH (11). Only 25% of the activity was recovered after treatment at pH 4 for 5 min. Its poor stability might explain why enzymic browning is not a problem in citrus processing and why ascorbic acid, an effective donor for the *p*-dihydroxyphenols, is stable in the presence of the many phenolic compounds in citrus juices.

Carboxyesterase. Carboxyesterase is relatively stable in orange juice. After 2 hr at 25°C, about one-half of the activity of the freshly reamed juice remained (13). Triacetin was the preferred substrate but other acetates, including ethyl, linalyl, terpinyl and octyl acetates and ethyl butyrate were reactive. At the pH of juice (3.5), the activity of a partially purified enzyme was about 15% of its activity, which was optimum, at pH 7; but even at 15% efficiency, the enzyme hydrolyzed about 15 ppm of ethylbutyrate in 1 hr at saturating level of substrate. The rate of ester hydrolysis in juice has not been measured, but it is probably much less than the calculated potential rate.

The change in aroma of orange juice at 25°C was determined by difference taste test (13). A panel of tasters detected a difference in aroma between freshly reamed juice, and juice held for 2 and 5 hr at 25°C after reaming. When freshly reamed juice was tasted immediately after addition of purified orange flavedo esterase or commercial esterase, no difference was detected by the taste panel. But 1 and 2 hr after addition of the esterase, the taste panel determined that the juice was different, at the 0.1% level of significance, from the untreated juice. These tests suggest that esterase-catalyzed hydrolyses, which occur primarily during the first few hours in freshly-prepared juice, contribute to change in aroma.

Pyruvic Decarboxylase. Acetaldehyde accumulates in orange juice from the reductive decarboxylation of pyruvic acid. At 30°C, 1.4 ppm acetaldehyde was formed in orange juice in 4 hr by the reaction catalyzed by pyruvic decarboxylase (14). Pyruvic acid-dependent accumulation of acetaldehyde in fresh orange juice was demonstrated by the addition of sodium pyruvate (Table VI).

Table V
Substrate Specificity of Orange Diphenol Oxidase (11).

Substrate	IUPAC Nomenclature	μliters O_2/ min/ mg protein
o-Dihydroxyl		
DOPA	3-(3,4-Dihydroxyphenyl)alanine	7.0
Pyrogallol*	1,2,3-Trihydroxybenzene	4.6
Catechol**	1,2-Dihydroxybenzene	4.0
p-Dihydroxyl		
p-Hydroquinone	1,4-Dihydroxybenzene	4.5
o-Hydroxyl-methoxy		
Homovanillic Acid	4-Hydroxy-3-Methoxyphenylacetic Acid	1.4
3-Methoxytyramine	4-Hydroxy-3-Methoxyphenylethylamine	0.34
Ferulic Acid***	4-Hydroxy-3-Methoxycinnamic Acid	0.14
Feruloyl-putrescine***	N-(4-Aminobutyl)-4-hydroxy-3-methoxycinnamide	0.21
Hesperetin	3',5,7-Trihydroxy-4-methoxyflavonone	0.56
Hesperidin	Hesperetin-7-rutinoside	0.56
o-Dihydroxyl		
Catechin	3,5,7,3',4'-Pentahydroxyflavan	2.7
Chlorogenic Acid	3-(3,4-Dihydroxycinnamoyl)quinic acid	0.81
Eriodictyol	3',4',5,7-Tetrahydroxyflavonone	1.1
Rutin	Quercetin-3-rutinoside	0.56
Quercetin	5,7,3',4'-Tetrahydroxyflavonol	1.3
Quercetrin	Quercetin-3-rhamnoside	1.3

*Assayed at pH 5.6 instead of pH 7.

**Catechol 1 x 10^{-8} M with ascorbate 1 x 10^{-4} M and EDTA 1 x 10^{-4} M.

***Ferulic acid or feruloylputrescine 3 x 10^{-5} M, ascorbate 1 x 10^{-4} M and EDTA 1 x 10^{-4} M.

Pyruvic acid is the most reactive substrate for the enzyme, 2-ketobutyric acid is only one-third as reactive, and higher homologues of 2-keto acids (5-7 carbons) are less than 20% as reactive. The enzyme is slowly deactivated in orange juice after a rapid 80% deactivation during the first hour after reaming of the juice. Acetaldehyde is a precursor of acetoin and diacetyl so that its accumulation could increase the formation of more diacetyl during storage of juice (14). Diacetyl produces a "cardboard" off-flavor in citrus products at 0.25 ppm (15).

Table VI
Pyruvate Dependent Accumulation of Acetaldehyde (14).

Sodium pyruvate μmoles/ml	Acetaldehyde nmoles/ml
0	95
1	120
5	330
10	485
15	485

Orange juice was incubated with sodium pyruvate for 2 hr at 30°C.

Pectinesterase. Citrus pectin is a polymer of 1,4 linked, α-D-galactopyranosyluronic acid units with about 65% of the available carboxyl groups methylated (degree of esterification, DE). Pectin is part of the stable colloidal system that gives citrus juices the characteristic turbid appearance. The colloid is "broken", or the juice is clarified, by pectinesterase (PE), a hydrolase that demethylates pectin to pectic acids and methanol. Demethylation proceeds block-wise, starting at a methoxyl group adjacent to a free carboxyl group. When the average DE of the pectin reaches a critical stage, the juice clarifies. Baker (16) found the critical DE for orange juice to be 28% and recorded partial clarification at 35% DE. The destabilization of citrus juices by PE is prevented commercially by pasteurization at 92°C to inactivate the enzyme.

Use of Enzymes in Citrus Processing.

Polygalacturonase (E.C. 3.2.1.15), naringinase, and limonoate dehydrogenase have been shown useful in improving the quality and quantity of citrus products by stabilizing colloid, decreasing viscosity, and reducing bitterness.

Polygalacturonase. Heat inactivation of PE is not the only method of stabilizing orange juice against clarification. Orange juice was stabilized by treatment with polygalacturonase (PG) (Table VII).

Table VII
Effect of PG on Turbidity of OJ (17).

PG Treatment	Turbidity (g/l bentonite) after 4° storage				
ppm Klerzyme*	0	12d	20d	28d	40d
0	1.6	0.4	0.3	0.3	0.3
50	1.6	0.9	1.0	1.1	1.1
200	1.6	1.1	1.3	1.5	1.5
400	1.6	1.4	1.6	1.6	1.6

*Commercial pectinase.

Both 200 and 400 ppm of a commercial pectinase sustained turbidity for more than 40 days at 4°C. The effects of PG on the distribution of soluble and insoluble pectins, insoluble pectates and oligogalacturonates show that 61% of the pectin in orange juice was hydrolyzed to soluble oligogalacturonates in 8 days at 4°C (Table VIII).

Table VIII
Effect of PG on Pectic Substances of Orange Juice (17).

Pectic substances	Control mg/l AGA**	PG-Treated mg/l AGA**	PG-Treated % of control
Insoluble pectins	556	141	25
Soluble pectins	118	103	87
Insoluble pectates	185	95	51
Total	859	339	39
Oligogalacturonates*		520	61

*By difference (control total minus treated total; % of control total.
**AGA = anhydrogalacturonic acid.

These data were used to explain stabilization of orange juice by PG. The intermediate states of demethylated pectin are susceptible to depolymerization. Before the critical state of 28% DE is reached, PG hydrolyzes the α-1,4-D-galacturonide links of the intermediates to short chain pectins, which are demethylated by PE to colloid stable oligogalacturonates.

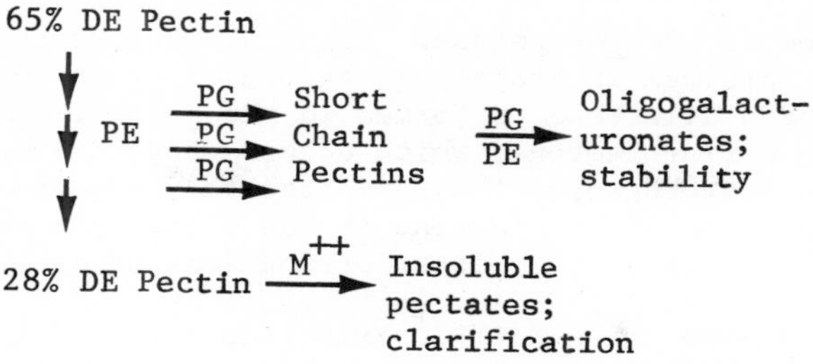

Depolymerization of pectin by PG results in reduction of viscosity and pulp volume--physical changes that aid processing of citrus juices (18). Reduced viscosity permits more efficient evaporation to high solids concentrates; reduced pulp volume increases the yield of finished juice. Commercially, PG is used to increase yield of pulp-wash liquids from orange and grapefruit juice pulps and to facilitate the evaporation of these liquids to high solids concentrates; but it is presently not used in juice processing (19).

Also PG can be used to reduce the viscosity of clarified juices for evaporation to syrups. Citrus juices are rapidly clarified by added polygalacturonic acid or pectins with low DE (20). Orange juice clarified with PG was concentrated to a clear, 90°Brix syrup. This syrup is presently being tested as a preservative for the formulation of a low pulp, orange juice concentrate that can be stored at room temperature (21).

Naringinase. Naringin, the 7-rhamnoside-β-glucoside of 4',5,7-trihydroxyflavonone is a bitter component of grapefruit. Most of the compound is found in the albedo and core of the fruit (2 to 4% by weight of wet tissue); but at times, the content in juice is as high as 0.1%. Naringinase hydrolyzes naringin to prunin (naringenin-7,β-glucoside) and rhamnose, reducing the bitterness.

The enzyme was used to debitter grapefruit juice (22), grapefruit concentrate (23), and grapefruit pulp (24). At present, a limited amount of grapefruit concentrate is debitterized commercially.

Use of naringinase to debitterize albedo and core of fresh intact grapefruit is under development at the Citrus and Subtropical Products Laboratory in Winter Haven, Florida (25). A naringinase solution, free from PG and cellulase, is vacuum infused into flavedo-shaved intact grapefruit. After 1 hr at 50°C, the albedo is less bitter. Solutions containing naringinase, protein concentrates, vitamins, minerals, flavor, and coloring compounds, sweeteners, gelling compounds, and bacteriostats have been infused into seedless grapefruit in this manner. The product is flavorful with a pleasant colorful appearance, and, when eaten entirely, (albedo and core as well as the juice sections) would supply food fiber as well as nutrients.

Limonoate Dehydrogenase. Limonin bitterness is recognized as a quality problem in grapefruit and navel orange juices. This triterpenoid dilactone is formed in the juice from the non-bitter limonoate A-ring lactone, which occurs naturally in fruit tissues. Limonoate dehydrogenase (LD) oxidizes the hydroxyl group on carbon 17 of limonoate A-ring lactone to the 17-dehydro compound, which cannot lactonize to the bitter D-ring lactone in acidic media (26).

The enzyme is an $NAD(P)^+$ oxidoreductase which is activated by $NAD(P)^+$. It was isolated from cultures of Pseudomonas grown on media containing limonoate as sole carbon source (27). Although activity was optimum at pH 8, LD decreased the limonin content of navel orange juice (pH 3.5 to 4) to an acceptable level (28) in 1 hr. Commercial use of LD must await further investigations on growth of the organism and its production.

Literature Cited

1. Vines, M. H., Edwards, G. J. and Grierson, W. Proc. Fla. State Hortic. Soc. (1965) 78:198-202.
2. Bryan, W. L., personal communication.
3. Moshonas, M. G., Shaw, P. E. and Sims, D. A. J. Food Sci. (1976) 41:809-811.
4. Bruemmer, J. H., and Roe, B. Proc. Fla. State Hortic. Soc. (1969) 82:212-215.
5. Bruemmer, J. H. and Roe, B. Proc. Fla. State Hortic. Soc. (1970) 83:290-294.
6. Lea, C. H. and Swoboda, P. A. Chem. & Ind. (1958) 1289-1290.
7. Kirchner, J. G. and Miller, J. M. J. Agric. Food Chem. (1957) 5:283-291.
8. Bruemmer, J. H. and Roe, B. J. Agric. Food Chem. (1971) 19:266-268.
9. Bruemmer, J. H. and Roe, B. Phytochem. (1971) 10:255-259.
10. Bruemmer, J. H., Roe, B., Bowen, E. R. and Buslig, B. J. Food Sci. (1976) 41:186-189.
11. Bruemmer, J. H. and Roe, B. J. Food Sci. (1970) 35:116-119.
12. Horowitz, R. M. in "The Orange" (334) Univ. of Calif., Berkeley, 1968.
13. Bruemmer, J. H. and Roe, B. Proc. Fla. State Hortic. Soc. (1975) 88:300-303.
14. Roe, B. and Bruemmer, J. H. J. Agric. Food Chem. (1974) 22:285-288.
15. Beisel, C. G., Dean, R. W., Kichel, R. L., Rowell, K. M., Nagel, C. W., Vaughn, R. H. J. Food Res. (1954) 19:633-643.
16. Baker, R. A. Proc. Fla. State Hortic. Soc. (1976) 89:0-0.
17. Baker, R. A. and Bruemmer, J. H. J. Agric. Food Chem. (1972) 20:1169-1173.
18. Baker, R. A. and Bruemmer, J. H. Proc. Fla. State Hortic. Soc. (1971) 84:197-200.

19. Braddock, R. J. and Kesterson, J. W. J. Food Sci. (1976) 41:82-85.
20. Baker, R. A., J. Food Sci. (1976) 41:1198-1200.
21. Bruemmer, J. H. Citrus Chem. & Techn. Conf. (1976) USDA Citrus & Subtropical Products Laboratory, Winter Haven, Florida.
22. Ting, S. V. J. Agric. Food Chem. (1958) 6:546-549.
23. Olsen, R. W. and Hill, E. C. Proc. Fla. State Hortic. Soc. (1964) 77:321-325.
24. Griffiths, F. P. and Lime, B. J. Food Technol. (1959) 13: 430-433.
25. Roe, B. and Bruemmer, J. H. Proc. Fla. State Hortic. Soc. (1976) (in press).
26. Hasegawa, S., Bennett, R. D., Maier, V. P. and King, A.D. Jr. J. Agric. Food Chem. (1972) 20:1031.
27. Hasegawa, S., Maier, V. P. and King, A. D. Jr., J. Agric Food Chem. (1974) 22:523.
28. Brewster, L. C., Hasegawa, S. and Maier, V. P. J. Agric. Food Chem. (1976) 24:21-24.

2

Use of Enzymes in the Manufacture of Black Tea and Instant Tea

GARY W. SANDERSON and PHILIP COGGON

Thomas J. Lipton, Inc., 800 Sylvan Ave., Englewood Cliffs, N.J. 07632

Black tea is manufactured from the rapidly growing shoot tips of the tea plant (Camellia sinensis, (L), O. Kuntze). These shoot tips are collectively known as the tea flush, and they are comprised of immature plant parts, namely, the terminal bud, the first 2 or 3 leaves, and the included stem piece. They also contain both the enzymes and the substrates that will react during the black tea manufacturing process. The enzymically catalyzed reactions involving tea flush constituents are essential to the preparation of black tea: These reactions are described further in the following discussion.

The black tea manufacturing process itself begins as soon as the tea flush is plucked (harvested), and it is carried out on, or very near, the plantation where the tea flush is grown. The entire process, from plucking to finished black tea product, requires about 8 to 24 hours. The process is outlined in Table 1, and more details can be found in reviews by Harler (1), Eden (2), Hainsworth (3), and Sanderson (4). The enzymes known to be involved in black tea manufacture are listed in Table 2 (part A) and they are discussed in the text below.

Instant tea may be prepared from either black tea or directly from fresh green tea flush. Processes for the manufacture of instant tea have been summarized by Pintauro (5), and by Sanderson (6). A few enzymic processes for use in manufacturing instant tea have been described in the patent literature. The enzymes involved are listed in Table 2 (part B) and their role in instant tea manufacture is discussed in the following text.

Enzymes Involved in Black Tea Manufacture (Table 2, Part A)

Tea Catechol Oxidase. This enzyme is certainly the most important single enzyme in black tea manufacture. As soon as the withered tea flush is macerated (Table 1), an enzymic oxidation of the tea flavanols (I-IV) is initiated. Tea catechol oxidase is the catalyst for this oxidation and the process is called "tea fermentation".

Table 1: Outline of Black Tea Manufacturing Process.

Steps of Process	Comments
1. Growth of tea leaves	Tea plants (Camellia sinensis, [L.] O. Kuntze) are cultivated to encourage the rapid growth of shoot tips collectively called the flush.
2. Harvest of tea leaves (i.e. plucking)	The flush is picked once every 2 to 4 weeks during the growing season.
3. Withering	The plucked flush is çaused partially to dessicate from about 80% moisture down to about 65% moisture in 4 to 18 hours.
4. Macerating	The withered tea flush is macerated to cause the endogenous enzymes and tea flavanols to come into contact.
5. Fermenting	The macerated tea flush is allowed to stand for 1 to 4 hours while it oxidizes (Fig. 1).
6. Firing	The fermented tea flush is dried to about 2% moisture in about 20 minutes using hot air at 70° to 95°C.
7. Sorting	The fired black tea is sifted into grades according to size of tea particles.
8. Storing/Shipping	Fired black tea may be consumed immediately but it usually takes 2 to 4 months for it to travel from point of origin to point of consumption.
9. Conversion to instant tea	Most instant tea is prepared from black tea in the country in which it will be sold.

Table 2: Enzymes with Known Functions in Tea Manufacture.

Name of Enzyme	Function
A. BLACK TEA MANUFACTURE (These enzymes are endogenous to fresh green tea leaves)	
1. Catechol Oxidase	Oxidation of tea flavanols resulting in formation of tea pigments (Fig. 1) and indirectly in formation of black tea aroma (Fig. 2).
2. Peroxidase	Removes any peroxide formed during tea fermentation (Fig. 1). Can cause oxidation of tea flavanols if hydrogen peroxide is added.
3. Tea Metalloprotein	Causes oxidation of unsaturated fatty acids, leading to formation of black tea aroma constituents.
4. Pectin Methyl Esterase	Demethoxylates tea pectins resulting in formation of hard, glossy covering on black tea particles when well made.
5. Alcohol Dehydrogenase	Supposed to play a role in the formation of some of the alcohols in black tea aroma by causing reduction of the corresponding carbonyls.
6. Transaminase	Supposed to play a role in the biosynthetic transformation of amino acids to compounds that are important in black tea aroma.
7. Peptidase	Acts to form amino acids from leaf proteins especially during withering (Table 1): Amino acids are one type of black tea aroma precursor (Fig. 2).
8. Phenylalanine Ammonia Lyase	Involved in the biosynthesis of tea flavanols.
9. 5-Dehydroshikimate Dehydrogenase	Involved in the biosynthesis of tea flavanols.

(Table 2, part B, continued next page)

Table 2: Enzymes with Known Functions in Tea Manufacture. (cont.)

Name of Enzyme	Function
B. INSTANT TEA MANUFACTURE (These enzymes are from non-tea sources. They have been added to tea preparations for specific purposes).	
1. Tannase	Catalyzes hydrolysis of galloyl groups from tea polyphenolic compounds to produce cold water soluble tea solids.
2. Polyphenol Oxidase	Causes "tea fermentation" (same as A.1. above).
3. Peroxidase	Causes "tea fermentation" when H_2O_2 is added.
4. Pectinase	Causes breakdown of water extracted tea pectins to reduce foam forming tendency of instant tea powders.
5. Cellulase	Breaks down cell wall material in tea leaf to increase yield of instant tea solids.

Tea catechol oxidase is an endogenous component of tea flush (7-9), and it appears to be soluble in the cytoplasm of intact fresh green tea leaves (10, 11). The enzyme does not react with the tea flavanols (I-IV), which are present in the cell vacuoles, until the tea leaf tissues are disrupted causing the cell contents to be mixed. The "Maceration" step of the Black Tea Manufacturing Process (Table 1) brings about mixing of the tea flush cell contents thereby initiating "Tea Fermentation". The primary reaction occurring during tea fermentation is the oxidation of the tea flavanols to their reactive oxidized intermediates (V-VIII) which then form the black tea pigments (IX-XVI): These reactions are outlined in Figure 1. These same oxidized tea flavanols (V-VIII) also act as oxidizing agents to cause the oxidative degradation of other tea flush constituents (Figure 2): Reactions of this type are important in the formation of black tea aroma. The biochemistry of tea fermentation is discussed further in reviews by Sanderson (4, 12), Sanderson and Graham (13) and Reymond (14).

Tea catechol oxidase is probably soluble in the intact tea leaf (10,11,15,16), but it is readily insolubilized by interaction with the tea flavanols (17). The result is that tea catechol oxidase is immobilized within the insoluble portion of the tea flush during the maceration step of black tea manufacture. One of the secrets of black tea manufacture is to obtain conditions that will bring the tea flavanols to the active site of the enzymes since the reverse cannot take place. In fact, withering, the step of black tea manufacturing just preceeding maceration, is critical in that it must be carried far enough to destroy the semipermeable nature of the cell membranes and yet not so far as to render the flush so dry that the tea flavanols are no longer in solution. Further, maceration of the tea flush must be of the kind that will cause mixing of the cell contents without reducing the tissues to a pulp thereby preventing the free entry of oxygen into the "fermenting" macerated tea flush.

The levels of activity of tea catechol oxidase in tea flush appear to vary during the growing season (18-20) and during the black tea manufacturing process (15, 17). The cause of these changes is not known but they must be important since the enzyme is essential to the black tea manufacturing process.

Credit for elucidation of the true nature of the tea catechol oxidase enzyme goes to Sreerangachar (21-24). The work, and the controversy, that led to the identification and characterization of tea catechol oxidase is a most important part of the history of tea chemistry but one should consult Roberts (25), or Sanderson (4), for more details on this subject. The enzyme is known to contain copper (24). The enzyme has a remarkable specificity for the ortho-dihydroxy group on the B-ring of the tea flavanols (I-IV): Besides the tea flavanols, catechol (XVII) and pyrogallol (XVIII) will serve as substrates for this enzyme but gallic acid (XIX) and chlorogenic acid (XX) do not (26). Finally, tea catechol oxidase is susceptible to precipitation by

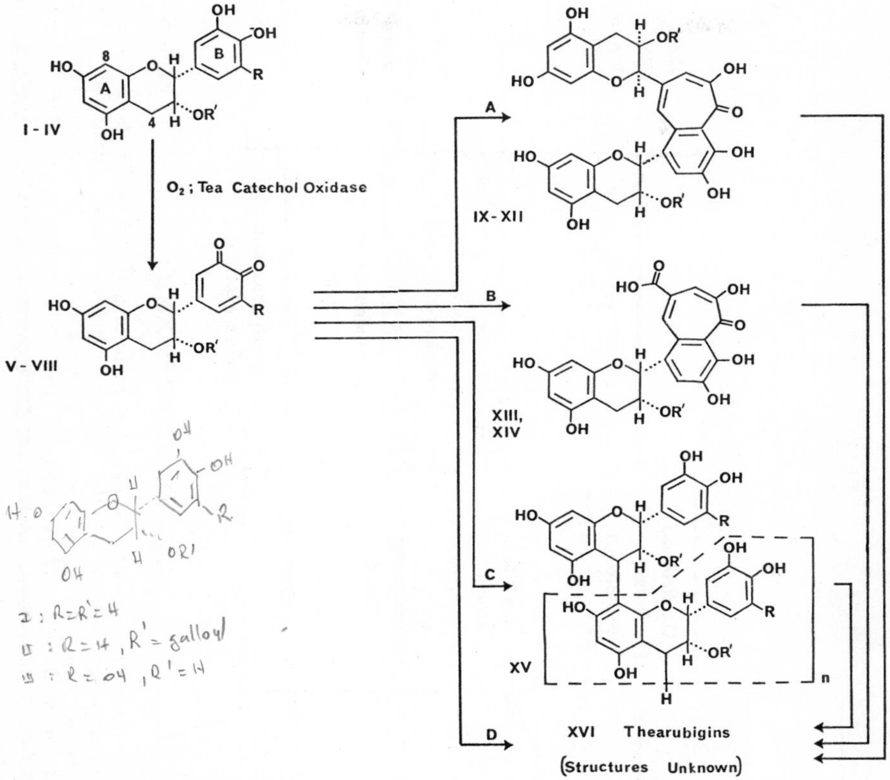

Figure 1. Oxidation pathway for tea flavanols during fermentation. The tea flavanols (I: R = R′ = H; II: R = H, R′ = galloyl; III: R = OH, R′ = H; and IV: R = OH, R′ = galloyl) are enzymically oxidized to their respective intermediates (V–VIII) which undergo the following condensation reactions: (A) I or II plus III or IV → IX–XII; (B) I or II plus gallic acid (XIX) → XIII or XIV; (C) I, II, III, or IV → XV (by 4,8 linkages); and (D) I, II, III, or IV → XVI (by unknown linkages).

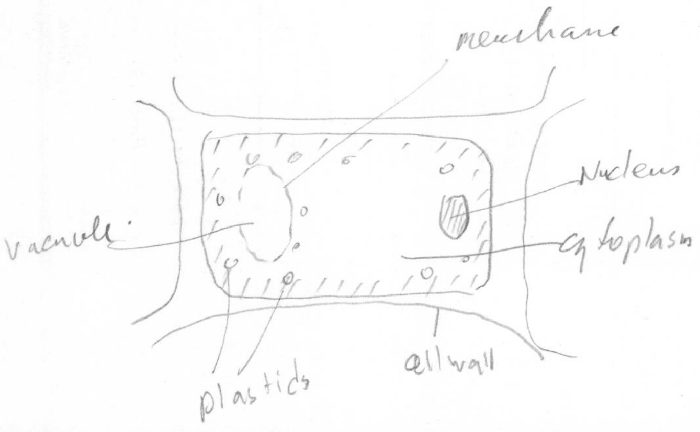

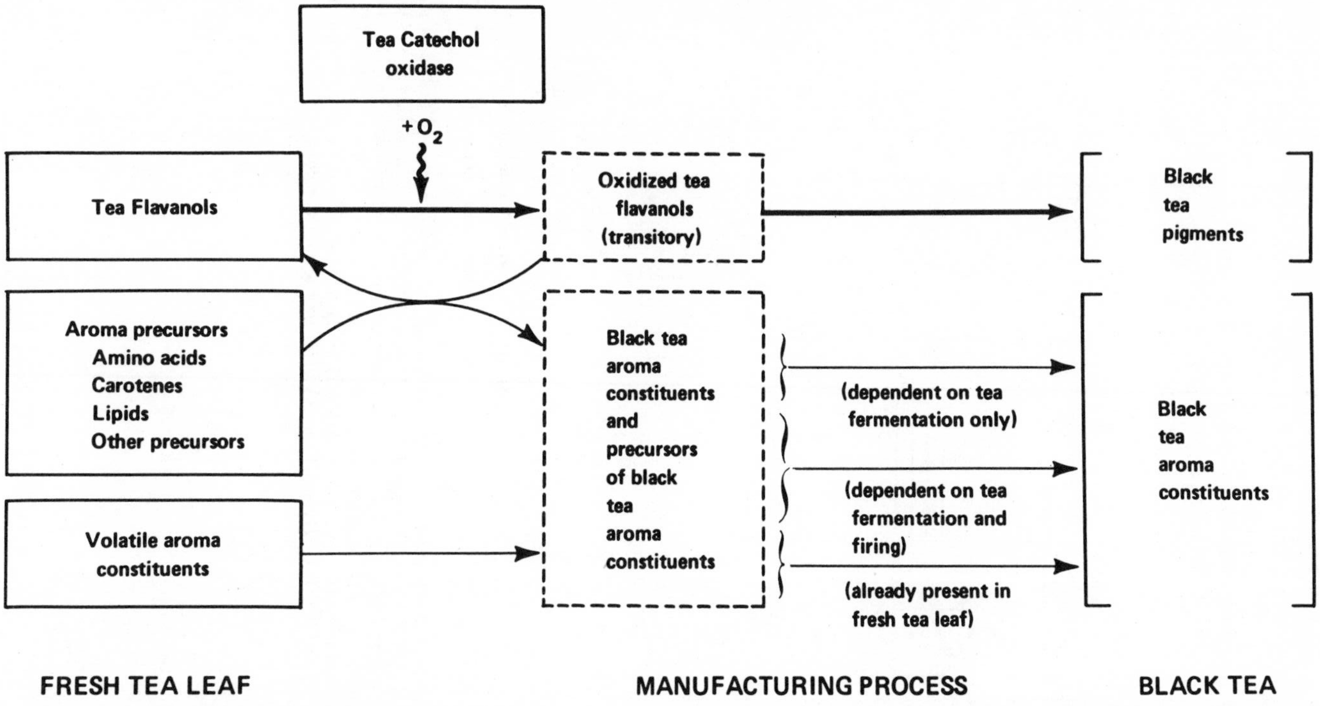

Figure 2. Proposed scheme for formation of black tea aroma. Materials present in fresh (or withered) tea flush are enclosed in rectangles. Materials present in dashed enclosures are largely transitory in that they are undergoing change during the black tea manufacturing process. Materials present in brackets are those present in the finished black tea product. The amount of any of the above materials present in the black tea produced would be dependent on the conditions of manufacture.

the tea flavanols (27), but the activity of this enzyme only decreases slowly when the enzyme is exposed to tea flavanols. It is clear that the tea catechol oxidase enzyme is well suited to catalyze the reactions of "tea fermentation."

XVII

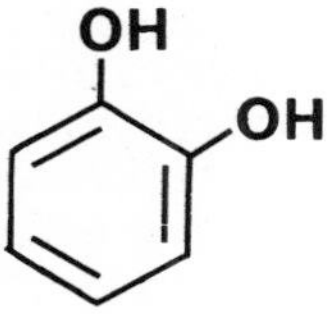

XVIII

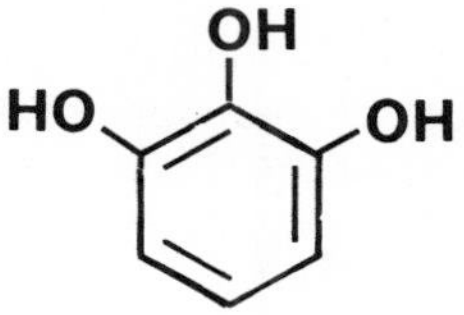

XIX

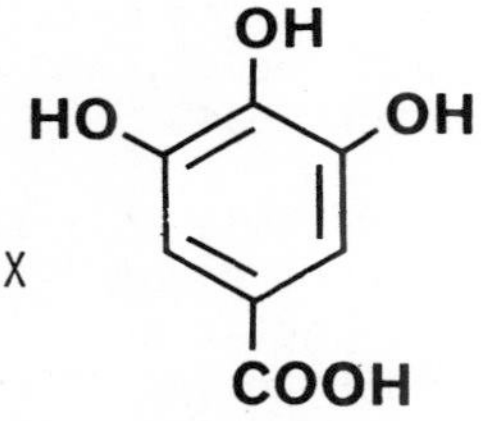

XX

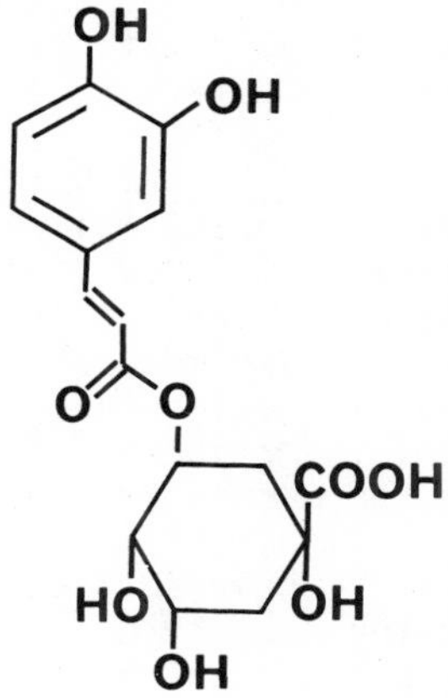

For additional information on work done on the extraction, purification and partial characterization of tea catechol oxidase, see papers by Takeo (28), Takeo and Uritani (29) and Gregory and Bendall (30). The work that establishes the role of tea catechol oxidase as the catalyst for the oxidation of the tea flavanols during tea fermentation has been reviewed by Roberts (25), Bokuchava and Skobeleva (31), and Sanderson (4).

Peroxidase. The presence of peroxidase in tea flush was established as early as 1938 by Roberts and Sarma (8, 32). The enzyme is readily extracted from tea flush (26), and, like tea catechol oxidase, retains its activity even after exposure to the tea flavanols. The tea peroxidase enzyme has not been characterized except that it is reported to be composed of 6 or 7 isoenzymes (33, 34). No role has yet been described for this enzyme in black tea manufacture, but it will catalyze the oxidation of tea flavanols if hydrogen peroxide is present (8, 32, 35, 36).

Tea Metalloprotein. Linoleic and linolenic acids act as black tea aroma precursors in that they are oxidized during fermentation to hexanal and 2(E)-hexenal, respectively (37, 38). Recently, a metalloprotein present in tea leaves has been shown (39) to be a non-enzymic factor that is responsible for this oxidation. Lipid changes in the finished, fired black tea have been implicated (40) in quality changes that occur during storage. Lipolysis has been shown to take place in teas held under high humidity conditions. On the other hand, lipid oxidation was found to be most severe in teas stored under dry conditions at elevated temperatures. The non-enzymic nature of lipolysis and lipid oxidation was demonstrated when tea heated at 110°C for 15 min. showed the same amount of change in the lipids during subsequent storage as unheated control samples of tea. These results suggest that non-enzymic factors, such as the metalloprotein catalyzed oxidation of unsaturated fatty acids, are important in determining the character of black teas.

Pectin Methyl Esterase. The activity of this enzyme in tea leaves has been implicated (41, 42) in the control of the rate of fermentation by causing the formation of pectic acid from methylated pectin. This hydrolysis was shown to lead to reduced oxygen uptake when an anaerobic period preceeded fermentation. It was suggested (42) that the demethylation causes some degree of gel formation and this subsequently impedes oxygen diffusion into the macerated tea flush, thereby inhibiting tea fermentation. Also, well made black tea particles have a dry hard pectin layer on their surface: This pectin coating is supposed to protect the quality of teas by acting as a barrier to water and oxygen, both of which promote reactions that are involved in loss of tea flavor and the formation of off-flavors.

Alcohol Dehydrogenase. This tea enzyme, which has been studied extensively by Hatanaka and his co-workers (43-48), may be determining the amount of the leaf alcohols 3(E)-hexenol, and 2(Z)-hexenol, and the corresponding aldehydes, that are present in finished tea products. These volatile compounds are important black tea aroma constituents (12, 49), which appear to enhance the flavor quality at certain levels (50) but which are associated with teas of poor quality when present at high levels (51,52).

There appear to be two types of alcohol dehydrogenase in tea flush (46). The major one requires the co-enzyme nicotinamide adenine dinucleotide (NAD) and the minor one needs nicotinamide adenine dinucleotide phosphate (NADP). This means that the level of these coenzymes, and their oxidation state during tea fermentation, should be another factor that determines the level of leaf alcohols, and their corresponding aldehydes, in black tea.

Transaminase. Wickremasinghe *et al.* (53, 54) have identified transaminase activity in macerated tea flush. Wickremasinghe (54) suggests that this enzyme is involved in the biosynthesis of terpenoid compounds that are important elements in the aroma of good quality black teas. Many factors other than transaminase, some of which are affected by weather, are also involved in the biosynthetic systems that were proposed (54).

Peptidase. There is an increase in the free amino acid content of tea leaves during the withering period (about 18 hr.) which follows plucking. A peptidase enzyme has been shown to be present in tea flush (55). This enzyme is probably important in black tea manufacture because it determines, in part, the level of free amino acids in tea flush, and free amino acids play a role in aroma formation (Fig. 2) (12, 13).

Phenylalanine Ammonia Lyase. Tea leaves have been shown (56) to contain phenylalanine ammonia lyase. This activity is highest in the young leaves and decreases as the leaves grow larger. A positive correlation was observed between the activity of this enzyme and catechin content as might be expected from an enzyme that is known to be involved in flavanol biosynthesis.

5-Dehydroshikimate Reductase. This enzyme has been extracted from tea leaves in an active soluble state (57). The activity of this enzyme was particularly high in tea flush. This is as would be expected of an enzyme which is known to be involved in flavanol biosynthesis and which is endogenous in plant material containing an extraordinarily high level of polyphenolic compounds.

Enzymes Involved In Instant Tea Manufacture (Table 2, Part B).

Tannase. Cold water solubility is an important characteristic of instant tea products since most instant tea is used to prepare iced tea. The presence of "tea cream", a cold water insoluble precipitate that forms naturally in brewed tea bever when they are allowed to stand for a few hours at below ab 40°C, is therefore a major problem in instant tea manuf It is known (58-61) that tea cream is a hydrogen bo formed between the polyphenolic substances of bl and caffeine (XXI). It has been found (62, 6 of tea solids with the enzyme tannase causes cream solids to become soluble in cold wat

Tannase is a fungal enzyme obtained *Aspergillus sp.* (64-66), and it is specif of esters of gallic acid (XIX) (66-68). The largest part of the tea flavanols (I-IV) is esterified with gallic acid (i.e. II and IV) and these galloyl flavanols are transformed into galloyl oxidation products (IX-XVI where R′=3,4,5-trihydroxybenzoyl) during the tea fermentation process (see Figure 1). Further, it is known that the preponderance of polyphenolic substances present in tea cream are esterified with gallic acid (59).

XXI

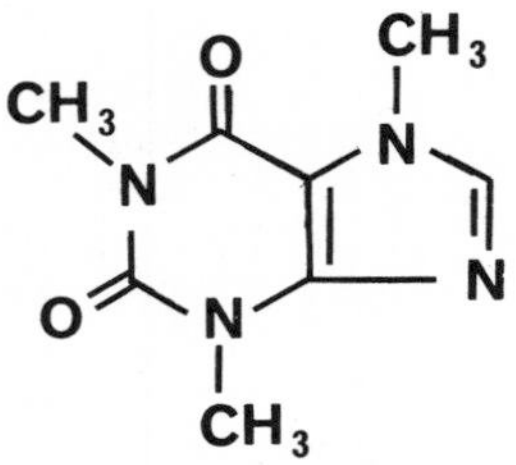

Some work has been done to prepare immobilized tannase (69, 70). These studies indicate that tannase can be successfully immobilized using virtually all of the usual methods and support materials available today. A process for the preparation of cold water soluble instant tea directly from fresh green tea flush, which uses tannase in a "pre-conversion" treatment, has been described by Sanderson and Coggon (71).

Polyphenol Oxidases. It was pointed out in the section on the role of tea catechol oxidase in black tea manufacture that the enzyme was central to the process by which green tea solids are converted to black tea solids. Seltzer, et al. (72) were the first to describe a process for the conversion of solids extracted from green tea to black instant tea solids utilizing the catechol oxidase enzyme from other tea leaves. In this process green tea extract is obtained from any green tea leaves and macerated fresh green tea leaves containing active tea catechol oxidase is added to catalyze the conversion.

Oxygenation of aqueous tea leaf homogenates has been described (73,74) as a means of preparing instant tea products directly from fresh tea flush. This kind of process makes use of the catechol oxidase endogenous to the tea flush to effect the tea fermentation process.

Polyphenol oxidases extracted from a number of different plant materials (ex. potato peelings; callus tissues of beans, sycamore, and apple plants; and certain molds) have been shown (75) to be capable of catalyzing tea fermentation in aqueous extracts of green tea flush. Unfortunately, the cost of these polyphenol oxidases is probably prohibitive and their stability is questionable.

Peroxidase. A process to prepare instant tea directly from fresh tea flush has been described (35,36) which utilizes the peroxidase enzyme that is endogenous to tea flush to catalyze the tea conversion process. These processes depend on adding hydrogen peroxide (H_2O_2) to stirred homogenates of fresh tea flush. Of course, the added hydrogen peroxide acts as the oxidizing agent in the tea fermentation process and the tea peroxidase serves as an essential catalyst for the oxidation of the tea flavanols (I-IV).

Pectinase. This enzyme has been described as being useful for improving the foam forming tendency of instant tea powders (76). It is presumed that pectinase treatment of tea extracts destroys some of the natural pectins present which lowers the ability of the solids present to form stable films.

Cellulases. Fungal enzymes which fall into this general classification have been used to release useful soluble tea solids which are normally bound to the tea leaves (77,78).

Summary

The manufacture of black tea is essentially the mechanical manipulation of fresh green tea flush to cause a conversion of the green tea solids to black tea solids: This conversion is catalyzed by enzymes that are endogenous to the tea flush (Table 2, Part A). The enzymology and biochemistry of the oxidation of the tea flavanols to black tea pigments (Figure 1) appears to be understood at least in outline. On the other hand, the enzymology and the chemistry of the formation of black tea flavor, i.e. taste and aroma (Figure 2), is only beginning to be understood. The difficulty of extracting enzymes from tissues rich in polyphenolic compounds (10, 79-81) such as tea flush has undoubtedly been an obstacle to more rapid development of knowledge in this field. The manufacture of instant tea offers many interesting possibilities for the use of enzymes (Table 2, Part B). It will be interesting to watch future developments in this field.

Literature Cited

1. Harler, C.R. (1963), "Tea Manufacture", Oxford Univ. Press, London, England, 126 pp.
2. Eden, T. (1965), "Tea", 2nd Ed., Longmans, Green and Co., Ltd., London, England.
3. Hainsworth, E. (1969), "Encyclopedia of Chemical Technology", 2nd. Ed., Editor, A. Standen, pp. 743-755, Wiley (Interscience), New York.
4. Sanderson, G.W. (1972), "Structural and Functional Aspects of Phytochemistry", Runeckles, V.C., and Tso, T.C., Ed., Academic Press, New York, pp 247-316.

5. Pintaro, N. (1970), "Soluble Tea Processes", Park Ridge, N.J., Noyes Data Corp.
6. Sanderson, G.W. (1972), World Coffee & Tea, April, 1972, pp 54-57.
7. Bamber, M.K. and Wright, H., 1902, "Year Book of the Planters Association for 1901-1902", Planters Association, Colombo, Ceylon.
8. Roberts, E.A.H., and Sarma, S.N. (1938), Biochem. J. 32: 1819-1828.
9. Sreerangachar, H.B. (1939), Current Sci., 8: 13.
10. Sanderson, G.W. (1964), Biochim. Biophys. Acta. 92: 622-624.
11. Buzun, G.A., Dzhemukhadze, K.M., and Mileshko, L.F. (1970), Prikl. Biokhim. Mikrobiol. 6: 345-347.
12. Sanderson, G.W. (1975), "Geruch- und Geschmackstoffe", Drawert, F. (editor), Verlag Hans Carl, Nurenburg, W. Germany, pp 65-97.
13. Sanderson, G.W., and Graham, H.N. (1973), J. Agric. Food Chem. 21: 576-585.
14. Reymond, E. (1976) (ACS Centennial Symposium, New York City, April, 1976, to be published).
15. Takeo, T. (1966), Agr. Biol. Chem. 30: 931-934.
16. Buzun, G.A., Dzhemukhadze, K.M., and Mileshko, L.F. (1966), Prikl. Biokhim. Mikrobiol. 2 (5): 609-11.
17. Sanderson, G.W. (1964), J. Sci. Fd. Agric. 15: 634-639.
18. Sanderson, G.W. (1964), Tea Quarterly 35: 101-110.
19. Sanderson, G.W. and Kanapathipillai, P. (1964), Tea Quarterly 35: 222-229.
20. Takeo, T., and Baker, J.E. (1973), Agr. Biol. Chem. 12: 21-24.
21. Sreerangachar, H.B. (1943), Biochem. J. 37: 653-655.
22. Sreerangachar, H.B. (1943), Biochem. J. 37: 656-660.
23. Sreerangachar, H.B. (1943), Biochem. J. 37: 661-667.
24. Sreerangachar, H.B. (1943), Biochem. J. 37: 667-674.
25. Roberts, E.A.H.(1962), "Chemistry of Flavanoid Compounds," Geissman, T.A., ed., pp 468-512, London, Pergamon Press.
26. Coggon, P., Moss, G.A., and Sanderson, G.W. (1973), Phytochem. 12: 1947-1955.
27. Sanderson, G.W. (1965), Biochem. J. 95: 24-25.
28. Takeo, T. (1965), Agr. Biol. Chem. 29: 558.
29. Takeo, T., and Uritani, I. (1966), Agr. Biol. Chem. 30: 155-163.
30. Gregory, R.P.F., and Bendall, D.S. (1966), Biochem. J. 101: 569-581.
31. Bokuchava, M.A., and Skobeleva, N.I. (1969), Advan. Food Res. 17: 215-280.
32. Roberts, E.A.H. (1939), Biochem. J. 33: 836-852.
33. Takeo, T., and Kato, Y. (1971), Plant and Cell Physiol. 12: 217-223.
34. Tirimanna, A.S.L. (1972), J. Chromatogr. 65: 587-588.

35. Fairley, C.J., and Swaine, D. (1973), British Patent No. 1,318,035, published May 23, 1975.
36. Fairley, C.J., and Swaine, D. (1975), U.S. Patent No. 3,903,306, published Sept. 2, 1975.
37. Gonzalez, J.G., Coggon, P. and Sanderson, G.W. (1972), J. Food Sci. 37: 797-798.
38. Saijyo, R., and Takeo, T. (1972), Plant and Cell Physiol. 13: 991-998.
39. Coggon, P., Romanczyk, L.J., and Sanderson, G.W. (1976), J. Agric. Fd. Chem. (in press).
40. Stagg, G.V. (1974), J. Sci. Fd. Agric. 25: 1015-1034.
41. Lamb, J., and Ramaswamy, M.S. (1958), J. Sci. Fd. Agric. 9: 46-51.
42. Lamb, J., and Ramaswamy, M.S. (1958), J. Sci. Fd. Agric. 9: 51-56.
43. Hatanaka, A. and Harada, T. (1972), Agr. Biol. Chem. 36: 2033-2035.
44. Hatanaka, A. and Harada, T. (1973), Phytochem. 12: 2341-2346.
45. Kajiwara, T., Harada, T., and Hatanka, A., Agr. Biol. Chem. 39: 243-247.
46. Sekiya, J., Kawasaki, W., Kajiwara, T., and Hatanka, A. (1975), Agr. Biol. Chem. 39: 1677-1678.
47. Hatanaka, A., Sekiya, J., and Kajiwara, T. (1976), Phytochem. 15: 487-488.
48. Sekiya, J., Numa, S., Kajiwara, T., and Hatanaka, A. (1976), Agr. Biol. Chem. 40: 185-190.
49. Yamanishi, T. (1975), Nippon Nogei Kagaku Kaishi 49: R1-R9.
50. Tenco Brooke Bond Ltd. (1973), British Patent No. 1,306,017.
51. Yamanishi, T., Wickremasinghe, R.L., and Perera, K.P.W.C. (1968), Tea Quarterly 39: 81.
52. Gianturco, M.A., Biggers, R.E., and Ridling, B.H. (1974), J. Agric. Food Sci. 22: 758-764.
53. Wickremasinghe, R.L., Perera, B.P.M., and deSilva, U.L.L. (1969), Tea Quarterly 40: 26-30.
54. Wickremasinghe, R.L. (1974), Phytochem. 13: 2057-2063.
55. Sanderson, G.W., and Roberts, G.R. (1964), Biochem. J. 93: 419-423.
56. Iwasa, K. (1974), Nippon Nogei Kagaku Kaishi 48: 445.
57. Sanderson, G.W. (1966), Biochem. J. 98: 248-252.
58. Roberts, E.A.H. (1963), J. Sci. Food Agr. 14: 700-705.
59. Wickremasinghe, R.L. and Perera, K.P.W.C. (1966), Tea Quarterly 37: 131-133.
60. Smith, R.F. (1968), J. Sci. Fd. Agric. 19: 530-535.
61. Rutter, P., and Stainsby, G. (1975), J. Sci. Fd. Agric. 26: 455-463.
62. Tenco Brooke Bond Ltd. (1971), British Patent No. 1,249,932.
63. Takino, Y. (1976), U.S. Patent No. 3,959,497, issued 5/25/76.
64. Yamada, K., Iibuchi, S., and Mirada, Y. (1967), Hakko Kagaku Zasshi 45: 233-240.

65. Iibuchi, S., Minoda, Y., and Yamada, K. (1968), Agr. Biol. Chem. 32: 803-809.
66. Iibuchi, S., Minoda, Y., and Yamada, K. (1972), Agr. Biol. Chem. 36: 1553-1562.
67. Dyckerhoff, H., and R. Armbruster (1933), Hoppe Seyler's Zeitschrift fur Physiologische Chemie, 219: 38-56.
68. Haslam, E., Haworth, R.D., Jones, K., and Rogers, H.J., J. Chem. Soc. (1961), 1829-1835.
69. Weetall, H.H. and Detar, C.C. (1974), Biotech. Bioengineer. 16: 1095-1102.
70. Coggon, P., Graham, H.N., and Sanderson, G.W. (1975), British Patent No. 1,380,135.
71. Sanderson, G.W., and Coggon, P. (1974), U.S. Patent No. 3,872,266.
72. Seltzer, E., Harriman, A.J., and Henderson, R.W. (1961), U.S. Patent No. 2,975,057.
73. Millin, D.J. (1970), British Patent No. 1,204,585, published Sept. 9, 1970.
74. Millin, D.J. (1972), U.S. Patent No. 3,649,297, published March 14, 1972.
75. Fairley, C.J., and Swaine, D. (1970), Australian Patent No. 21460/70 assigned to Tenco Brooke Bond Ltd.
76. Sanderson, G.W., and Simpson, W.S. (1974), U.S. Patent No. 3,787,582.
77. Misawa, Y., Matubara, M., and Inzuka, T. (1968), Nippon Shokuhim Kogyo Gakkaishi 15: 306-309.
78. Toyama, N., and Owatashi, H. (1966), J. Fermentation Technol., (Japan) 44: 830-834.
79. Loomis, W.D. (1974), "Methods in Enzymology, Vol. 31, Biomembranes", (Fleischer, S., and Packer, L., editors). Academic Press, New York, pp. 528-544.
80. Anderson, J.W. (1968), Phytochem. 7: 1973-1988.
81. Lam, T.H., and Shaw, M. (1970), Biochem. Biophys. Res. Comm. 39: 965-968.

3

Coffee Enzymes and Coffee Quality

HENRIQUE V. AMORIM and VERA L. AMORIM

Department of Chemistry, Sector of Biochemistry, Esc. Sup. Agric., "Luiz de Queiroz," University of São Paulo, 13400 Piracicaba, Sp, Brazil

Coffee beverage is the infusion of the roasted coffee seed. It is the most consumed stimulant beverage in the world and is second in international trade. The coffee tree is a perennial plant and grows in tropical and sub-tropical regions. The price of coffee beans (bean, is a worldwide used name, but the correct form should be coffee seed) depends on the quality. It is almost impossible to define quality because what is a good coffee for one, could be a bad coffee for others. However, it is important to take into consideration the acceptance of the product by the majority of the consumers. It will be considered in this paper that the best coffee quality is the one that is accepted as being high quality by the majority of people involved in the coffee industry.

Coffea arabica L. for example, is considered the most flavorful and is also the most expensive. It accounts for 3/4 of the world consumption. Following this comes *C. Canephora*, Pierre ex Froehner, known in the international trade by Robusta, and accounts for 1/4. A small percentage comes from *C. Liberica*, Hiern. Within each specie, coffee is also classified by its quality, which includes physical aspects of raw bean (color, size, shape and impurities) and the flavor quality of the beverage (acidity, body and aroma) after roasting. The quality of the coffee within a given specie is dictated by the region where the plant is grown, by the cultural practices and by harvesting, processing and storage conditions.

In spite of the importance of coffee quality to the price of the commercial bean, little has been done on the chemical and enzymological aspects as coffee seed. Tea is probably the most studied stimulant beverage. The flavor precursors of black tea and several of the chemical components in the

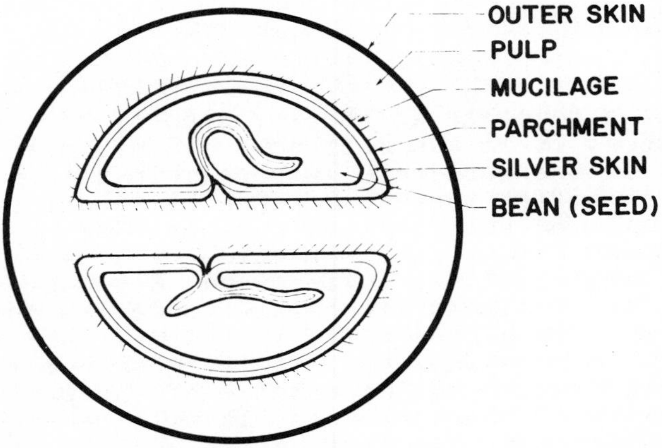

Figure 1. Cross section of a ripe coffee fruit

finished infusion are well known and their effect on quality are fairly well established (1, 2). Cocoa fermentation studies have shown that chocolate flavor precursors develop during this process (3, 4). However, the whole mechanism and reactions are not completly understood. The activity of polyphenol oxidase and hydrolytic enzymes are recognized to play an important role on flavor precursors formation of these two commodities.

Coffee is perhaps the least studied with regard to the green bean and flavor precursors. A few papers indicate that trigonellin, free sugars and peptides are responsible for the coffee aroma (5, 6). On the other hand, coffee aroma after roasting and infusion, has been extensively investigated, and more than 400 volatiles have been identified (7). Not much is known about coffee taste and the compounds which affect it (6). In one aspect, coffee is completely different from tea and cocoa. For the formation of flavor precursors, tea and cocoa have to be subjected to a kind of fermentation, which affects the chemical composition of the whole leaf (tea) or cotyledon (cocoa), while on coffee, with the exception of mucilage removal, it seems that any kind of fermentation or change in chemical composition of the seed after maturity depreciates the flavor quality. It is not intended to make a complete review on coffee processing. Instead, full attention will be given to enzymatic activities and the possibility of these enzymes affecting the quality of coffee.

Coffee Fruit, Harvesting and Processing

A cross section of a coffee cherry is shown in Figure 1. This is a schematic drawing of a ripened fruit. The outer skin can be red or yellow-orange depending on the variety. To attain the best coffee quality the fruit should be harvested in the fully ripe stage regardless of specie or variety. There are two main types of coffee processing: The wet ("washed") process and the dry ("natural") process. In the wet process the coffee cherry is harvested and transported to a processing plant where the cherry is squeezed in a pulping machine and the outer skin and pulp are removed. The mucilagenous and slipery layer is then removed from the bean by a natural fermentation (bacteria, yeast and mold), alkali or added pectinolytic enzymes (8). The beans are washed and dryed to 10-12 per cent moisture. In the "natural" process the cherry is havested and dried to about 12 per cent moisture in the whole fruit. The outer layers are then removed by hulling.

A detailed description of each processing step can be seen in SIVETZ and FOOTE's book (8). Figure 2 shows the main steps of the wet and natural processes.

WASHED COFFEE	NATURAL COFFEE
Harvesting	Harvesting
Transportation	Transportation
Storage	Storage
Sorting	Sorting
Pulping	Drying
Mucilage Removal	(Storage)
Washing	Hulling
Drying	Storage
(Storage)	Removal of Impurities
Hulling	Roasting
Storage	Grinding
Removal of Impurities	Infusion (coffee beverage)
Roasting	
Grinding	
Infusion (coffee beverage)	

Figure 2. Main steps of coffee-handling process

Enzyme Activities in Different Coffee Species and Varieties

The first report of enzyme detection in green coffee beans was made by HERNDLHOFER in Brazil in 1932 (9). Lipase, protease, amylase, catalase and a "trace" of peroxidase were the enzymes detected. Polyphenol oxidase could not be confirmed in the dried seeds. However, the first to try to correlate enzyme activity with the quality of coffee was WILBAUX in 1938 (10). He was sucessfull in measuring the activity of polyphenol oxidase and peroxidase in the dried seeds using various substrates. In his extensive work, WILBAUX compared the activity of lipase, catalase, polyphenol oxidase and peroxidase in seven different coffee species and correlated them with the organoleptic properties. Although he did not try to correlate these differences in enzymatic activities with the quality of the beverage (because other components such as caffeine and oil also differed among the species), this pioneering work was not followed by other enzymatic studies. Recently, the higher activity of polyphenol oxidase of Robusta coffee, in relation to Arabica species found by WILBAUX (10), was confirmed in 1972 by OLIVEIRA (11) and VALENCIA (12).

Among the different varieties of C. arabica in Brazil, no difference was found in the quality of the beverage if the coffee is processed in the same way. The activity of polyphenol oxidase is also very similar.

Table I shows the activity of polyphenol oxidase in four different species and three different varieties.

TABLE I. Relative activity of polyphenol oxidase in coffee seeds (DOPA as substrate) of different species and varieties (recalculated from OLIVEIRA, 11).

COFFEE	Relative PPO activity %
C. dewevrei De Wild et Durant	
var. dewevrei	100
C. canephora Pierre ex Froehner	
var. Bukobensis	58.2
C. liberica Hiern	23.5
C. arabica Linneu	
var. Bourbon amarelo	10.5
var. Mundo novo	13.0
var. Catuai amarelo	12.2

The quality of the beverage of three C. arabica coffees is the same, it is characteristic of the specie and indistinguishable among them. However, it is very different from C. canephora, C. liberica and C. dewevrei which have their characteristic flavor, but is not much appreciated.

Recently, PAYNE et al. (13) utilized malate dehydrogenase and general protein banding patterns to differentiate coffee species, varieties and cultivars with some success.

Although variations in different protein patterns and enzyme activities may contribute to different flavors after the beans have been roasted, coffees of different species also have different contents of caffeine, chlorogenic acid, trigonelline, fats, and possibly other components (14). For this reason, it seems reasonable to speculate that the difference in flavors among species is given by their different contents of several chemical components. However, the effect of enzymatic activities on flavor can not be excluded.

Coffee Enzymes on Harvesting, Transport and Storage

Natural Coffees. In Brazil, coffee is classified with respect to quality of the beverage, by the following grades: (from the best to the poor) Soft (mild), Almost Soft, Hard (astringent), Slightly Rio and Rio (medicinal).

When the coffee cherry is harvested in the fully ripened stage, the moisture content is about 65 per cent and this kind of fruit can produce the best beverage. Immature beans and black beans if not separated from ripe beans, depreciate the final liquor. These defective beans will not be considered in this paper.

In regions where all the fruits reach maturity at almost the same time, the harvesting of only ripe cherries becomes impossible. The early mature beans maybe over-ripe in the tree or fall dow to the ground. If the weather is dry, the quality of the beverage of these beans may be fair to good. However, if the climate is humid, hydrolytic and oxidative reactions caused by the fruit enzymes and microorganisms deteriorate the quality of the resulting beverage.

One experiment utilizing fruits collected in the same tree, but harvesting in different stages of maturity, and dried in different ways showed that the quality of the beverage and the activity of polyphenol oxidase differ, depending upon the process used (Table II). The poorer the liquor quality, the lower was the activity of polyphenol oxidase.

TABLE II. Relative activity of polyphenol oxidase of green coffee and quality of the beverage of cherries harvested and prepared in different ways under wet climate conditions.

Treatment	Relative PPO act. %	Beverage Quality
Ripe cherry, pulped and sun dried	100	Soft
Over ripe, dried in the tree		
Sample (1)	46.9	Slightly Rio
Sample (2)	35.1	Slightly Rio
Dried in the ground 15 days	16.9	Rio

FRANCO, TEIXEIRA, AMORIM, 1971 (unpublished).

Rio flavor also develops when the bean is drying in the sun and gets wet by rain. After the pulp is removed, no Rio flavor develops. Working with coffee harvested in different regions, AMORIM and SILVA (15) found a positive correlation between the activity of polyphenol oxidase (DOPA as substrate) of the green bean and quality of the beverage of C. arabica. Several papers confirmed their results (11, 16, 17, 18, 19). Polyphenol oxidase, or o-diphenol oxidase, is a copper enzyme (EC, 1.10.3.1.) which oxidizes o-diphenols to quinones.

Figure 3 shows the results obtained by OLIVEIRA (11) who used 7 coffee samples of each kind of beverage and compared them with the activity of polyphenol oxidase. The correlation was significant at the 1 per cent level. The activity of peroxidase and catalase were also studied by OLIVEIRA (11). However, only peroxidase gave a significant result. The Rio coffee had lower activity in relation to the others.

Localization of polyphenol oxidase in coffee seed has been unsucessful up to now. However, peroxidase was localized (p-phenylenediamine + H_2O_2) (20) chiefly in the outer layers of the seed and in the embryo (Figure 4). The intensity of oxidation of p-phenylenediamine in Soft coffees is much greater than in Rio coffees. In general, Rio coffee is light green or yellow-brown, even when it is of the current year's crop. The localization of more intense peroxidase activity in the outer layers of the endosperm increases the chance of it being associated with the discoloration generally found in Rio coffees.

By using alkaline gel electrophoresis (21, 22) of crude extracts without dialysis, several polyphenol oxidases and one peroxidase were found in green coffee beans. However, the number of PPO bands depends on the substrate used (23) (Figure 5). It was not possible to differentiate Soft from Rio coffees by the intensity of the bands, because of great variations. In Brazil, where most of the commercial beans are "natural coffee", these different qualities result from different methods of harvesting, handling and drying. If one harvests only ripened fruit or collects green, ripe and over-ripe fruits, but classifies them before drying, the ripe fruit always give a good beverage (Soft). The classification of the Hard coffee in Brazil is a puzzle. The characteristic of Hard coffee is its astringency. However, some coffees have no astringency, no Rio flavor and also lack the full taste and aroma of the

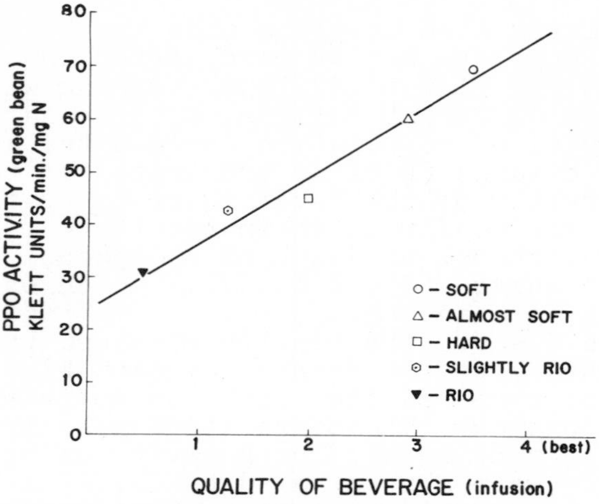

Figure 3. Activity of polyphenol oxidase in raw coffee and the quality of the beverage after roasting and infusion. Each point represents an average of seven different samples (11).

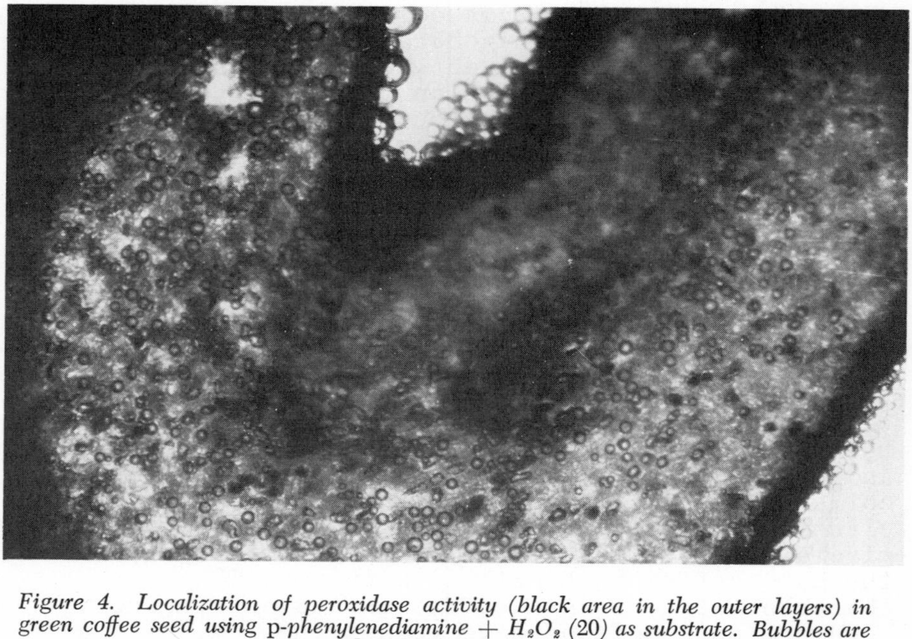

Figure 4. Localization of peroxidase activity (black area in the outer layers) in green coffee seed using p*-phenylenediamine + H_2O_2 (20) as substrate. Bubbles are O_2 produced by catalase activity.*

Soft coffee. These coffees are then classified by the professional tasters as Hard, considering that they occupy an intermediate position in the quality scale (A.A.TEIXEIRA, personal communication). The literature shows that many kinds of defective beans, drying temperatures and attack by fungi may give rise to this peculiar flavor (24).

Although the activity of polyphenol oxidase seems to be correlated with the quality of Brazilian "natural" coffees, other estimates of chemical components of green coffee were made. Soluble carbohydrates, polysaccharides (25), chlorogenic acids, total soluble phenolics, hydrolyzable phenols (26), soluble proteins (27), eletrophoretic patterns of soluble proteins (24) and a multiple linear regression analysis of all data were calculated (28). The most significant results showed that Rio coffee had less hydrolyzable phenols and less soluble proteins (TCA precipitable). The different patterns of soluble proteins in agar gel eletrophoresis were explained by the fact that, in the extraction procedure, chlorogenic acid was more oxidized by polyphenol oxidase (higher in the Soft coffees) and it binds with proteins changing their electrical charge (24, 29).

One tentative explanation for the lower activity of polyphenol oxidase found in Rio coffees is this: in any step during harvesting and processing the enzyme has a chance to come into contact with the substrates. The oxidized phenolics are very reative and bind covalently with the enzyme, inhibiting its activity (30).

Concerning the lesser amount of TCA-precipitable protein reported earlier in Rio coffee (27), recent results confirmed this fact and shed more light on the protein differences between Soft and Rio coffees (21). It was found that, although the total soluble nitrogen is apparently the same between these two coffees, SDS (sodium dodecyl sulfate) gel eletrophoresis of the water-soluble proteins (treated with 2-mercaptoethanol) shows a different pattern between the proteins extracted from 5 different samples of Rio coffee and 4 samples of Soft coffee (Figure 6). SDS gel eletrophoresis separates most of the proteins according to their molecular weights (31). If this holds true for coffee proteins, the gel patterns show that Soft coffee has some proteins of high molecular weight (~64,000 Daltons) that are absent in Rio coffee. On the other hand, Rio coffee shows some bands of low molecular weight that are

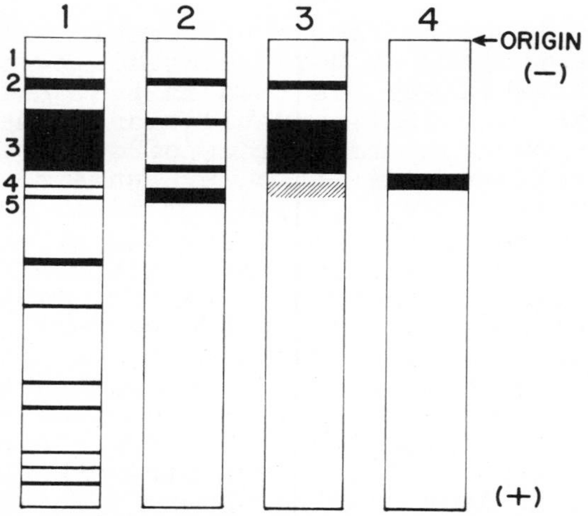

Figure 5. Alkaline polyacrylamide gel electrophoresis ($T = 7\%$) of seed water-soluble proteins: protein banding pattern (1); polyphenol oxidase, DOPA (2), and caffeci acid coupled with m-*phenylenediamine (3) as substrates; and peroxidase,* o-*dianizidine + H_2O_2 in 10% acetic acid (4)*

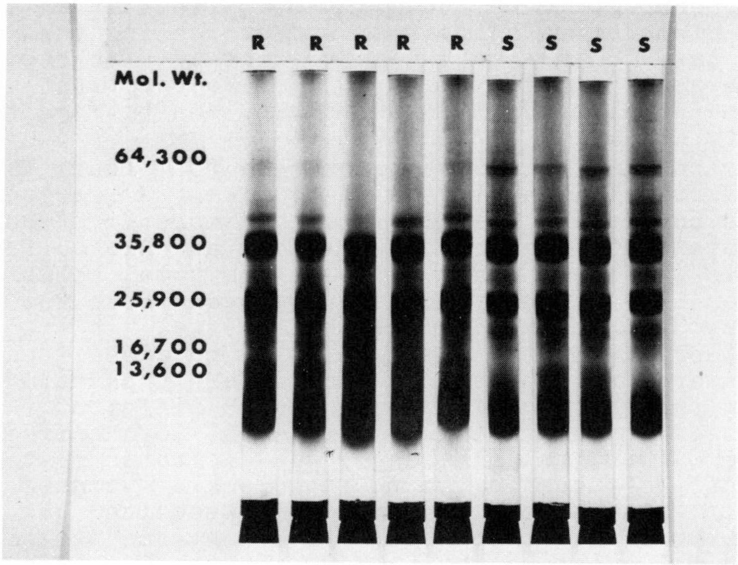

Figure 6. SDS polyacrylamide gel electrophoresis ($T = 10\%$) of green coffee, water-soluble proteins (treated with 2-mercaptoethanol) of Soft and Rio samples

absent in the Soft coffee. Furthermore, the bands of small polypeptides (~9,000 Daltons) are much more intense in the Rio coffee than in Soft coffees. This suggests that the coffee beans which give Rio flavor undergo hydrolysis during processing and/or modification of the tertiary structure of their proteins. However, at the present time one cannot be sure that these polypeptides are the cause of the Rio flavor, because other hydrolytic reactions also take place, probably at the same time: for example, the total extractable lipids (chloroform/methanol) were lower in Rio coffee in relation to Soft coffee as shown in Table III.

TABLE III. Extractable lipids (chloroform/methanol) of Arabica coffees which differ in the quality of the beverage and age.

Rio		Soft	
age (yrs)	% lipids	age (yrs)	% lipids
7	14.84	3	16.28
3	13.51	3	14.98
3	13.32	3	15.44
2	13.45	2	15.12
mean	13.78*	mean	15.45*

*Statistical significant at 1% level.

The density, thickness of cell wall and volume of cell walls are lower in Rio coffee in comparison with Soft coffee (32), as shown in Table IV.

TABLE IV. Physical measurements of the whole bean and the hard (outer) endosperm of Arabica coffee beans which differ in quality of the beverage, (the symbols in parentheses represent the number of measurements of the cell wall)

Coffee Sample	Wt. 500 beans (g)	Density	Cell wall thickness (mean) µm	Volume of cell wall %
Soft-1	71.03	1.085	6.2 (972)	50.7
Soft-2	57.69	1.014	5.6 (822)	49.7
Rio-1	58.21	0.967	5.0 (731)	37.5
Rio-2	49.24	0.761	4.3 (475)	35.6

AMORIM, SMUCKER, PFISTER (32)

Although all samples were from the same coffee variety (C. arabica L., var. Mundo Novo), the age of the beans were different. Soft-1, Soft-2 and Rio-1 were kept in sealed cans for three years and Rio-2 was 7 years old. (In a later section we will discuss the effect of storage conditions on physical and chemical aspects of raw coffee).

Coffee beans contain several glycosidases (33), and three α-galactosidases were recently purified by affinity chromatography (34). BARHAM and coworkers (35) found only one α-galactosidase (they did not mention the coffee specie) and found a molecular weight of 26,300. SHADAKSHARASWAMY and RAMACHANDRA (36, 37) observed that the activity of α-galactosidase and β-fructofuranosidase incresed when the seed was soaked in water. This might partly explain the thinner cell wall in Rio coffees. However, differences in soluble carbohydrates between Rio and Soft coffee (25) were not found, probably because the simple sugars were metabolized, giving rise to CO_2. This accounts for part of the dry weight loss in Rio coffees.

It may be inferred from these chemical and physical analyses, that the Rio flavor is associated with hydrolytic and oxidative reactions that take place during harvesting and subsequent processing, while the bean is still wet and covered with the pulp.

Washed Coffees. The physiological stage of fruit development upon harvesting which gives the best beverage for "natural" coffees, also applies to washed coffees. The ripe cherry, if well processed, gives the best beverage.

After harvesting, the coffee cherry should be "pulped" as soon as possible. The time required to initiate the reactions which lead to "off" flavors depends on the temperature of the environment and the storage conditions. If the whole beans are kept under a stream of water, the time from harvesting to pulping may take 48 hours without affecting cup quality.

ARCILA and VALENCIA (38), in Colombia, studied several factors which affect the quality of the beverage of washed coffees and tried to correlate them with activity of polyphenol oxidase of the dried bean. The enzyme activity is higher in coffee pulped immediately; that is within a 12-hour interval from harvesting to pulping. It drops in 24 to 36 hours and again in 48 to 72 hours. Statistical analysis of the data showed that enzymatic activities were different and inversely correlated with acidity, the higher the activity, the lower the acidity. No difference was found in the body of the beverage, although the 72-hour treatment yielded a very poor aroma. By considering a series of different conditions and using multiple linear regression analysis, these same authors found that the activity of PPO was negatively correlated with acidity and positively correlated with the body of the beverage (38).

It is believed that upon harvesting, hydrolytic and oxidative reactions in the pulp and mucilage begin. If the cherry is injured, attack by microorganisms speeds up these reactions. We have been able to detect activity of pectin esterase, polygalacturonase, α-galactosidase, peroxidase and polyphenol oxidase in the mucilage of uninjured ripe fruits (unpublished results). Polyacrylamide gel electrophoresis (39) of mucilage water-soluble proteins indicated that there are 2 polyphenol oxidases and four protein bands (Figure 7). The contribution to "off" flavors developed before pulping by microorganisms and by degradative and oxidative reactions catalyzed by the enzymes of the pulp/mucilage have not yet been stablished. However, WOOTON (40) states that brown pigments formed in the pulp and mucilage layers depreciate the quality of the beverage, as well physical aspects of the bean, if it diffuses through the parchment and reaches the bean.

Enzymes in Coffee Fermentation

Mucilage removal after the bean is pulped is a very important step in washed coffee processing. If the mucilage is not removed, the drying process becomes difficult. On the other hand, the mucilage is easily attacked by microorganisms which add "off" flavor into the bean and taint the beverage.

The chemical composition of the mucilage has been partialy resolved. COLEMAN and coworkers (41) determined that pectic acid is one the polysaccharides of the mucilage; it is a polymer of polygalacturonic acid (77.6% d. wt. basis), with a small amount of arabinose, galactose, xylose and rhamnose. Further work carried out by CORRÊA and his group (42, 43, 44) showed that the pectin is a linear polymer containing (1-4) linked residues of D-galactose. They also isolated a highly branched galactoaraban in which L-arabinofuranose residues linked (1-3) and (1-5) apparently occur in the exterior chain, with a core of a branched galactan composed of (1-6) and (1-3) linked D-galactopyranose residues.

Reducing sugars, sucrose, caffeine, chlorogenic acids (45) and amino acids are also found in the mucilage (8). The water-soluble protein patterns are shown in Figure 7. Although the mucilage has no

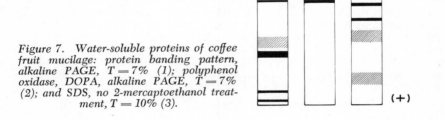

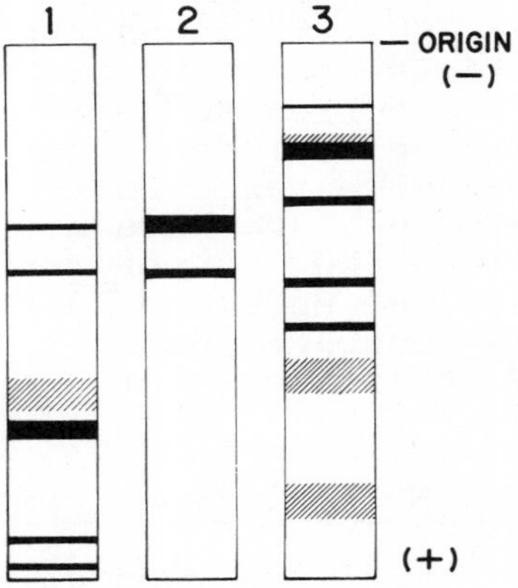

Figure 7. Water-soluble proteins of coffee fruit mucilage: protein banding pattern, alkaline PAGE, T = 7% (1); polyphenol oxidase, DOPA, alkaline PAGE, T = 7% (2); and SDS, no 2-mercaptoethanol treatment, T = 10% (3).

celluar structure (46) it has several hydrolytic and oxidative enzymes, such as pectinesterase (47), galacturonase, α-galactosidase, peroxidase and polyphenol oxidase (catechol and DOPA as substrate) (AMORIM, TEIXEIRA, OLIVEIRA, COSTA - in preparation). However, these hydrolytic enzymes of the mucilage seem to be unable to remove the mucilage when the seed is kept in microorganism free environment (48). CARBONNEL and VILANOVA (46) explain this fact by the equilibrium between the substrate and product of an enzymatic reaction. Apparently, the fruits do not have enough enzyme to completely catalyze solubilization of the mucilage; conclusive proof for this is still lacking.

The fermentation process which includes the removal of the mucilagenous layer adhering to the parchment, is achieved first by the endogenous mucilage enzymes, followed by bacterial, fungi and yeast fermentation (50). WOSIAK and ZANCAN (51) isolated *Penicillium sp*, *Fusarium sp*, *Cladosporium sp* and *Aspergillus sp* from Brazilian coffee fruits and all mold strains produced in the culture medium enzymes able to break down pectic acid and galactoaraban from coffee pulp and mucilage. A decrease in pH is caused by exposure of the carboxyl groups of pectin (46), and by the production of acetic and lactic acids (40). There is also a decrease in carbohydrates during fermentation (47) and a concurrent production of ethanol (52). For low-grown coffee, the fermentation is fast and ethanol production is very high in the begining, but then it disappears. For high-grown coffees, ethanol increases steadlly but does not reach 10% of that by low-grown coffee production. Acetic acid production is steady and reaches 15% of all volatiles. Propionic and butyric acid are also produced, but they increase only after 20 hours fermentation (52). Propionic and butyric acids give a bad flavor to the coffee beverage. For this reason, fermentation time should be kept as short as possible. On the other hand, if the coffee is exposed to the liquor (degradation and oxidized products of mucilage) for too long, these products may pass through the parchment, cause a poor appearance of the raw coffee (brown silver skin and center cut) and possibly taint the beverage characteristics (40). These oxidized products are possibly chlorogenic acids which are oxidized by polyphenol oxidase and peroxidase and link covalently with amino acids and polypeptides (29). To avoid these browning reactions, WOOTON (40) added a reducing agent, sodium sulfite (1%), with apparent sucess, but the fermentation was delayed; probably because of inhibition of microorganisms.

SANINT and VALENCIA (53) found that long fermentation times of more than 24 hours, depreciate quality of the beverage and lower activity of polyphenol oxidase of the dried bean.

To avoid a long fermentation time, one can add several enzymes that are commercially available, but these are not used on a large scale because of the added cost. Benefax (54) is a material produced from molds and contains 10% of a useful crude enzyme preparation (8). Recently, Ciba-Geigy S.A. marketed ULTRAZIM 100, which is used by orange juice industries. It has been used to remove coffee mucilage with success in Kenya (55) and Guatemala (56), and in Brazil we observed a drastic reduction in time when 4 mg of the commercial ULTRAZIM 100 was mixed with 2 kg of pulped coffee, as demonstrated in Table V.

TABLE V. Fermentation time of pulped coffee under different conditions (temperature 18-20°C).

Treatment	Hours	pH end	pH initial
Natural, dry	16	4.02	5.64
Natural, under water	18	5.29	5.75
ULTRAZIM 100 (4 mg - 100 ml/2 kg)	9	5.17	5.69
Natural + Metabisulfite 500 ppm - 100 ml/2 kg	26	4.54	5.67

AMORIM, TEIXEIRA, COSTA, OLIVEIRA, 1976 (in preparation)

This commercial enzyme may be reused for at least 3 consecutive fermentations by recycling the decanted liquid. To avoid browning of the silver skin, 250 ppm of metabisulfite may be added per kg of pulped coffee. No inhibition of the enzyme was noted. Furthermore, metabisulfite may react with acetaldehyde and other carbonyl compounds, avoiding seed contamination and affecting the quality of the beverage (57).

It is important to note that, after fermentation, the beans should be washed to remove mucilage and polysaccharide degradation products (free sugars). Otherwise, these free sugars bind to the silver skin in the region of the center cut and, after the roasting process, the center cut becomes brown (58). This brown center cut indicates that the bean was not well processed.

Possible Role of Enzymes During the Drying Process

The drying process is an important step in coffee preparation. If drying is not done properly it may change the physical aspect of the bean and the quality of the beverage. FERRAZ and VEIGA in 1954 (59), were the first to call attention to the importance of drying temperature of natural coffee and the quality of the beverage. The best temperature was found to be 45°C. Drying between 50 to 60°C had the worst effects on the quality of the beverage. Temperatures above 60°C and up to 90°C gave a fairly good beverage for natural coffees. They speculate that temperatures around 55°C activate enzymes which in turn produce compounds deleterious to the coffee flavor. On the other hand, other temperatures activate (or inactivate) enzymes which can produce compounds of good or bad flavors. Although no enzymatic activities were measured, the speculation of FERRAZ and VEIGA (59) should be taken into consideration, because temperature may affect bean enzymatic activities.

In Colombia, ARCILA and VALENCIA (38) studied the effect of drying temperature on the activity of polyphenol oxidase and on the quality of the beverage. The best temperatures for acidity, body and aroma were found to be 40, 50 and 80°C. Temperatures of 30, 60 and 70°C gave a poor beverage, with respect to these three organoleptic characteristics. The activity of polyphenol oxidase generally decreased when temperature increased, with one exception at 40°C. The enzyme activity at 40°C was lower than that at 30 and 50°C. From the data presented, it would appear that there is still no direct relationship between quality, temperature and enzymatic activities.

In the section on roasting the resistance of some coffee enzymes to high temperatures, probably because of the dry state of the bean, will be discussed. However, it does seem that both enzymatic and nonenzymatic reaction might be involved in changing the chemical composition during the drying process, but this aspect needs further investigation.

Role of Enzymes During Coffee Storage

Coffee storage in humid and warm regions is a serious problem, because the coffee bean becomes white, or yellow to brown, depending on the degree of moisture in the air and the time of storage. This change in color is accompanied by a decrease in flavor quality.

MULTON and his group (60, 61, 62) have done a series of studies on coffee storage in different environment conditions, analyzing microorganisms, water absorption, and enzymatic activities, and comparing them with the quality of the beverage. Their major findings may be summarized as follows: above 75% relative humidity, the coffee bean absorbs a significant amount of water and the increase in moisture content is positively correlated to fungal growth, although the number of bacteria decreases. The quality of the beverage is completely changed and after 30 days at 95% relative humidity, the coffee beans are completely rotten. At 95% relative humidity there is also an increase of lipase activity, with a concurrent increase in free fatty acids. In addition, ribonuclease and protease activities seem to increase under these conditions. The dry weight loss may reach 25% in 200 days if the relative humidity of the air is above 90%. In Brazil, JORDÃO et al. (63) also found an increase in free fatty acids during three years storage of green coffee. The peroxide value increased only after 2 years storage. In Africa, ESTEVES (64) studying acidity in the oil of Robusta and Arabica coffees also found an increase in titratable acidity with increasing time of storage in both types of coffees.

PEREIRA (65) in Portugal was the first to report a decrease in polyphenol oxidase activity in green coffee with storage time. His results were confirmed later by OLIVEIRA (11), VALENCIA (12) and AMORIM et al. (19).

Figure 8 shows the polyphenol oxidase activities obtained by OLIVEIRA (11) with different coffee species after different times of storage. The value for Arabicas represents the average of four different varieties grown in different regions; all gave similar patterns.

Stored green coffees that give different qualities of beverage also show decreases in polyphenol oxidase activity and total carbonyls with time of storage (Table VI).

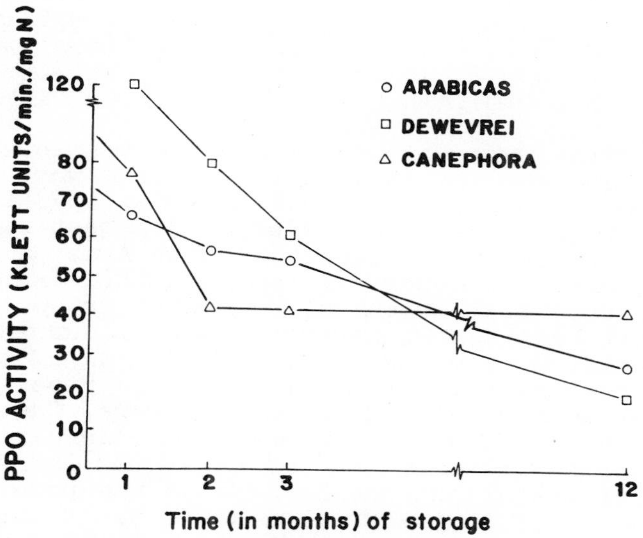

Figure 8. Polyphenol oxidase activity during raw coffee bean storage (11)

TABLE VI. Total carbonyls in coffee oil and polyphenol oxidase (PPO) activity of Arabica green coffee beans (each symbol is the average value of four samples of different coffees).

Coffee Sample	Total carbonyls umol/g oil	PPO Activity Abs. 10 min/g powder
Soft (stored 1 yr)	113.7 a	1.72 a
Rio (stored 1 yr)	88.8 b	0.72 b
Soft (stored 2 yr)	93.0 b	0.97 c
Rio (stored 2 yr)	87.6 b	0.51 d

Different symbols (columns) mean significance at 5% level.
AMORIM et al. (19).

The higher amount of total carbonyls found in the best coffees agrees in part with the work of CALLE (66) in Colombia and GOPAL et al. (67) in India, who found more aldehydes in the best coffees.

In a preliminary experiment with Arabica coffees, MELO and coworkers (68) and TEIXEIRA et al. (69) observed that green coffee stored in sealed cans or plastic bags for 21 months were still green, whereas coffee stored for the same period of time in bags made of paper, cotton or vegetable fiber were white-yellow and yielded a poor beverage. These bleached beans showed concurrent lower densities because of an increase in size (swelling) (Table VII). The activities of polyphenol oxidase and peroxidase were also lower in the spoiled coffees. It is interesting to note that the beans kept in cans and plastic had practically a constant moisture content (~10%), but coffee stored in paper, cotton and vegetable fiber containers changed moisture contents from 9 to 13%, depending upon the season.

It is well known that, upon storage green coffee gradually becomes white. The discoloration starts in the outer layers of cells and goes towards the center. The color of green coffee of a new crop, if well processed, is dark green or greenish-blue. With increased time of storage it becomes light green and white. Depending upon the environmental conditions (humidity and temperature), the bean may change to yellow or brown.

This change in color of green coffee with storage was extensively studied by BACCHI (70) in Brazil. By using two varieties of C. arabica, different methods of processing (wet and natural), different pulping and hulling processes, and different storage conditions, BACCHI concluded that

TABLE VII. Physical and chemical changes of green coffee stored in different containers during 21 months (all estimations were made with four replications).

Container type	Dry wt. 1000 beans	Density	Soluble N % d.wt.	PPO* Activity	PEROX.* Activity	Color	Quality of beverage
Can	120.5	1.124	1.46	100	100	green	Almost Soft
Plastic	120.9	1.118	1.32	96	105	green	Almost Soft
Paper	118.7	0.875	1.28	53	81	yellowish	Hard
Cotton	118.8	0.905	1.28	51	88	yellowish	Hard
Vegetable fiber	119.6	0.901	1.35	52	83	yellowish	Hard
l.s.d. 5%	1.9	0.05	0.28				

*Relative activity in percentage relative to activity in sealed cans. (substrate: PPO = DOPA; PER = o-dianizidine + H_2O_2).

MELO et al. (68).

the most important factor which causes the bean to become white is mechanical injury, from harvesting to hulling, regardless of whether the coffee is processed by the wet or the natural method. The environmental factor which enhances most of the change in color is the moisture of the air.

The chemical mechanism of coffee discoloration has not yet been establhished. However, from the data available in the literature, and by comparison with discoloration in vegetables and fruits (71), it seems that green coffee bean discoloration is due to enzymatic oxidation of phenolic compounds by polyphenol oxidase and peroxidase, with a concurrent oxidation of ascorbic acid (72). The ascorbic acid oxidation might be coupled with the reduction of quinones produced by the action of PPO and PER on phenolics (71). In a general manner, discoloration in fruits is followed by a decrease in phenolics, ascorbic acid and polyphenol oxidase activity (71, 73). In coffee, AMORIM et al. (26) found less hydrolyzable phenolics in Rio coffee than in Soft coffee, but the soluble chlorogenic acids (A.O.A.C. procedure (74)) content in the best coffee was significantly lower, in comparison with the other qualities (26). Chlorogenic acid determination in green coffee should be reinvestigated because different types of coffee may give different results, depending upon the method used (75). There is a possibility that phenolics bound to cell wall are responsible for the discoloration reactions in coffee.

It is well documented in the literature, that injury causes discoloration or/and browning reactions in plant tissues. Coffee beans seem to be no exception. However, better methods for analysis should be devised in order to explain many of the chemical and physical changes which occur in coffee deterioration.

Enzyme Activities and Protein Solubility During the Roasting Process

Coffee aroma and taste develop only after the bean is roasted. The roasting temperature is 240°C or higher and it requires 3 to 15 min., depending upon the temperature, load, machinery, etc. Coffee aroma begins to form when the bean reaches 180-190°C (14, 76, 77). The principal precursor of coffee aroma seems to be trigonelline, sucrose, fructose, glucose, free amino acids and peptides (5). Carboxylic and phenolic acids play an important role in the taste of the beverage (6, 77).

The activities of several enzymes in the green bean must play an important role in modifying the quality of the final beverage, as has been discussed in this paper. In several processing steps there are occasions to modify the chemical composition of the bean. It is not probable that enzyme activities, per se, play a major role in coffee flavor during the roasting process, because the coffee bean is dehydrated and the high temperatures would inactivate all enzymes. However, due to the fact that 20% of the activity of polyphenol oxidase could be retained after 5 hours exposure of green coffee beans to 125°C (MELO and AMORIM, 1972, unpublished), it was necessary to look at the activity of several hydrolytic (β-galactosidase, β-glucosidase, acid phosphatase, (77)) and oxidative (polyphenol oxidase, peroxidase) enzymes during the roasting process.

Figure 9 shows the relative activities of several enzymes plus the water-soluble proteins (TCA precipitatable) during roasting. As can be seen, the degree of inactivation of the enzymes with increasing temperature, differs depending on the enzyme. At 5 min. (170°C), the powder has no coffee aroma. Instead, an odor reminiscent of roasted peanut aroma was observed. At 8 min. (187°C), a clearly defined coffee aroma was detected, which increases at 12 and 15 min. of roasting. The insolubilization of proteins coincides with the inactivation of enzymes and coffee aroma formation. Between the 5th to 8th minute, the free amino acids, peptides, trigonelline and free sugars should be reacting to produce the coffee aroma.

If the free amino acids and polypeptides are important in coffee aroma formation (5, 76), the ratio of these amino acids, as well as their total amounts should be important to the flavor of the final beverage. Also, the amino acid composition of the proteins and perhaps their tertiary structures also play a role in coffee flavor. It is well known that basic and sulfur amino acids are destroyed by the roasting process (79, 80, 81). Furthermore, POKORNY et al. (81) observed that free sulfur and basic amino acids in green coffee decrease with storage. In Brazil, DOMONT et al. (82), comparing the behavior of amino acids during the roasting of Soft and Rio coffees, observed that Rio coffees had 25% more free amino acids than Soft coffees, and the speed of pyrolysis of serine, threonine and chiefly

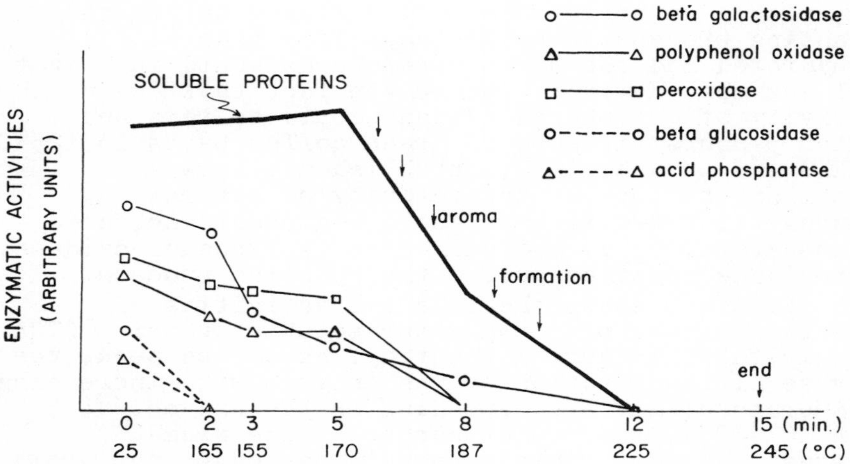

Figure 9. Water-soluble proteins and activity of β-galactosidase (p-nitrophenyl-β-D-galactopyranoside), polyphenol oxidase (DOPA), peroxidase (pyrogallol + H_2O_2), β-glucosidase (p-nitrophenyl-β-D-glucopyranoside), and acid phosphatase (p-nitrophenyl phosphate) during the roasting process. The temperature indicated is that of the air surrounding the bean.

arginine, is much fastly in Rio than in Soft coffee samples. These results agree in part with the observation that Rio coffee has more low molecular weight polypeptides than Soft coffees, as seen in Figure 6. Thus, there is a possibility that these polypeptides may affect the precursors of aroma formation.

Conclusion

From a review of the literature and the data presented in this report, there is only one enzyme that is certain to depreciate coffee quality. This is the polyphenol oxidase of the pulp and mucilage (two enzymes detected), which oxidizes chlorogenic acid and other phenolics to quinones. These quinones then form complexes with amino acids and polypeptides having a brown color, that bind to the center cut of the silver skin. This brown color of the center cut and silver skin depreciate the quality of raw coffee and its appearance after roasting. However, whether these pigments positively affect the quality of the beverage, is still disputed.

Commercial pectinolytic enzymes may be used in conjunction with reducing agents to speed up mucilage removal and avoid browning reactions.

Although polyphenol oxidase and peroxidase activities in many instances are correlated with quality of the beverage, proof of the mechanism and the compounds which take part in these changes and affect cup quality is still lacking. Because of the fact that peroxidase is located chiefly in the outer layers of the seed, it seems probable that this enzyme is associated with coffee discoloration upon storage.

Considering all hydrolytic reactions catalysed by coffee seed enzymes, the lipases and other enzymes of lipid metabolism should be important, because oil is the aroma carrier. Also, the protein hydrolases deserve attention because they affect a known important coffee aroma precursor: the amino acids and polypeptides. By affecting the ratio of these aroma (and probably taste) precursors, it is reasonable to believe that both aroma and taste should be altered.

From the available literature on coffee chemistry it seems that the characteristic flavor of each coffee specie is given chiefly by the inherited chemical composition of the bean. However, the variation in quality (physical and organoleptic) whithin each specie, seems to be associated with variations in chemical composition, caused by the action of hydrolytic and oxidative enzymes of the bean and/or microorganisms. The deleterious effects of enzymes on the quality of the resulting beverage may be prevented by using more suitable methods for coffee processing and storage.

Acknowledgment

Research was carried out with grants from the Instituto Brasileiro do Café, Conselho Nacional do Desenvolvimento Científico e Tecnológico e Fundação de Amparo a Pesquisa do Estado de São Paulo. We are grateful to Dr. A. A. Teixeira and Dr. C. M. Franco for helpful discussions and to Dr. Robert L. Ory for reviewing the manuscript.

Literature Cited

1. Sanderson, G.W. "Structural and Functional Aspects of Phytochemistry". V. C. Runeckless and T. C. Tso Editors. vol. 5, p. 247-316. Academic Press, New York and London, 1972.
2. Sanderson,G.W., Ranadive, A.S., Eisenberg, L.S., Farrel, F.J., Simons, R., Manley, C. H., Coggon, P., "Phenolic,Sulfur, and Nitrogen Compounds in Food Flavors". ACS Symposium Series, 26, p. 14-46 (1976).
3. Biehl, B., Ann. Technol. Agric. (1972) 21: 435-455.
4. Purr, A., Ann. Technol. Agric. (1972) 21: 457-472.
5. Russwurm, H., 4ème Colloque International sur la Chemie du Cafè Vert, Torrefies and leur Derivès. (ASIC), p. 103-109, Amsterdam, (1969).
6. Feldman, J.R., Ryder, W. S., Kung, J. T., J. Agr. Food Chem. (1969) 17: 733-739.
7. Vitzthum, O. G., "Kaffee und coffein". p. 1-77. Springer-Verlag, Berlin, Heildelberg, New York, 1975.
8. Sivetz, M., H. E. Foote, "Coffee Processing Technology", vol. I, p. 48-99. The Avi Publishing Company, Inc. Westport, Connecticut, 1963.
9. Herndlhofer, E., Biochem. Z. (1932). 255: 230-246.
10. Wilbaux, R., Repport annual. 2ème partie, p. 3-45. Congo, Inst. Nat. Et. Agron. 1938.
11. Oliveira, J.C., Relacão da atividade da polifenol oxidase, peroxidase e catalase dos grãos de café e a qualidade da bebida. Doctor Thesis. Esc. Sup. Agr. "Luiz de Queiroz". University of São Paulo. (1972), Piracicaba, SP, Brazil, 80 pages.

12. Valencia, G.A., Cenicafé (Colombia) (1972) 23: 3-18.
13. Payne, R.C., Oliveira, A.R., Fairbrothers, D.E., Biochem. Systemat. (1973) 1: 59-61.
14. Streuli, H., 6ème Colloque International sur la Chemie des Cafés (ASIC) p. 61-72, Bogotá, Colombia. 1973.
15. Amorim, H.V., Silva, D.M., Nature (1968) 219: 381-382.
16. Rotemberg, B., Iachan, A., Rev. Bras. Tecnol. (1971) 2: 67-69.
17. Rotemberg, B., Iachan, A., Rev. Bras. Tecnol. (1972) 3: 155-159.
18. Gopal, N.H., Annual Detailed Technical Report, Coffee Board (India) (1974-75) 28: 109-139.
19. Amorim, H.V., Legendre, M.G., Amorim, Vera L., St. Angelo, A., Ory, R.L., Turrialba (1976) 26: 193-195.
20. Raa, J., Physiol. Plant. (1973) 28: 132-133.
21. Amorim, H.V., Josephson, R.V., J. Food Sci. (1975) 40: 1179-1184.
22. Haard, N.F., Tobin, C.L., J. Food Sci. (1971) 36: 854-857.
23. Van Loon, L.C., Phytochem. (1971) 10: 503-507.
24. Amorim, H.V., Teixeira, A.A., M.Melo, V.F. Cruz, E. Malavolta, Turrialba (1975) 25: 18-24.
25. Amorim, H.V., Teixeira, A.A., Breviglieri, O., Cruz, V.F., Malavolta, E., Turrialba (1974) 24: 214-216.
26. Amorim, H.V., Teixeira, A.A., Guercio, M.A., Cruz, V.F., Malavolta, E., Turrialba (1974) 24: 217-221.
27. Amorim, H.V., Teixeira, A.A., Melo, M., Cruz, V. F., Malavolta, E., Turrialba (1974) 24: 304-308.
28. Amorim, H.V., Cruz, V.F., Teixeira, A.A., Malavolta, E., Turrialba (1975) 25: 25-28.
29. Melo, M., Amorim, H.V., Turrialba (1975) 25: 243-248.
30. Loomis, W.D., Battaile, J., Phytochem. (1966) 5: 423-438.
31. Weber, K., Pringe, J.R., Osborn, M., "Methods Enzymol." vol. 26, Part C. Ed. Hirs, C. H. W. and Timasheff, S. N. p. 3-27. Academic Press, New York, London (1972).
32. Amorim, H.V., Smucker, R., Pfister, R., Turrialba (1976) 26: 24-27.
33. Courtois, J.E., Petek, F., "Methods Enzymol." vol. 8 Ed. Neufeld E., Ginsburg, V. p. 565-571. Academic Press, New York and London (1966).

34. Harpaz, N., Flowers, H.M., Sharon, N., Biochem. Biophys. Acta (1974) 341: 213-221.
35. Barham, D., Dey, P.M., Griffiths, D., Pridham, J. B., Phytochem. (1971) 10: 1959-1963.
36. Shadaksharaswamy, M., Ramachandra, G., Phytochem. (1968) 7: 715-719.
37. Shadaksharaswamy, M., Ramachandra, G., Enzymologia (1968) 35: 93-99.
38. Arcila, P.J., Valencia, A.G., Cenicafé (Colombia) (1975) 26: 55-71.
39. Davis, B.J., Ann. New York Acad. Sci. (1964) 121: 404-427.
40. Wooton, A.E., Report C. R. 12. East African Research Organization (1963).
41. Coleman, R.J., Lenney, J.F., Coscia, A.T., Di Carlo, F.J., Arch. Biochem. Biophys. (1955) 59: 157-164.
42. Corrêa, J.B.C., Fontana, J.D., An. Acad. bras. Ciênc. (1971) 43: 803-812.
43. Corrêa, J.B.C., Odebrecht, S., Fontana, J.D., An. Acad. bras. Ciênc. (1974) 46: 349-356.
44. Corrêa, J.B.C., Coelho, E.O., Fontana, J.D., An. Acad. bras. Ciênc. (1974) 46: 357-360.
45. Franco, C.M., J. Agronomia (Brasil) (1939) 2: 131-139.
46. Carbonnel, R.J., Vilanova, T.M., El café de El Salvador (1952) 22: 411-556.
47. Rolz, C., Menchu, J.F., Espinosa, R., Garcia--Prendes, A., 5ème Colloque International sur la Chemie des Cafès Verts, Torréfiés et leur Dérivés, Lisbonne (ASIC) p. 259-268, Lisbonne (1971).
48. Franco, C.M., Bol. Supta. Serv. Café, São Paulo (1944) 19: 250-256.
49. Franco, C.M., Bragantia (1960) 19: 621-626.
50. Calle, V.H., Cenicafé (Colombia) (1965) 16: 3-11.
51. Wosiaki, G., Zancan, G.T., Arq. Biol. Tecnol. (1973) 16: 129-134.
52. Menchu, J.F., Rolz, C., Café, Cacao, The (1973) 17: 53-61.
53. Sanint, O.B., Valencia, G.A., Cenicafe (Colombia) (1970) 21: 59-71.
54. Johnston, W.R., Foote, H.E., U.S.A. patent 2, 526, 873 (10/24/1950).
55. Arunga, R.O., Kenya Coffee (1973) 38: 354-357.
56. Kupferschmied, B., Revista Cafetalera (1975) 42.
57. Rodriguez, D.B., Frank, H.A., Yamamoto, H.Y., J. Sci. Fd. Agric. (1969) 20: 15-19.
58. Teixeira, A.A., Ferraz, M.B., A Rural (1963) Maio: 28-29.

59. Ferraz, M.B., Veiga, A.A., Bol. Supta. Serv. Café (1954) 29: 5-6.
60. Multon, J.L., Poisson, J., Cahagnier, B., Hahn,D., Barel, M., Santos, A.C., Café, Cacao, The (1974) 18: 121-132.
61. Santos, A.C., Hahn, D., Cahagnier, B., Drapron,R., Guilbot, A., Lefebvre, J., Multon, J.L., Poisson, J., Trentesaux, E., 5ème Colloque International sur la Chemie des Cafés (ASIC), 304-315, Lisbonne (1971).
62. Poisson, J., Cahagnier, B., Multon, J.L., Hahn,D., Santos, A.C., 7ème Colloque International sur la Chemie des Cafés (ASIC), Hamburg (1975) (in press).
63. Jordão, B.A., Garrutti, R.S., Angelucci, E., Tango, J.S., Tosello, Y., Colet. Inst. Tec. Alim. (1969-70) 3: 253-281.
64. Esteves, A.B., Estud. Agron. (Lisboa) (1960) 1: 297-317.
65. Pereira, M.J., Estud. Agron. (Lisboa) (1962) 3: 153-156.
66. Calle, H.V., Cenicafe (Colombia) (1963) 14: 187-194.
67. Gopal, N.H., Venkataramana, D., Ratna, N.G.N., Indian Coffee (1976) 40: 29-33.
68. Melo, M., Fazuolli, L.C., Teixeira, A.A., Amorim, H.V., Resumos Soc. Bras. Progr. Ciênc. Doc. 133-6.3, p. 846., (Brasília) (1976).
69. Teixeira, A.A., Fazuoli, L.C., Carvalho, A., Resumos Soc. Bras. Progr. Ciênc. Doc. 72-601, p. 777 (Brasília) (1976).
70. Bacchi, O., Bragantia (1962) 21: 467-484.
71. Ponting, J.D., Joslyn, M.A., Arch. Biochem. (1948) 19: 47-63.
72. Vasudeva, N., Gopal, N.H., J. Coffee Res. (1974) 4: 25-28.
73. Weaweaver, C., Charley, H., J. Food. Sci. (1974) 39: 1200-1202.
74. Official Methods of Analysis of The A.O.A.C. p. 217. 10th Edition, Washington (1965).
75. Amorim, H.V., Guercio , M.A., Cortez, J.G., Malavolta, E., An. Esc. Sup. Agr. "Luiz de Queiroz", USP (1973) 30: 281-291.
76. Baltes, W., 7ème Colloque International sur la Chemie du Cafè (ASIC), Hamburg (1975) (in press).
77. Reymond, D., 171st A.C.S. meeting, New York (1976).
78. Berndt, W., Meier-Cabell, E., 7ème Colloque International sur la Chemie des Cafés (ASIC), Hamburg (1975) (in press).

79. Thaler, H., Gaigl, R., Z. Lebensm. Untersuch. Forsch. (1963) 120: 357-363.
80. Oliveira, J.S., 5eme Colloque International sur la Chemie des Cafés (ASIC), p. 115-119, Lisbonne (1971).
81. Pokorny, J., Con, N.H., Bulantova, H., Janiced,G., Nahrung (1974) 18: 799-805.
82. Domont, G.B., Guimarães, V., Solewicz, E., Perrone, J.C., An. Acad. brasil. Ciênc. (1968) 40: 259 R.

4

Enzymatic Modification of Milk Flavor

W. F. SHIPE

Department of Food Science, Cornell University, Ithaca, N.Y. 14853

Until recently enzymatic action in milk has been regarded as undesirable. So most of the attention has been given to inactivating the enzymes, rather than using them. But research in the last few years indicate that enzymatic modification of milk may be feasible in some cases.

Enzymes reported to affect milk flavor are either oxidative or hydrolytic. Table I shows the oxidative enzymes and their real or potential effect.

Table I. Oxidative Enzymes Reported to Have Flavor Impact

1.	"Oleinase"	Enhances oxidation?
2.	Xanthine oxidase	Enhances oxidation?
3.	Lactoperoxidase	Enhances oxidation (nonenzymatically).
4.	Sulfhydryl oxidase	Reduces cooked flavor.
5.	Hexose oxidases	Increases acidity.
6.	Lactose dehydrogenase	Increases acidity.

In 1931 Kende (1) reported that an enzyme which he called oleinase was involved in oxidized flavor development. This conclusion was based primarily on the fact that oxidized flavor development was inhibited by heat treatment. It was assumed that the inhibition was a result of the inactivation of "oleinase". It was not recognized at that time that heat treatment tends to inhibit oxidation, presumably by producing active sulfhydryl groups. The existence of an oleinase was never confirmed but people are still searching for it.

It has been reported that xanthine oxidase contributes to spontaneous oxidized flavor development (2). Evidence has been presented which shows a good correlation between xanthine oxidase activity and production of TBA reactive components. Furthermore, thermal and chemical treatments known to inactivate xanthine oxidase also inhibited the development of oxidized flavor. However, there is no direct evidence to show that milk contains a substrate that is acted on by xanthine oxidase. Recently, it has been postulated (3) that xanthine oxidase produces superoxide anion in milk which may undergo non-enzymatic dismutation to form singlet oxygen which could catalyze lipid oxidation. Furthermore, it was postulated that the superoxide dismutase could inhibit this oxidation by catalyzing the conversion of superoxide anion to triplet oxygen and hydrogen peroxide. Superoxide anion has also been detected (4) in milk serum that had been exposed to fluorescent light. Regardless of the source of the superoxide anion, superoxide dismutase could possibly contribute to the milk antioxidant system. However, more research is needed to determine if either xanthine oxidase or superoxide dismutase have any impact on milk flavor. Incidentally, if xanthine oxidase does catalyze oxidation the enzyme can be inactivated by heating to 94°C following normal pasteurization. Milk plants with steam injection-vacuum processing equipment can easily produce xanthine oxidase free milk with only a slight cooked flavor.

Eriksson (5) reported that lactoperoxidase can catalyze oxidation by virtual of the heme iron that it contains. In other words, it is a non-enzymatic catalyst. I mention this because there may be other enzymes that have significant non-enzymatic effects. Consequently, when we use enzymes we must not ignore possible non-enzymatic effects of enzymes.

In 1967 Kiermeier and Petz (6) in Germany isolated a sulfhydryl oxidase from milk. This enzyme oxidizes sulfhydryl groups to disulphides. Sulfhydryl oxidase studies have also been conducted at North Carolina State and Cornell. Table II shows the effect of this enzyme on sulfhydryl groups and cooked flavor as reported by Siewright (7). As will be noted this enzyme can significantly lower cooked flavor. It has been suggested that sulfhydryl oxidase might be used to eliminate the strong cooked flavor in milk that is produced by ultra high temperature (UHT)

Table II. Effect of Sulfhydryl Oxidase on Sulfhydryl Concentration and Cooked Flavor.

Sample	Sulfhydryl Level[a]	Flavor Intensity[b]
Control	0.18	3.4
1% enzyme solution	0.09	1.6
2% enzyme solution	0.05	1.1

[a] Expressed as optical density. Ellman - DTNB Method
[b] Flavor intensity: 1 = slight → 4 = very strong cooked

treatment. However, one needs to determine whether the sulfhydryl oxidase treatment will make the milk more susceptible to oxidation by eliminating sulfhydryl groups which have anti-oxygenic properties. Since UHT treatment is usually given to prolong storage life and the risk of oxidized flavor reaching the detectable level increases with age, one should avoid lowering the oxidative stability of the UHT milk. To preserve the anti-oxygenic benefit and still eliminate cooked flavor, one might wait to add the sulfhydryl oxidase until after storage, i.e. just before consumption.

Rand (8) has studied two methods for conversion of lactose in milk to acid. In one method he first hydrolyzed the lactose to glucose and galactose with lactase and then converted the hexoses to the corresponding aldobionic acids with hexose oxidases. In the second method he used lactose dehydrogenase to produce lactobionic delta lactone which subsequently hydrolyzed to lactobionic acid. He reported that this latter method was less efficient than the first method. The rate of acid production with hexose oxidases was increased by the addition of catalase and H_2O_2 which supplied oxygen for regenerating the reduced enzymes. The enzymatic process changes the flavor and also gets rid of the lactose which is an additional advantage for those with lactose intolerance.

Glucose oxidase-catalase systems also have potential value as oxygen scavengers in products such as milk powders, dry ice cream mixes, coffee and active dry yeast (9). Since these products do not contain enough moisture for the enzymatic reaction it is necessary to provide water. This problem has been solved by putting glucose, glucose oxidase-catalase and sufficient moisture in a plastic envelope. This envelope should readily transmit gaseous oxygen but should restrict moisture passage. It has been reported that these enzymatic oxygen scavenger packets produced lower levels of residual oxygen than vacuum treatment followed by flushing with nitrogen. Although these packages have had only limited commercial use they do illustrate an ingenious approach to flavor control.

Table III lists the hydrolytic enzymes that have a flavor

Table III. Hydrolytic Enzymes with Flavor Impact

1.	Lipase	Causes hydrolytic rancidity.
2.	Lactase	Increases sweetness.
3.	Milk proteases	Increases bitterness?
4.	Phospholipase C	Inhibits oxidation.
5.	Phospholipase D	Inhibits oxidation.
6.	Trypsin	Inhibits oxidation.

impact. Of course, lipase heads the list because it is the naturally occurring enzymes that appears to have the greatest flavor impact. In fact two survey taken in the past year in New York City indicates that hydrolytic rancidity was the principal off-flavor in commercial milk samples. Therefore, the dairymen needs to know how to control lipolysis. Control in this case primarily involves keeping the enzyme away from the substrate. At the time of milk secretion the fat globule is fairly well protected from the lipase by the fat globule membrane. This membrane can be altered by thermal or mechanical abuse. Obviously the alteration is increased by increasing the abuse. The mere cooling of some milk from 37 to 5°C is enough to cause lipolysis presumably by lipase located in the membrane. Heating cooled milk up to 30°C and recooling to 5°C causes even greater increases in lipolysis. The effect of this so-called temperature activation is shown in Table IV. Apparently, these temperature changes causes disruption

Table IV. Effect of "Activation" Temperature on Lipase Action

Temperature °C	15	20	25	30	35	40
Acid degree values[a]	1.0	1.3	2.8	3.9	2.2	0.9

[a]As determined by method of Thomas, Nielsen and Olson (10).

of the fat globule membrane so as to increase the contact between fat and lipase. Mechanical agitation such as simple stirring or pumping can also expose the fat to the enzyme. The magnitude of this effect is dependent on the temperature as is shown in Table V. Obviously, the effect of mechanical abuse is less if the milk is kept cool. Vigorous agitation in a Waring Blendor or homogenizer produces even more drastic increases. The increase produced

Table V. Effect of Temperature and Shaking on Lipolysis

	Temperature	
	5°C	26°C
% Increase in lipolysis on shaking[a]	17	370

[a]As determined by Kurkovsky and Sharp (11).

by homogenization exceeds the amount of increase of fat globule surface area. Apparently, part of the increase is due to changes in the nature of the fat globule membrane. Fortunately, most of the lipase in milk is inactivated by pasteurization which is done either just before or after homogenization. Most lipolysis can be avoided if milk is not subjected to thermal or mechanical abuse and is pasteurized soon after production. However, bulk handling of milk and transporting it over as much as 300 or 400 miles makes it difficult to follow simple control measures. Furthermore, pasteurization does not completely inactivate all of the lipase (12) and in the case of homogenized milk, this can cause flavor problems. Consequently, other methods of controlling lipolysis are being sought. Lipase can be inactivated by adding H_2O_2, but such additions are not legal. Exposure to fluorescent light can inhibit lipolysis but this would be a questionable procedure because of the risk of producing light induced off-flavors. Shahani and Chandan (13) observed that non-fat milk solids inhibited purified milk lipase, presumably by absorbing on the surface of the substrate. The addition of solids would also increase the nutritive value of the product and thus would have a double benefit.

Additional evidence on the effect of additives was revealed in a recent study (14) on the effect of lecithin, casein and trypsin on lipolysis (Table VI). In this study, emulsions of 2% milk fat were stabilized with 0.5% casein + 0.5% bovine lecithin,

Table VI. Interaction of Lipase and Trypsin in Different Emulsions of Milk Fat

Additives	Free Fatty Acids (μeg/ml)[1] Without Trypsin	With Trypsin
Lecithin	1.18[a]	---
Lecithin + Casein	3.40[b]	2.50[b]
Casein	7.44[c]	15.51[d]
Casein + Lecithin	0.65[a]	0.87[a]

[1]Numbers with same superscript are not significantly different ($P < 0.05$)

or 0.5% of each. The emulsions were then incubated at 21°C for 60 min. with lipase or lipase and trypsin. Casein enhanced lipolysis when added alone or after lecithin, but it appeared to inhibit when added prior to lecithin. Since milk lipase is known to associate with casein, the casein absorbed on the fat may "attract" lipase to the fat surface. The addition of lecithin after the casein may block the absorption of the lipase. The fact that trypsin enhances lipolysis when casein alone was used, suggest that the unhydrolyzed casein does protect the fat to some extent. Regardless of the explanation of these results, they clearly illustrate that lipolysis is affected by the material that is absorbed on the fat globule.

So far I have been talking about inactivation or inhibiting lipolysis but in many products free fatty acids are desirable. Even good fresh milk contains some free fatty acids. But by comparison good butter contains 6 to 7 times more free fatty acids, cheddar cheese 4 to 5 times more and blue cheese 50 to 100 times more. Lipases play a dual role in blue cheese flavor production by releasing volatile fatty acids, some of which contribute directly to flavor and some serve as precursors of methyl ketones. The methyl ketones are major contributors to blue cheese flavor. Richardson and Nelson (15) have indicated that cheddar cheeses were organoleptically preferred when gastric lipase preparations were included in their manufacture. So lipolysis is often desirable if we can control the amount and kinds of fatty acids released. Milk and pancreatic lipases are fairly non-specific i.e. the ratio of fatty acids in the hydrolysate and the unhydrolyzed fat are essentially the same. By contrast pre-gastric (i.e. oral) lipases from calves and kids and fungal sources preferentially release high proportions of short chain fatty acids. But even the fungal enzymes exhibit considerable differences as is shown in Table VII. Arnold *et al*. (16) have reviewed the differences in specificity of various lipases and how different lipases can be used to produce lipolyzed milk fat with specific flavor characteristics. They pointed out that selectively hydrolyzed fats are added to a variety of foods including cereal products, margarines, salad dressings, coffee whiteners, soups, snack foods and confectionary products such as milk chocolates, creams and

Table VII. Relative Release of Free Fatty Acids by Two Fungal Lipases

Source of Lipase	Total free fatty acids (μ Equiv.)	Short chain fatty acids (%)
Penicillium roqueforti	110	38
Achromobacter	96	2

fudges.

Lactase is the next hydrolytic enzyme on the list but since this topic is covered in the next chapter, I will only take time to point out a unique use of this enzyme. A commercial company is currently producing packets of lactase which bears the fanciful name "Lacti-Aid". The consumer is instructed to add the Lact-Aid to a quart of milk and to store it in the refrigerator for 24 hours. During this time the lactose will be hydrolyzed and the sweetness will be increased. Certainly, the concept of having the consumer carry out the enzymatic treatment increases the potential use of enzymes. Since consumers are already using enzymatic meat tenderizers, home use of enzymes may become quite commonplace. Perhaps, we should consider preparing enzyme packets for use by the farmers. For example, we might be able to eliminate the lipase problem by having the farmer add the required amount of H_2O_2 to inactivate the lipase and then have him add sufficient catalase to get rid of the excess H_2O_2. Similarily, appropriate amounts of sulfhydryl oxidase might be packaged for use in bulk-dispensed milk such as is used in cafeterias. The sulfhydryl oxidase could be added far enough in advance of serving time to allow the enzyme to eliminate cooked flavor.

Milk has been known to contain proteases for some time, but there is no direct evidence to indicate that they contribute to flavor in fluid milk. Some scientists believe that they contribute to the flavor of raw milk cheeses. In fact, a recent report (17) indicated that milk proteases survived pasteurization and therefore, was also important in pasteurized products. Similar proteases from bacterial origin are known to produce bitter flavors, and perhaps in some cases bitter flavor in milk may be produced by them. This might particularly be true in milk that has been stored for several days.

The last three hydrolytic enzymes on the list have been shown to inhibit oxidation. The action of these enzymes are of especial interest because they illustrate how rather limited enzymatic action can have a significant flavor impact. Furthermore, it is worth noting that these three enzymes catalyze three different hydrolytic reactions yet all inhibit oxidation. This immediately raises two questions: First, why should hydrolysis inhibit oxidation? and second, do they have some common inhibitory mechanism?

Before trying to answer these two question, lets consider the individual enzymes, starting with phospholipases C and D.

If phosphotidylcholine is the substrate, phospholipase C yields a diglyceride plus phosphorycholine whereas D yields phosphatidic acid and choline. Skukla and Tobias (18) claimed that the phosphatidic acid produced by phospholipase D was responsible for its anti-oxygenic effect. On the basis of our results we have assumed that the anti-oxygenic effect of phospholipase C was due to a rearrangement of one or more of the fat globule membrane components. Presumably this rearrangement decreases the susceptibility of the lipids to oxidation by reducing the exposure of the substrate to pro-oxidants such as copper or iron. The inhibition of copper induced oxidation by phospholipase C, as reported by Young (19), is shown in Table VIII. Whereas, phospholipase C inhibits oxidation, Chrisope and Marshall (20) reported that it enhances lipolysis. Therefore, this enzyme should not be used in raw milk unless one is interested in increasing the free fatty acid content.

Table VIII. Effect of Phospholipase C Action[a] on Development of TBA Reacting Components[b]

	Cow Number		
Sample Identification	326	362	377
Control Milk	.02	.03	.04
Milk + 0.05 ppm Copper	.07	.09	.09
Milk + Copper + Enzyme	.03	.04	.04

[a]Incubated for 1 hr at 37°C, followed by thermal inactivation and storage for 3 days at 5°C.
[b]TBA values expressed as absorbance at 532 nm. Samples with TBA values greater than 0.04 usually have detectable oxidized flavors.

The use of trypsin to inhibit oxidation is perhaps the most indirect method for controlling milk flavor. Ever since the use of trypsin for this purpose was first reported (21) in 1939, scientists have been trying to elucidate the mechanism of its action. Since preliminary studies had indicated that tryptic action affected the fat globule membrane we isolated the membrane using the method of Bailie and Morton (22). Our results indicated that approximately 25% of the membrane material was released into the serum phase by the enzymatic action. Furthermore, SDS-polyacrylamide electrophoretic separations revealed (23) that the membrane material had been partially hydrolyzed by the trypsin. This suggests that the trypsin effect may be due to both a release of membrane material and protein hydrolysis.

It has been suggested that protein hydrolysis inhibits oxidation by increasing the copper binding capacity of the milk. Results of our studies (24) indicated that tryptic action does increase the copper binding capacity of the milk. Although there was some variability in the amount bound, all samples showed significant increases. Heat inactivated trypsin did not cause any increase in copper binding thus indicating that the effect was due to hydrolytic action rather than direct binding to the enzyme.

Results by Babish (25) showed that tryptic action reduced oxygen uptake in samples containing added copper (Table IX). The linearity of the data reveals the lack of an induction period which suggests a heme catalyzed oxidation. This theory is

Table IX. Effect of Trypsin Treatment on Oxygen Intake

	Time (hours)				
	0	12	24	36	48
Control Milk	6.4	---	---	---	5.2
Milk + Copper	6.2	5.5	4.0	3.4	2.7
Milk + Trypsin + Copper	5.7	5.4	4.9	4.7	4.4

supported by the observation that both histidine and antimycin A inhibited oxidation in milk. These two substances are capable of complexing with heme protein. Tryptic action may also affect the catalytic action of heme iron. Regardless of the mechanism of action trypsin does inhibit oxidized flavor development. The results of a typical experiment is shown in Table X. Our results (23) also indicate that trypsin treatment increased Vitamin A stability. This indicates a fringe benefit of the enzymatic treatment.

Table X. Effect of Trypsin Treatment on the Oxidative Stability of milk

Sample	TBA Values[a]	
	Trial 1	Trial 2
Control Milk	0.03	0.03
Milk + Copper	1.00	0.70
Milk + Trypsin + Copper	0.03	0.02

[a]Values represent absorbance at 532 nm, 3 days after treatment

The foregoing results were obtained with soluble trypsin but we have also used immobilized trypsin (26,27). Inasmuch as we only want to hydrolyze a small fraction of the protein in this treatment, the immobilized enzyme has the advantage of enabling us to carefully control the extent of hydrolysis. We have bound trypsin to both porous glass and tygon tubing. Although porous glass has much greater surface area for binding enzymes, the tygon tubing is easier to sanitize and does not plug up. Incidentally, we have used 10% ethanol to sanitize the reactor or tubing after each use. By bubbling air or nitrogen into the flow stream through the tygon tubing, turbulence is increased and better substrate-enzyme contact is obtained (26). The tygon tubing still does not provide as much contact but it is less likely to be plugged by the fat in whole milk. Skim milk is less apt to plug a column reactor and whey passes through quite readily. So it is more feasible to treat these two products with immobilized enzymes than whole milk. Swaisgood and associates (28,29) at North Carolina State have used immobilized sulfhydryl oxidase, rennin and a fungal protease to treat skim milk. A number of workers have prepared immobilized lactase for treating skim milk and whey.

In conclusion, enzymes can be used to improve milk flavor in three ways. Namely -

1. By producing or increasing desirable flavors such as sweetness.

2. By removing off-flavors such as strong cooked flavor.

3. By preventing development of off-flavors, such as oxidized flavor. In most cases oxidized flavor development is not a critical problem at the present time. However, it could become one if we adopt wholesale fortification of milk with iron or increase the polyunsaturated fatty acid content by feeding encapulated unsaturated oils.

Whether any of these treatment are used will depend on the seriousness of the flavor problem and the cost of the treatment.

Literature Cited

1. Kende S. Proc. IX Intern. Dairy Congr. (1931), Subject 3, paper 137.
2. Aurand, L. W. and Woods, A. E. J. Dairy Sci. (1959) 42, 1111.
3. Hicks, C. L., Korycka-Dahl, M. and Richardson, T. J. Dairy Sci. (1975) 58, 796 (Abstr.).
4. Korycka-Dahl, M. and Richardson, T. Abstr. 71st Ann. Mtg. Amer. Dairy Sci. Assn. (1976) p54,D49.
5. Eriksson, C. E. J. Dairy Sci. (1970) 53, 1649 (Abstr.).
6. Kiermeier, F. and Petz, E. Z. Lebensmittelunters u-Forsch. (1967) 132, 342.
7. Sievwright, C. A. M.S. Thesis, Cornell Univ., Ithaca, NY (1970).
8. Rand, A. G., Jr. and Hourigan, J. A. J. Dairy Sci. (1975) 58, 1144.
9. Reed, G. "Enzymes in Food Processing" pp386-390. Academic Press, New York (1966).
10. Thomas, E. L., Nielsen, A. J. and Olson, J. C., Jr. Amer. Milk Rev. (1955) 17, 50.
11. Krukovsky, V. N. and Sharp, P. F. J. Dairy Sci. (1938) 21, 671.
12. Harper, W. J. and Gould, I. A. XV Intern. Dairy Congr. (1959) I455.
13. Shahani, K. M. and Chandan, R. C. J. Dairy Sci. (1963) 46, 597.
14. Marshall, R. T. and Charoen, C. Abstr. 71st Ann. Mtg. Amer. Dairy Sci. Assn. (1976) p51,D40.
15. Richardson, G. H. and Nelson, J. H. J. Dairy Sci. (1967) 50, 1061.
16. Arnold, R. G., Shahani, K. M. and Dwivedi, B. K. J. Dairy Sci. (1975) 58, 1127.

17. Noomen, A. Netherlands Milk Dairy J. (1975) 29, 153.
18. Shukla, T. P. and Tobias, J. J. Dairy Sci. (1970) 53, 637.
19. Young, M. H. M.S. Thesis, Cornell Univ., Ithaca, NY (1969).
20. Chrisope, G. L. and Marshall,, R. T. J. Dairy Sci. (1975) 58, 794 (Abstr.).
21. Anderson, E. O. Milk Dealer (1939) 29(3)32.
22. Bailie, M. J. and Morton, R. K. Biochem. J. (1958) 69,35.
23. Gregory, J. F. and Shipe, W. F. J. Dairy Sci. (1975) 58, 1263.
24. Lim, Diana and Shipe, W. F. J. Dairy Sci. (1972) 55, 753.
25. Babish, J. E. M.S. Thesis, Cornell Univ., Ithaca, NY. (1974).
26. Senyk, G. F., Lee, E. C., Shipe, W. F., Hood, L. F. and Downes, T. W. J. Food Sci. (1975) 40, 288.
27. Lee, E. C., Senyk, G. F. and Shipe, W. F. J. Dairy Sci. (1975) 58, 473.
28. Brown, R. J., Poe, L. B., Kasper, G. A. and Swaisgood, H. E. Abstr. 71st Ann. Mtg., Amer. Dairy Sci. Assn. (1976) p49,D35.
29. Janolino, V. G. and Swaisgood, H. E. Abstr. 71st Ann. Mtg. Amer. Dairy Sci. Assn. (1976) p50,D37.

5

Use of Lactase in the Manufacture of Dairy Products

J. H. WOYCHIK and V. H. HOLSINGER

Eastern Regional Research Center, Agricultural Research Service, U.S. Department of Agriculture, Philadelphia, Pa. 19118

The potential for lactase application in the manufacture of dairy products has been recognized for many years. Lactase (β-D-galactosidase) hydrolyzes milk lactose into its constituent monosaccharides, glucose and galactose. Chemical and physical changes that occur as a result of lactose hydrolysis provide the rationale for its application. The principal changes are reduced lactose content, increased carbohydrate solubility, increased sweetness, higher osmotic pressure, reduced viscosities, and more readily fermentable sugar. Enzymatic hydrolysis of lactose in dairy foods would improve product quality and provide low-lactose products for the lactose intolerant segment of the population.

Available Lactases

Efforts to utilize lactases for the manufacture of hydrolyzed lactose (HL) products have been restricted primarily by the lack of suitable, commercially available, enzymes. Lactases are found in plants, animals and microorganisms (Table I), but only microbial sources can be used commercially. The pH optima of the microbial lactases are quite varied. The two enzymes of commercial value are isolated from the fungus, *Aspergillus niger* (1), and the yeast, *Saccharomyces lactis* (2), and differ widely in their properties, particularly in pH optimum (Table II). The *S. lactis* enzyme (pH optimum of 6.8-7.0) is ideally suited for the hydrolysis of lactose in milk and sweet whey (pH 6.6 and 6.1, respectively); lack of stability below pH 6.0 precludes its application to acid whey (pH 4.5). The *A. niger* lactase (pH optimum 4.0-4.5) has good pH stability, but the fall-off in relative activity at pH values above 4.5 limits its usefulness primarily to acid whey. Both of these lactases are available commercially in purified form.

Table I
Lactase Distribution

Plants

Phaseolus vulgaris, kefir grains, almonds, tips of wild roses, and seeds of soybeans, alfalfa and coffee.

Animal

Intestinal brush border (pH optimum 5.5-6.0).

Microbial

Bacteria (pH optimum 6.5-7.5)

Fungi (pH optimum 2.5-4.5)

Yeasts (pH optimum 6.0-7.0)

Table II
Properties of *A. niger* and *S. lactis* Lactases

	A. niger	*S. lactis*
pH Optimum	4.0-4.5	6.8-7.0
Temperature Optimum	55°C	35°C
pH Stability	3.0-7.0	6.0-8.5
Half-life (Days) at 50°C		
Whole Acid Whey	8	
5% Lactose	100	

Purity, availability, and cost of lactases are important considerations in any full-scale enzymatic process for lactose hydrolysis. Consequently, although lactose hydrolysis can be achieved most simply by the addition of soluble enzyme to milk or whey, immobilized enzyme technology has been evaluated with lactases in an effort to improve the economics of lactose hydrolysis (3,4,5,6). While efforts have been made to develop a satisfactory immobilized process for the *S. lactis* lactase (2), its stability following immobilization has not been sufficient to warrant its use. Problems associated with protein adsorption to a variety of enzyme support systems and maintaining acceptable column sanitation levels working with nutritive substrates, such as milk or sweet whey, further limit its usefulness. It would appear that batch treatment with soluble enzyme would be the method of choice for HL dairy products made with *S. lactis* lactase.

In contrast, the *A. niger* lactase has proven to be adaptable to use in immobilized forms and has been bound to a wide variety of solid supports (3,4,7,8,9). Depending on the substrates used and operating conditions, operational half-lives from 8 to 100 days have been obtained. Although some operational problems still exist, there is little doubt that the *A. niger* lactase can be used successfully in immobilized systems.

Applications

The hydrolysis of lactose in whole or skim milk, best accomplished by batch treatment with *S. lactis* lactase, can be achieved by incubation with 300 ppm lactase at 32°C for 2.5 hr (10), or at 4°C for 16 hr (11). Hydrolysis levels of 70-90% are obtained and the milk can be processed into a variety of products or used directly as a beverage. Some of the advantages and disadvantages in application of lactases in the manufacture of dairy products and processes follow.

Fluid and Dried Milk Products

Beverage Products. In recent years, numerous studies (12, 13,14) have established a pattern of lactose intolerance among non-Caucasian children and adults that is attributable to low intestinal lactase levels. Flatulence, cramps, diarrhea, and possibly a general impairment in normal digestive processes may accompany lactose intolerance. The inability of individuals to hydrolyze lactose can have serious implications in nutrition programs based on milk products.

The HL milk prepared in our laboratory for beverage use was evaluated in terms of its physical and organoleptic properties. The fact that HL milk is sweeter than normal milk posed some problems in defining its flavor in taste panel testing. Based on flavor scores (Table III) it was evident that a reciprocal relationship existed between the amount of lactose hydrolyzed in the product and its flavor score. It was clear that judges treated the marked sweetness as a "foreign" flavor. Additional taste panel studies showed that hydrolysis of 30, 60 and 90% lactose had the same effect on flavor of milk as the addition of 0.3, 0.6, and 0.9% sucrose. Paige et al. (15) reported that Negro adolescents found milk with 90% of its lactose hydrolyzed quite acceptable to drink, although the majority of the group surveyed judged it sweeter than control milk.

Table III
Flavor Scores of Hydrolyzed-Lactose Milk

% Lactose Hydrolyzed	Flavor Score
0	37.0
30	37.0
60	36.7
90	36.2

In addition to overcoming the problems of the lactose intolerant, hydrolysis of lactose in milk is advantageous for the preparation of milk concentrates. A highly attractive method of preserving whole milk is freezing a 3:1 concentrate. The products keep much longer than fluid milk under normal refrigeration and, when reconstituted, have a flavor virtually indistinguishable from fresh milk. However, concentrates prepared from normal milk have a tendency to thicken and coagulate on standing. This protein destabilization results from crystallization of lactose which has been brought to its saturation point in the concentration process. Early research (16) showed that lactose hydrolysis led to improvement in the physical stability of the concentrates during storage. However, hydrolysis of 90% of lactose alone did not increase storage stability by more than one month over the control (Figure 1). HL samples heat treated at 71°C for 30 min after canning showed only a moderate rise in viscosity after 9 months storage (lower curve). Organoleptic evaluation showed no difference in flavor score of the reconstituted concentrate with 90% of its lactose hydrolyzed and a fresh whole milk control with sucrose added. Similarly, lactose crystallization was avoided when skim milk concentrates were prepared from HL milk and stored at cold temperatures.

Dried Products. No problems were encountered in the manufacture of milk powder from HL whole milk. However, HL skim milk powder, with a major portion of its lactose hydrolyzed, had a tendency to stick to the hot metal surfaces of the dryer at temperatures significantly lower than regular skim milk sticks (60 *vs* 75°C) at comparable moisture levels. Thus, the surfaces of the powder collecting apparatus need to be held below 60°C to avoid the sticking problem.

Cultured Products. Cultured products have long been thought to contain low levels of lactose because of its fermentative utilization. However, while the lower levels of lactose may make these products more compatible to the lactose intolerant, the practice of fortifying yogurt with lactose results in residual levels as high as 3.3-5.7% (17). Gyuricsek and Thompson (18) reported that among yogurts prepared from lactase treated milk in which more than 90% of the lactose was hydrolyzed, the HL yogurts set more rapidly than controls did and had good acceptability. Advantages claimed for the application of lactase in yogurt manufacture include accelerated acid development, which reduces the required set-time, and a reduction of "acid" flavor by the glucose and galactose, which made the plain yogurt more acceptable to consumers.

Supplying starter culture organisms with glucose and galactose, rather than lactose, as an energy source can be expected to result in altered carbohydrate utilization patterns and metabolite production. O'Leary and Woychik (19,20) undertook a study to define the microbial and chemical changes that might occur in cultured dairy products manufactured from HL milk. They reported that typical sugar concentrations in the HL milk for yogurt are glucose, 2.6%; galactose, 2.3%; lactose, 2.0%. Decrease in pH accompanied fermentation in the control and HL milk; less time was required to reach pH 4.6 in the HL milk than in the control (Figure 2).

The faster acid development in the HL milk is a result of a higher initial rate of fermentation. These findings support the decreased set-times reported by Gyuricsek and Thompson (18). The growth curves of the mixed starter culture consisting of *Streptococcus thermophilis* and *Lactobacillus bulgaricus* shown in Figure 3 demonstrate that lactose hydrolysis has no effect on the symbiotic growth relationships of the starter organisms. The carbohydrate utilization patterns by the cultures in the two yogurts were different. The values in Table IV show that almost twice as much galactose was catabolized in the control milks as in HL milk. At the same time, the total amount of lactic acid produced by the starter organisms was greater in the HL milk. These findings reflect altered patterns of metabolite production resulting from the utilization of a greater proportion of the total available sugar in the form of glucose.

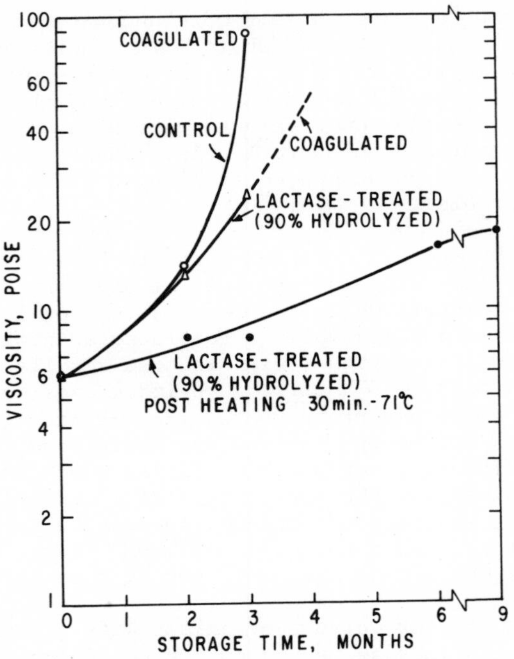

Figure 1. Effect of storage on viscosity of frozen 3:1 pasteurized whole milk concentrates

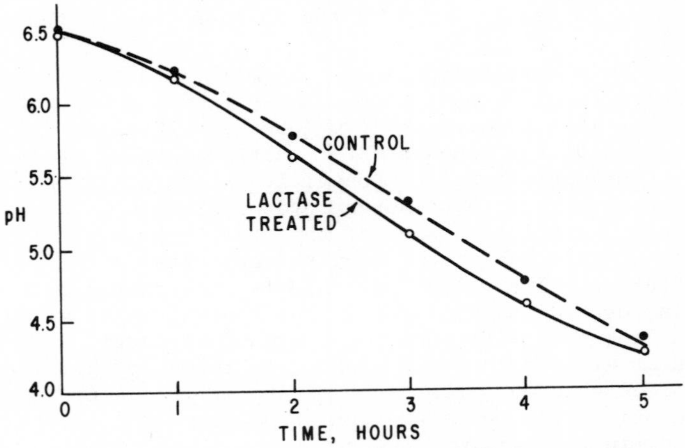

Figure 2. Change of pH with time during yogurt production

Table IV
Galactose Utilization in Control and HL Yogurts

	Control		HL	
	CHO (%)	Galactose (%)	CHO (%)	Galactose (%)
Initial	7.11	3.50	7.01	3.30
Final	5.23	2.73	5.20	2.95
% Utilized	1.88	0.77	1.81	0.35

The accelerated acid production by yogurt culture organisms in HL milk was also observed by Gyuricsek and Thompson (18) with cultures used in cottage cheese manufacture. Advantages of using HL milk reported (18) include reduced setting-time, less curd shatter and reduced loss of fines, a more uniform curd, and use of lower cook-out temperatures.

With the exception of improved starter cultures and more automated equipment, few modifications have been developed for improving the manufacture or accelerating the ripening of Cheddar cheese. Thompson and Brower (11) extended the use of HL milk to the manufacture of Cheddar cheese and found that lactose hydrolysis modified the process considerably. The faster acid development reduced renneting times and significantly reduced curd cooking and cheddaring times. With equivalent times in cure, the HL Cheddar was superior to control cheeses in flavor, body, and texture and more rapidly developed these characteristics closer to that of older control cheeses. The improved qualities were attributed to the effects of lactose hydrolysis. The body and texture changes may be the result of increased osmolarity accompanied by reduced casein solvation. The accelerated ripening may be attributable to increased levels of enzymes released into the cheese by higher bacterial populations when glucose is available as a growth substrate.

Whey Utilization

The nation's cheese industry annually produces over 32 billion pounds of whey, of which little more than half is utilized. In the case of lactose, this excess production over utilization represents approximately 600 million pounds which could conceivably be converted into salable products. Modification of whey by lactase hydrolysis could offer new avenues for whey utilization. Two of the most promising processes appear to be conversion to lactose-derived sirups and fermentative utilization.

Sirups. The low sweetness level of lactose relative to sucrose does not allow much direct competition; however, a considerable increase in sweetness results following hydrolysis of lactose. Relative sweetness is dependent on several factors, the principal one being concentration. Table V lists the sweetness of several sugars at 10% concentrations relative to sucrose.

Table V
Relative Sweetness of 10% Aqueous Sugar Solutions

Sugar	Sweetness (%)
Sucrose	100
Lactose	40
Glucose	75
Galactose	70

Thus, the availability of HL wheys increases the potential for producing lactose-derived sirups. HL sirups have been prepared by Guy (21) using both hydrolyzed wheys and lactose solutions. Deproteinized and demineralized sirups concentrated to 65% total solids exhibited good solubilities, showed no mold growth, and had good humectant properties. A ready application for these sirups was found in caramel manufacture where the HL sirup showed the same stabilizing effect against sucrose crystallization as did "invert" sugar. Further applications for these sirups await development by food technologists.

Fermentation. Utilization of whey as a fermentation substrate, especially for single cell proteins (22,23,24) and alcohol production (25), has been studied extensively. A major obstacle to broader utilization of whey as a fermentation medium has been the fact that relatively few organisms are able to utilize lactose. Several yeasts were evaluated for alcohol production in whey (26) and, although *Kluyveromyces fragilis* was found to be the most efficient, only 55% of the lactose was converted to alcohol. The low conversion level has been attributed

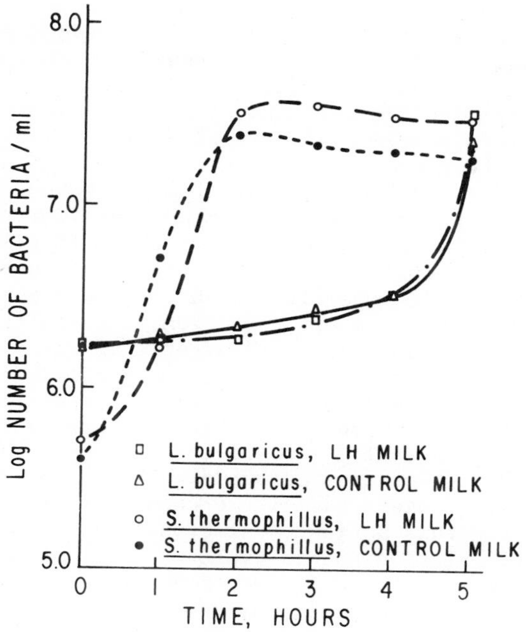

Figure 3. Growth of yogurt cultures in control and HL milks

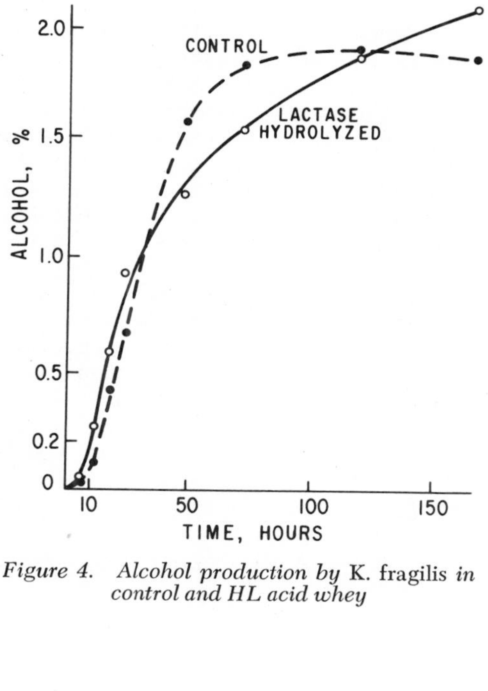

Figure 4. Alcohol production by K. fragilis *in control and HL acid whey*

to a possible inability of the yeast to tolerate high alcohol concentrations possibly because of sensitivity of the yeast lactase to alcohol. The availability of whey having its lactose hydrolyzed to its constituent monosaccharides permitted an evaluation of nonlactose fermenting organisms such as *Saccharomyces cerevisiae*, which tolerate high alcohol concentrations, for alcohol production. O'Leary *et al*. (*27*) evaluated HL wheys using *S. cerevisiae* and *K. fragilis* for comparison. Curves illustrating alcohol production by glucose-pregrown *K. fragilis* in control and HL wheys are shown in Figure 4. Ethanol production was more rapid in HL whey during the first 24 hr fermentation, but later decreased so that the rate of production was less than that in the control whey. The total yield of ethanol was similar in both cases and averaged 1.9%. Reducing sugar levels decreased at the same rate in the first 24 hr fermentation, but less utilization occured in HL wheys in the later stages of fermentation. The pattern of sugar utilization in the HL wheys (Figure 5) showed that although glucose disappeared rapidly during the first 24 hr fermentation, there was little change in the galactose concentration during this period. Galactose utilization began only after 24 hr fermentation and is typical of a diauxic pattern of sugar utilization. A longer fermentation period was required in the HL whey due to the diauxic phenomenon.

S. cerevisiae, a nonlactose fermenting yeast, is commonly used in wine and beer production because of its ability to withstand relatively high alcohol concentrations. Growth curves of *S. cerevisiae*, pregrown on glucose and on galactose, showed that similar cell populations existed in HL whey for the first 48 hr growth, but declined thereafter, except that those pregrown on galactose increased significantly after the intial decline. Alcohol production by these organisms in the HL wheys showed a similar pattern in that comparable levels of alcohol were attained after 48 hr (Figure 6). Alcohol production by the glucose pregrown *S. cerevisiae* began to decrease after 48 hr, whereas the galactose pregrown cells continued their alcohol production which reached a level twice that of the glucose-pregrown cells. Sugar utilization closely paralleled alcohol production. *S. cerevisiae* pregrown on glucose utilized glucose rapidly but did not utilize galactose. *S. cerevisiae* pregrown on galactose similarly utilized glucose but also utilized galactose after 96 hr. These studies indicated that although lactose hydrolysis in whey permits alcohol production by organisms unable to ferment lactose, the galactose released by hydrolysis is not efficiently utilized. In model system studies, the efficiency of converting sugar to alcohol by *K. fragilis* was as follows: glucose > lactose > galactose.

Conclusions

The dairy industry today has available existing technology for the application of enzymatic hydrolysis of lactose in milk

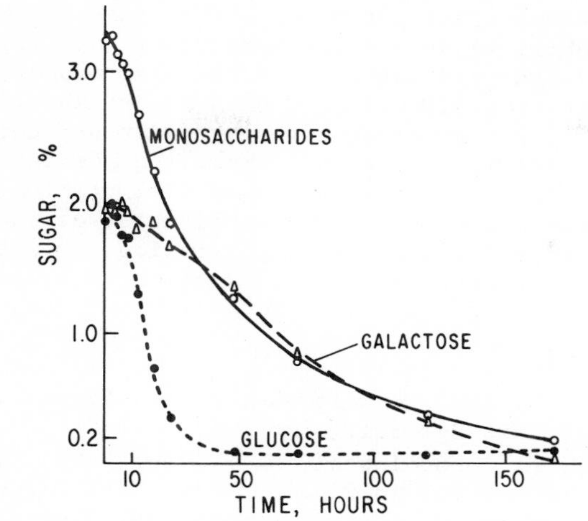

Figure 5. Change in sugar concentrations during fermentation of HL acid whey by K. fragilis

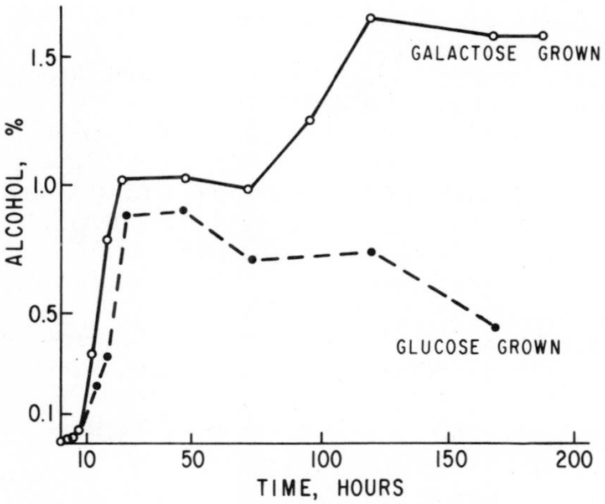

Figure 6. Alcohol production in HL acid whey by glucose-pregrown and galactose-pregrown S. cerevisiae

and milk products. Utilization of the lactase enzyme, in either "free" of immobilized systems, can result in quality improvements in a number of products and yield processing economies in others.

The application discussed in this report should be reviewed and assessed for potential benefits by dairy product manufacturers in their specific operations. Whey, in particular, with its ever increasing production, poses a special problem whose solution can be attained only through application of unconventional technology. It is believed that application of lactase offers a potential for innovative whey utilization.

Literature Cited

1. Woychik, J. H., and Wondolowski, M. V. Biochim. Biophys. Acta (1972) 289, 347-351.
2. Woychik, J. H., Wondolowski, M. V., and Dahl, K. J. In "Immobilized Enzymes in Food and Microbial Processes," Olson, A. C. and Cooney, C. L., editors, 41-49, Plenum Press, New York, N. Y. (1974).
3. Woychik, J. H., and Wondolowski, M. V. J. Milk Food Technol. (1973) 36, 31-33.
4. Wierzbicki, L. E., Edwards, V. H., and Kosikowski, F. V. Biotechnol. Bioeng. (1974) 16, 397-411.
5. Pitcher, W. H. Jr. Am. Dairy Rev. (1957) 37, 34B, 34D, 34E.
6. Pastore, M., Morisi, F., and Viglia, A. J. Dairy Sci. (1974) 57, 269-272.
7. Hustad, G. O., Richardson, T., and Olson, N. F. J. Dairy Sci. (1973) 56, 1111-1117.
8. Olson, A. C., and Stanley, W. L. J. Agri. Food Chem. (1973) 21, 440-445.
9. Hasselburger, F. X., Allen, B., Paruchuri, E. K., Charles, M., and Coughlin, R. W. Biochem. Biophys. Res. Commun. (1974) 57, 1054-1062.
10. Guy, E. J., Tamsma, A., Kontson, A., and Holsinger, V. H. Food Prod. Develop. (1974) 8, 50-60.
11. Thompson, M. P., and Brower, D. P. Cultured Dairy Prod. J. (1976) 11, 22-23.
12. Paige, D. M., Bayless, T. M., Ferry, G. C., and Graham, G. C. Johns Hopkins Med. J. (1971) 129, 163-169.
13. Rosensweig, N. S. J. Dairy Sci. (1969) 52, 585-587.
14. Lebenthal, E., Antonowicz, I., and Schwachman, H. Amer. J. Clin. Nutr. (1975) 28, 595-600.
15. Paige, D. M., Bayless, T. M., Huang, S., and Wexler, R. ACS Symposium Series, "Physiological Effects of Food Carbohydrates," (1974), 191.
16. Tumerman, L., Fram, H., and Cornely, K. W. J. Dairy Sci. (1954) 37, 830-839.
17. Anonymous. Cultured Dairy Prod. J. (1973) 8, 16-19.
18. Gyuricsek, D. M., and Thompson, M. P. Cultured Dairy Prod. J. (1976) 11, 12-13.

19. O'Leary, V. S., and Woychik, J. H. J. Food Sci. (1976) 41, 791-793.
20. O'Leary, V. S., and Woychik, J. App. Environmental Microbiol. (1976) 32, 89-94.
21. Guy, E. J. Abstract 71st Annual Meeting, Amer. Dairy Sci. Asso. (1976), D7.
22. Wasserman, A. E., Hopkins, W. J., and Porges, N. Sewage and Industrial Wastes (1958) 30, 913-920.
23. Wasserman, A. E., Hampson, J. W., Alvare, N. F., and Alvare, N. J. J. Dairy Sci. (1961) 44, 387-392.
24. Bernstein, S., Tzeng, C. H., and Sisson, D. First Chem. Congr. N. Amer. Cont. (1975) Abstract BMPC 68.
25. Rogosa, M., Browne, H. H., and Whittier, E. O. J. Dairy Sci. (1947) 30, 263-269.
26. O'Leary, V. S., Green, R., Sullivan, B. C. and Holsinger, V. H. First Chem. Congr. N. Amer. Cont. (1975) Abstract BMPC 158.
27. O'Leary, V.S., Sutton, C., Bencivengo, M., Sullivan, B., and Holsinger, V. H. Unpublished.

6

Immunochemical Approach to Questions Related to α- and β-Amylases in Barley and Wheat

J. DAUSSANT

C.N.R.S., Physiologie des Organes Végétaux, 92190 Meudon, France

I. Introduction

In food and beverage processing α- and β- amylases have an important function in two main areas : beer and bread making. In these industries, there is still a need for defining the source and the quality of barley, malt, and wheat.

Among the biochemical parameters used for characterizing these products, α- and β- amylases are of great interest : the activities are used for defining the seed quality and a better understanding of the isoenzymic contents seems promising for characterizing seeds. However, β- amylase activities and β- amylase zymograms are difficult to define when α- amylase is present.

The immunochemical characterization of α- and β- amylases which provided new information on barley and wheat amylases in physiological studies may also be useful in the above mentioned characterizations.

This article will present recently developed techniques :

1) for studying qualitatively and quantitatively β- amylases in presence of α- amylases, and

2) for identifying the origin of α- amylase sometimes present in small amounts in mature and apparently ungerminated wheat seeds.

II. Quantitative and Qualitative Techniques for β- Amylase Investigations in the Presence of α- Amylase

A. The problem. The characterization of amylases is particularly important for evaluating malt used in brewing. The activities of both α- and β- amylases and their ratio characterize malt ; the data are of major importance for the brewing process. Electrophoretic analysis of α- and β- amylase constituents may provide means of getting a deeper insight into the quality and the origin of the malt. However, the measurement and the detection of these activities remain approximate. In fact, α- amylase may be characterized in the presence of β- amylase by using starch preincubated with β- amylase. This procedure is used for quantitative as well as for qualitative techniques such as electrophoresis. Nevertheless, for quantitating β- amylase activity by measurement of reducing sugars that appear upon incubation of starch with the enzyme, the evaluation is approximate when α- amylase is present. In fact, α- amylase also produces reducing sugars by incubation with starch, and there is a synergistic effect between α- and β- amylases. On the other hand, in electrophoretic analysis of β- amylase constituents, some of them may be hidden by α- amylase components.

An inhibition of α- amylase and not of β- amylase could then be used : α- amylase requires Ca^{++} for its activity which is not the case for β- amylase. Thus, the treatment with polyphosphates, citrate or EDTA which inhibits the α- amylase activity (1) could solve the problem although α- amylases which may not require Ca^{++} have been reported (2). We shall describe here techniques involving the selective absorption of α- amylase constituents by using an immune serum specific for α- amylase ; these techniques permit qualitative as well as quantitative analysis of β- amylases in presence of α- amylases. For quantitation of β- amylase, we have developed a diffusion technique similar to a diffusion technique developed for α- amylase determination (3).

β- amylase and, to a smaller extent, α- amylase are damaged by the roasting process of malting ; both enzymes being altered in their activities. The results of previous immunochemical studies on barley and malt β- amylases indicated that malting, and in particular roasting, do not affect much the antigenic structure of the enzyme. An immunochemical measurement of the enzymatic protein contents in barley and malt extracts could thus be carried out (4). On the basis of immunochemical characterization of barley β- amylase, techniques for quantitating the enzyme are mentioned. Data concerning β- amylase activity, even in presence of α- amylase, and β - amylase protein content provide a means of obtaining β- amylase specific activity. This parameter could than serve for further characterizing effects of roasting in the malting process.

B. <u>Combined immunoabsorption and α - amylase determination in the same gel medium (5, 6).</u> α- amylase solutions were poured into wells cut in an agarose gel containing 1,5 % starch preincubated with β- amylase. The enzymes were allowed to diffuse for 24 h at 20°C and the gels were stained with an iodine solution. The α - amylase activity was reflected by white spots occuring on a red background. There is a linear relationship between the diameter of the reaction spots and the logarithm of the α- amylase concentration (3).

The immune serum anti α - amylase of germinated seeds was the one described in section III, B. α- amylase purified by an affinity technique (7) from extract of germinated barley seeds was used for immunization.

For immunoabsorption the wells were first filled with the immune serum and when the serum was sucked in by the gel they were refilled with the malt extract. During diffusion the α- amylase meets the antibodies which are all around the wells and precipitate, or are inhibited by an excess of antibodies.

Fig.1 shows the results of the immunoabsorption of α - amylase extracted from 7 days germinated barley seeds. Different dilutions of the immune serum were tested for the immunoabsorption of α- amylase from the malt extract. Dilution 3 of the immune serum absorbed practically all of the α- amylase in the malt extract. This dilution of the immune serum was then used for eliminating α- amylase from the malt extract, in order to measure its β- amylase activity. It is noteworthy that, in order to eliminate the small α- amylase-like activity present in the immune serum, the latter was heated at 50°C for 30 min before use.

C. Diffusion technique for β- amylase determination in agarose gel in the absence of α- amylase. This technique is similar to the one developed by Briggs (3) for α - amylase determination ; starch is used as substrate instead of limit dextrins (Fig.2 upper part). Different dilutions of barley extract were poured into

Fig.1 - Immunoabsorption of α- amylase from malt extract and evaluation of the remaining activity after absorption. This experiment shows that dilution 3 of the anti α- amylase immune serum absorbs practically all α- amylase present in the malt extract.

Upper and middle part : wells were cut in a 1.2 % agarose gel prepared in 0.2 M acetate buffer, pH 5.7, containing 1.5 % starch preincubated with β- amylase. Upper part : wells were filled with different concentrations of α- amylase (dilution series). In the initial malt extract, the α- amylase concentration was arbitrarily designated as 100. Middle part : For α- amylase immunoabsorption, different dilutions of the immune serum were first deposited in the wells. After these solutions were sucked in by the gel the initial malt extract (conc. 100 of the enzyme) was deposited in the wells. After 24 hr diffusion at 20°C, the gel was stained with iodine. Note that the reaction spot corresponding to the immunoabsorption of the malt extract with dilution 3 of the immune serum (IS/3) is much smaller than the reaction spot corresponding to concentration 0.2 of the malt α- amylase. Lower part of the figure : Relationships between diameter of reaction spot and logarithm of α- amylase concentration. The points represent the mean values of two measurements made on each of three reaction spots corresponding to one experiment. Absorption with dilutions 27 and 9 of the immune serum resulted in the absorption of only part of the initial activity (respectively about 40 % and 75 %). Dilution 3 of the immune serum absorbs more than 99.8 % of the initial activity.

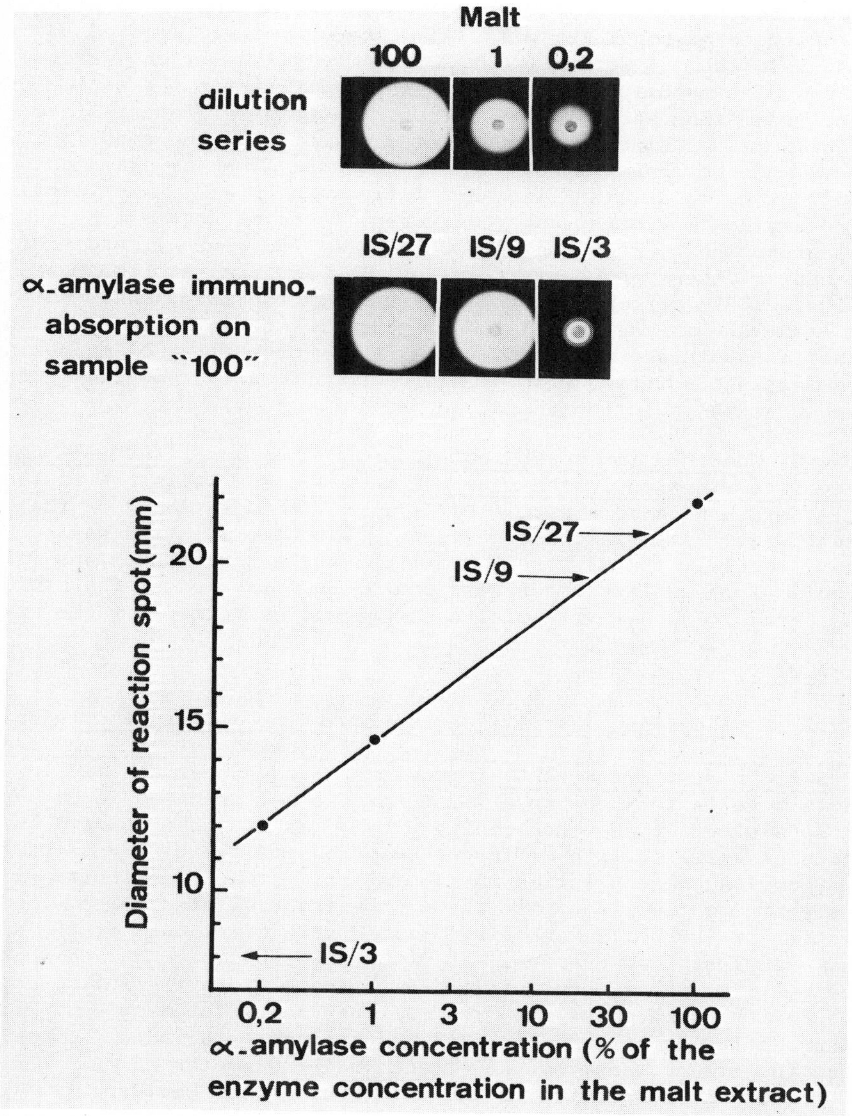
Malt
100
1
0,2
dilution series
IS/27
IS/9
IS/3
α-amylase immuno-absorption on sample "100"
Diameter of reaction spot (mm)
20
15
10
IS/27
IS/9
IS/3
0,2
1
3
10
30
100
α-amylase concentration (% of the enzyme concentration in the malt extract)

Figure 1

3 mm diameter wells cut in agarose gel, containing 0.8 % starch (see legend of Fig.2 for details). The enzyme was allowed to diffuse into the gel for 24 h. at 20°C. The gel was then stained with iodine solution. The β- amylase activity in the barley extract solutions was reflected by pink spots of different sizes, occurring on a blue background. The diameter of the reaction spots and the logarithm of the corresponding β - amylase concentrations correlate in a linear relationship (Fig.2, lower part). β - amylase may exist under different forms of aggregation which are disrupted by reducing agents (see (4) for review). In order to convert these aggregate forms of β - amylase into their basic units, 0.03 % mercapto ethanol was used in the solutions. When α - amylase is present, as is the case in the malt extracts, the reaction spots are white and there is no possibility of measuring β - amylase activity without having specifically removed α - amylase from the solution (Fig.2, upper part, Malt).

D. Combined immuno-absorption of α - amylase, and β - amylase determination in the same gel medium. The wells cut in the gel containing starch were first filled with dilution 3 of the heated anti α - amylase immune serum. When the solution was sucked in by the gel, the different dilutions of barley extract and dilution 2 of the malt extract were poured into the wells (Fig.2 middle part). The activity of the enzyme was reflected for the ex-

Fig.2 - β- amylase activity estimation using a diffusion technique which also involves the specific elimination of α - amylase present in malt extracts by immunoabsorption. Upper and middle parts : wells were cut in a 1.2 % agarose gel prepared in 0.1 M phosphate buffer pH 6 containing 0.8 % starch. Upper part : The wells were filled with different dilutions of the barley extract (β- amylase concentration in the initial extract was arbitrarily designated as 100), or with the malt extract diluted twice. Middle part : The wells were first filled with dilution 3 of the anti α - amylase immune serum. When the solutions were sucked in by the gel, the wells were refilled with the different dilutions of the barley extract, or with the malt extract diluted twice. After 24 hr diffusion at 20°C, the gel was stained with iodine. Note that the immune serum has no effect on the diameter of the reaction spots corresponding to the dilutions of the barley extract. With the malt extract, the absorption allows one to get a reaction spot corresponding to β- amylase and not to α- amylase. Lower part : Relationship between diameter of reaction spot and logarithm of β- amylase concentration in barley extracts. Each point represents the mean value of two measurements made on each of the three reaction spots corresponding to one experiment. β - amylase activity in the twice diluted malt extract corresponds to 60 % of the activity in the barley extract.

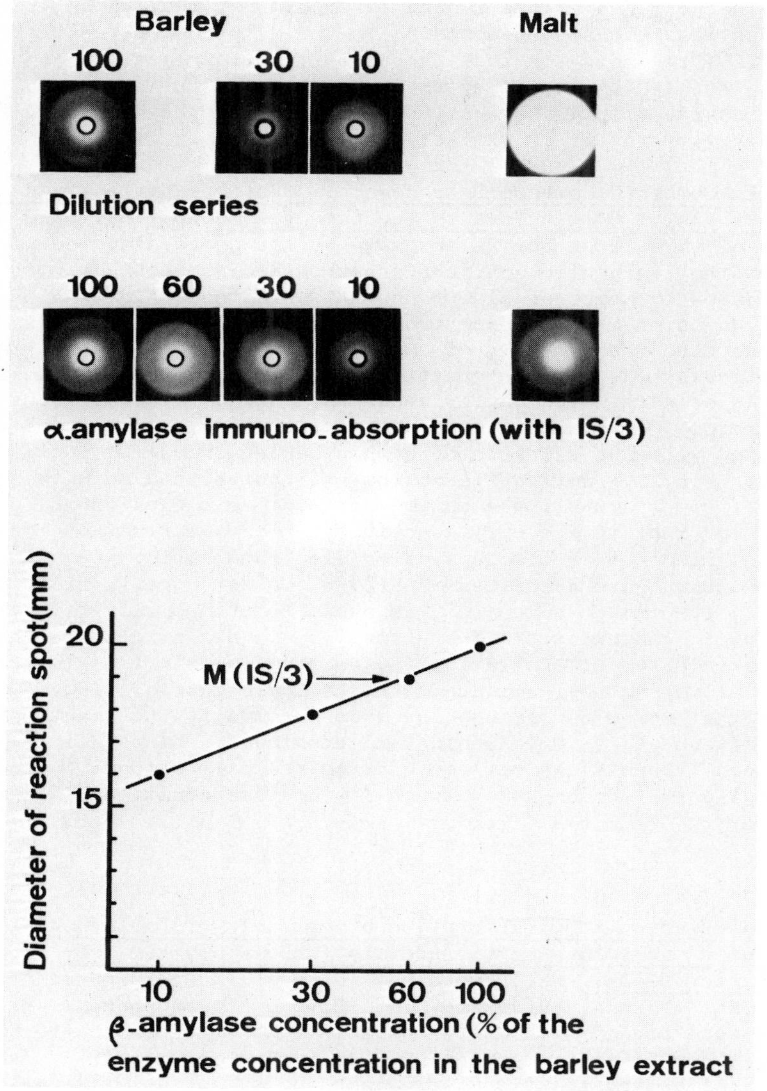
Barley
Malt
100
30
10
Dilution series
100
60
30
10
α.amylase immuno_absorption (with IS/3)
Diameter of reaction spot(mm)
20
15
M (IS/3)
10
30
60
100
β_amylase concentration (% of the
enzyme concentration in the barley extract

Figure 2

tracts of barley and malt by pink spots on a blue background.

The presence of the immune serum did not modify the size of the spot diameters corresponding to the dilutions of the barley extract. This indicates that the activity measurement of β- amylase was still possible under these conditions. β- amylase activity in the twice diluted malt extract was found to be nearly 60 % of the activity in barley extract.

E. Combined immunoelectroabsorption of α - amylase and electrophoresis of β - amylase in the same gel medium (8). The principle of this technique is the same as the preceeding technique, except that enzymatic antigens are brought into contact with the antibodies by means of electrophoresis, instead of diffusion. The technique aims at providing from a mixture of α- amylase and β - amylase, electropherograms of β - amylase alone by eliminating specifically α- amylase constituents : the immune serum anti α - amylase of germinated barley seeds was poured into wells cut in the agarose gel ; when it was sucked in by the gel, the wells were refilled with the extracts of either barley or malt. During electrophoresis, α- amylase meets the antibodies, and when these are in sufficient amount, the enzyme-antibody complexes form a precipitate or inhibit the enzyme activity. Characterization reactions were carried out on the gels after electrophoresis. Details of the technique are described in Fig.3. On the upper part of the figure, the immune serum (IS) is shown to absorb all α- amylase present in the extract. The absorption is specific since the serum taken before immunization (S) does not modify the electrophoretic pattern of α- amylase. On the lower part of figure 3 it is shown that serum and immune serum do not modify the electrophoretic pattern of β- amylase in barley extract. Since all α- amylase constituents can be removed from the malt extract by immunoabsorption, a comparison between β- amylase constituents in barley and malt can be made. A clearcut diffe-

Fig.3 - Electro-immunoabsorption of extracts of barley and barley malt using an immune serum specific for α - amylase from germinated barley seeds. Wells were cut in an 1.2 % agarose gel prepared in 0.025 M veronal buffer pH 8.6. Before electrophoresis the wells were filled with different solutions : veronal buffer (V), serum taken before immunization (S), immune serum anti α- amylase (IS). After the solutions were taken up by the gel, the wells were refilled with either the barley or the malt extract. After the solutions were taken up by the gel, the wells were filled with hot liquid agarose, and electrophoresis was carried out for about 5 hr at 3°C, starting with 6 V/cm. After electrophoresis, characterization reactions were carried out on the gels using limit dextrins for α- amylase characterization and starch for α- and β- amylase characterization.

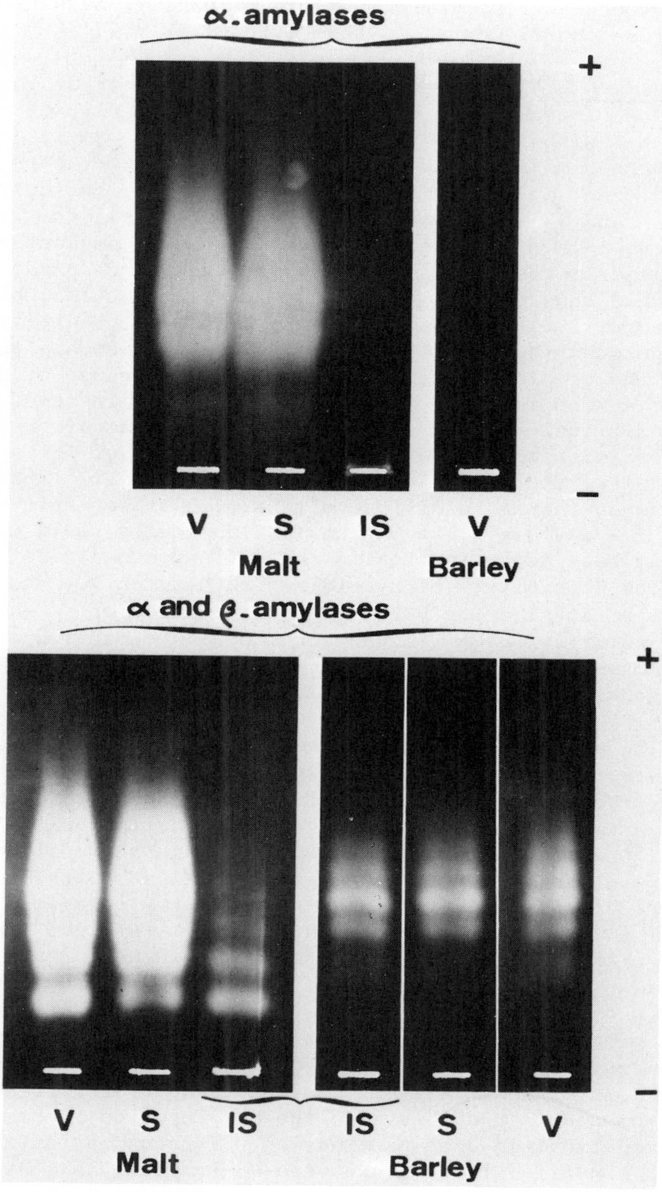
α. amylases
+
−
V
S
IS
V
Malt
Barley
α and β. amylases
+
−
V
S
IS
IS
S
V
Malt
Barley

Figure 3

rence in the electrophoretic mobility of barley and barley malt β - amylase constituents was thus evinced.

F. Remarks. 1) the non detectable reactivity of the anti α - amylase immune serum with β - amylase of barley extract indicates that with this immune serum, α - and β - amylases display no antigenic similarities. This confirms that they correspond to quite different molecular species. However, if in the malt extract, α - and β - amylases form hybrid molecules, the immunoabsorption of α - amylase could act on the β - amylase analyses. The complexes formed by α - amylases and the corresponding antibodies would thus contain those β - amylase molecules which are hybridized with α - amylases. If the complexes precipitate, in both techniques the β - amylase molecules would escape the analysis. If the complexes still remain soluble, because of a large excess of antibodies, the β - amylase contained in the complexes may be inhibited. In any case, the physico-chemical properties of the complexes are quite different from those of β - amylases. In particular, the complexes are of much larger size than β - amylase and must therefore diffuse with a much slower rate in the gel than β - amylase. The β - amylase molecules would then escape the β - amylase determination in the diffusion quantitative technique. The possibility of hybridization between α - and β - amylases must be studied further. Immunoabsorption of α - amylase using an anti β - amylase immune serum could be carried out ; a diminution of α - amylase activity would evidence that α - amylase is enclosed in β - amylase-antibody complexes.

2) The difference in electrophoretic mobilities between β - amylases from barley and malt shown by this technique confirms the preceding results obtained using immunoelectrophoretic analysis. β - amylases from barley and malt appeared antigenically identical using various immune sera, whereas one immune serum allowed the observance of antigenic similarities as well as antigenic differences (see (4) for review). The same observations were reported for wheat β - amylases. The causes of the β - amylase modifications were investigated by experiments combining in vivo labelled amino acids incorporation into proteins and immunochemical identification of β - amylase. It was found that the differences were due to a process modifying the enzymes existing in mature wheat upon germination rather than to a disappearance of these enzymes and synthesis of new β - amylase molecules during germination (9). This phenomenon is probably also the one involved in barley β - amylase modification upon germination. It remains to be seen whether each constituent of mature seed β - amylase is converted into a specific constituent in germinated seed β - amylase.

3) We described a technique for evaluating β - amylase activity in the presence of α - amylase by specifically removing α - amylase from the mixture. Immunochemical studies on barley

and malt β- amylases allowed one to envisage immunochemical quantitation of β - amylase proteins in protein mixtures, such as barley and malt extracts. Both evaluations provide means for estimating specific β - amylase activity in the extracts (see (4) for review) : β - amylase from barley and malt appeared to be antigenically identical using various immune sera, whereas with other immune sera, antigenic differences between these enzymes were detected. Only the first sort of immune serum could be used for comparing amounts of β - amylase proteins in barley and malt extracts. With the immunoabsorption technique, all β - amylase constituents of barley and malt were shown to react with the immune serum. Thus, all β- amylase constituents could actually be quantitated at once. Techniques involving a migration of the enzyme in a gel containing the immune serum could be applied. The migration may be promoted by diffusion (10) or by electrophoresis (11), the latter technique for quantitation of one α - amylase constituent is described in the next section.

The study of specific activity of β - amylase during malting and especially during kilning is thus possible for further characterization of malt.

III. Immunochemical Identification of α - Amylase in Wheat Seeds

A. The problem. For breadmaking, too high a level of α - amylase activity in wheat constitutes a major concern in Europe (12, 13). α - amylase activity in mature wheat often occurs in northern and western Europe as a consequence of the moist and temperate climate prevailing during the ripening period of the grain (14). This activity may result from the presence of sprouted grains in the crops, but not necessarily so ; the development of α - amylase need not be associated with sprouting (12). In fact, the correlation between visible sprouted grains and α - amylase activity is poor (15). It was observed that the enzyme level in absence of visible sprouting may already be much higher in sprouted seeds than in sound grains (16). An increase in α - amylase activity in wheat seeds during storage and without apparent sprouting was also reported (17).

Several sorts of α- amylases may be accounted for by the activity in sound and apparently ungerminated seeds : enzymes exogenous to the seeds (fungi, bacteria) or enzymes endogenous to the seeds. In the latter case, several isoenzymes may be involved and it is worthy to recall briefly the feature of α- amylase in developing, maturing and germinating wheat seeds. α- amylase activity present in developing seeds vanishes during maturation and reoccurs dramatically during germination (18, 19, 20). A set of α - amylases present in developing seeds appears to be electrophoretically and antigenically identical to α- amylases of germinated seeds (21, 22, 23, 24). Another set of α - amylases in developing seeds appears to be specific for the developing seeds (21) whereas the major part of the activity in germinating seeds seems to be due to enzymes not present in developing seeds (21, 22, 23, 24, 25). α- amylases occurring in cereals during germination are known to be synthesized in the aleurone layer, the enzymes found in developing seeds, however, have been localized in the pericarp (13, 19, 20, 26, 27, 28). It is worth noting that α- amylase found in endosperm and aleurone during late stages of development was associated with damaged areas of the kernel (26). One may speculate that amylase secreted in the aleurone layer may, in vivo slightly diffuse into the starchy endosperm. Thus, it could modify part of the starch, whereas the α- amylase located in the pericarp would have less opportunity to attack in vivo the starchy endosperm. The difference in the localization of the two types of α- amylases and the differences in the specific activity of these enzymes may explain some cases concerning triticale, where both falling number and activity levels were found to be high (29). The disappearance of α- amylase during maturation seems more probably due to a destruction of the enzyme than to an inactivation or to an insolubilization (30, 31).

Further evidence was provided, indicating that the greatest proportion of α- amylase which appears during germination and is identical to α- amylase of developing seeds, results more probably from in vivo synthesis than from a regeneration of preexisting enzymes (21, 30).

The problem concerning α- amylase activity in wheat is quite different in countries with a more arid climate compared to northwestern Europe, and flour must often be complemented with amylases of exogenous origin (14). Nevertheless, the sort of α- amylase added to flour (enzymes from germinated wheat or barley seeds, from fungi, or from bacteria) is not indifferent to the process of breadmaking and to the quality of the end product, due particularly to the different thermostability of the enzymes (17).

The characterization of α- amylase present in wheat seeds and flour would thus constitute an interesting complementary means of evaluating these products. I will now show how an immunochemical approach to the problem could be attempted on the basis of

Fig.4 - Electrophoretic and immunoelectrophoretic analyses of extracts of germinated wheat seeds (upper wells) and developing wheat seeds (lower wells). Electrophoresis was performed in 1.2 % agarose gel prepared in 0.025 M veronal buffer pH 8.6 at 4°C under 6 V/cm for 50 min. α- amylase characterization reaction was performed after electrophoresis or immunoelectrophoresis. Upper part : electropherogram. Middle part : immunoelectropherogram involving a serum specific for α- amylases of developing seeds. Lower part : immunoelectropherogram involving a serum specific for α - amylases of germinated seeds. D_1 D_2, G_1 G_2 designate α- amylase groups detected in extracts of developing and of germinating seeds, respectively.

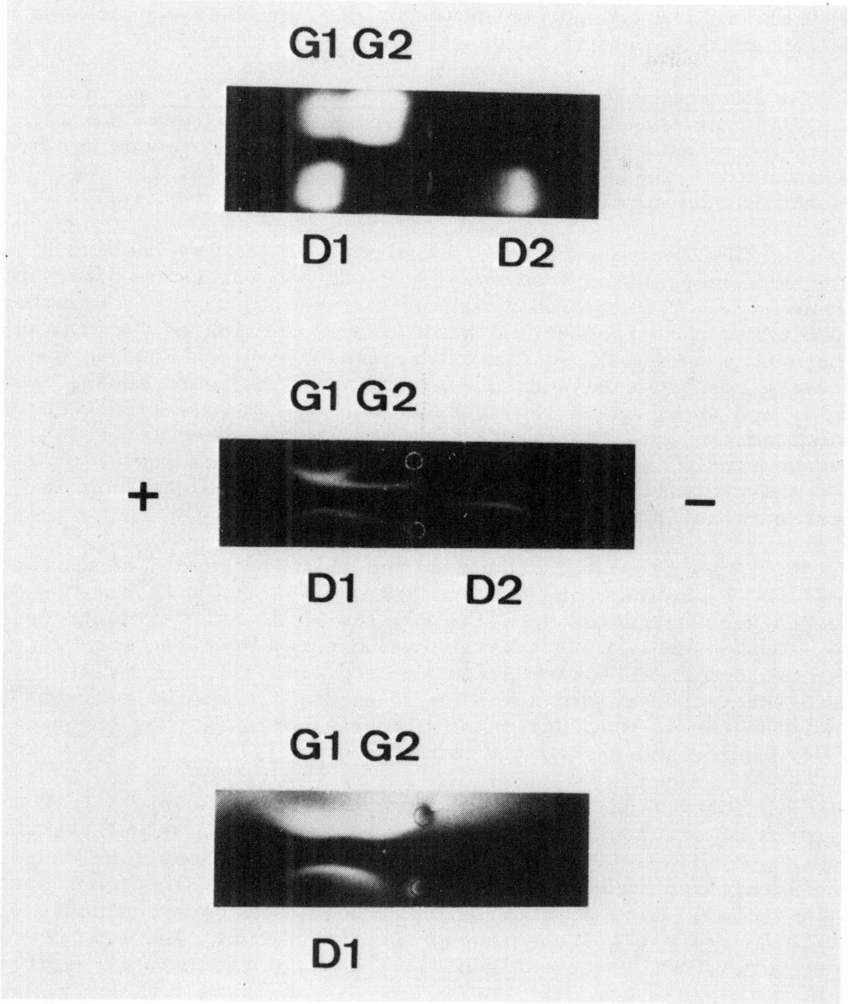

Figure 4

immunochemical data concerning wheat α- amylases and using a particular technique to be described.

B. Immunochemical characterization of wheat α- amylases. The characterization was carried out using an immune serum specific for α- amylases of developing seeds and another immune serum specific for α- amylases of germinating seeds (4, 21, 30). The results may be summarized as follows :

1. The immune sera were prepared by injecting rabbits with α - amylase purified from seed extracts according to Loyter and Schramm (7). Freezedried wheat seeds taken 15 days after anthesis and a mixture of freezedried barley seeds germinated for 3,5 and 7 days were used for protein extraction. Previous studies showing identity reactions between α- amylasesof germinated seeds, from barley and wheat using the anti-barley α- amylase immune serum, indicated the possibility of using the immune serum for wheat α - amylase studies (32). Later, the immune serum was shown to react with the proteins bearing most of the α- amylase activity in wheat germinated seeds (25).

2. Absorption techniques showed that the immune serum specific for α - amylase of developing seeds reacted with all α - amylase enzymes extracted from the developing seeds. As just reported, it also confirmed that the immune serum specific for α - amylase of germinated barley seeds reacted with all α- amylase enzymes extracted from germinated wheat seeds. Thus, the reagents permitted the identification of all α- amylase types endogenous to developing and germinating wheat seeds(21).

3. Identification with the immune serum specific for α - amylases of developing seeds is shown on Fig.4. By using immunoelectrophoretic analysis, two antigenic α- amylases (D_1 and D_2) were identified in extracts of developing seeds. They correspond to the two α- amylase groups, one anodic, the other cathodic, detected by agarose gel electrophoresis at pH 8.6. The two α- amylases were shown to be antigenically quite different. In extracts of germinating seeds, two antigenic α- amylases (G_1 and G_2) were detected, which corresponded to the two α- amylase groups evinced by agarose gel electrophoresis at pH 8.6. A partial identity reaction was suggested between these two constituents using different techniques. The partial identity reaction indicated that, while antibodies react with both G_1 and G_2 constituents, other antibodies react only with G_1. An identity reaction was observed between D_1 and G_1. This means that all antibodies which react with D_1 react also with G_1. (By using long duration agarose gel electrophoresis at pH 8.6, it was shown that D_1 and G_1 each contained three constituents differing by their electrophoretic mobility but all antigenically identical),(21, 30).

To summ up, the immune serum permitted the identification of one constituent specific for developing seeds (D_2) and another with electrophoretically and antigenically identical properties in developing and germinating seeds. Furthermore, certain antigenic relationships were suggested and antigenic differences were shown between G_1 and G_2, consequently between D_1 and G_2.

4. Identification with the immune serum specific for α - amylases of germinating seeds is shown on Fig.4. Using immunoelectrophoretic analysis, two groups of antigenic α - amylase (G_1 and G_2) were detected in extracts of germinating seeds. The antigens corresponded to the enzymatic groups detected by agarose electrophoresis. A partial identity reaction between the constituents was suggested : this indicated that, while antibodies reacted with G_1 and G_2, further antibodies reacted only with G_2. The partial identity reaction implies, in particular, that more antibodies react with G_2 than with G_1. The immune serum reacted with only one of the antigenic α - amylase groups in developing seeds, D_1. A reaction of identity was observed between D_1 and G_1. The results indicate, in particular, that the enzymic group which bears most of the α - amylase activity in germinating seeds (G_2) has an antigenic structure which does not exist in α - amylase of developing seeds (31,30).

C. <u>Principle of identification of α - amylases particular to germinating seeds (G_2)</u>. The above mentioned results indicate that the immune serum anti α - amylase from germinated seeds, contains antibodies specific for enzymes of the germinated seeds, G_1 and G_2, but some of these antibodies also react with one α - amylase of developing seeds, D_1. By means of an absorption technique which consists in letting the immune serum react first with α - amylase from developing seeds, the antibodies against D_1 could be eliminated and the immune serum could thus be made specific for G_2 only. Since it was shown that the immune serum did not react at all with fungal or bacterial α - amylases (6), the absorbed immune serum would constitute a reactive agent specific for α - amylases typical of germinated seeds only. A sensitive technique was developed on the basis of rocket immunoelectrophoresis (11) and enzymatic characterization reaction after rocket immunoelectrophoresis.

D. <u>Technique of identification</u>. In this technique, the antigens are forced to migrate into a gel containing antibodies. The antigens form complexes with the antibodies and these complexes continue to migrate becoming more and more rich in antibodies. When a certain ratio between antigens and antibodies is reached, the complex precipitates, forms peaks and does not move any further. For the same amount of antibodies, the surface of the peaks is in relationship with the amount of antigens analyzed (33).

In order to obtain a maximum of sensitivity, a minimal concentration of the immune serum in the gel was determined. An α -

amylase characterization reaction was performed on the gel after electrophoresis.

In these preliminary experiments designated for investigating seeds with low α- amylase activity, dilutions series of a diluted extract of 7 days germinated seeds was used. The highest activity of the dilution series evaluated according to the diffusion technique (3) described earlier for determination of malt α - amylase was 14.75 mm/24 h. This activity was designated as 100. (Such an activity was found for seeds displaying 4.5 units according the S K B technique). The concentration of the absorbed immune serum in the gel was 0.03 %. Under these conditions one precipitin peak was detected for the solutions with activity 100, 30, 10 respectively (Fig. 5A, g). It is worthy to note that the α - amylase characterization on the precipitin band increases the sensitivity of the technique ; the immunopeaks are not visible after protein amidoblack staining.

In order to increase the sensitivity of the technique even more, we combined rocket immunoelectrophoresis and line immunoelectrophoresis (34) with the enzymatic characterization reaction (Fig.5 B). A gel strip containing a dilution of the germinated seed extract was inserted between the gel containing the immune serum and the gel where the wells were cut. During electrophoresis, the α - amylase of the gel strip migrates as a continuous.

Fig.5 - Immunochemical technique for identifying the α- amylase antigen characteristic for germinated seeds in extracts of apparently ungerminated seeds. The technique was performed in 1.2 % agarose prepared in 0.025 M veronal buffer pH 8.6. Electrophoresis was carried out at 4°C for 5 hrs starting with 6 V/cm. α - amylase characterization reaction was performed on the gel after electrophoresis.

A : Rocket immunoelectrophoresis (see text).

B and C : Combination of rocket and line immunoelectrophoresis (see text).

g : increasing dilution series of extract of 7 days germinated wheat seeds.

d : dilutions of extracts of wheat seeds taken 15 days after anthesis.

b : dilutions of bacterial α- amylase solution.

Cl, Ca, Ma : extracts of three apparently ungerminated seed samples.

Activities : activity in each solution was evaluated by a diffusion technique (3). Activities are expressed in percent of the activity in the first dilution of the germinated seed extract (see text).

Note that the shape of the wells in C permits to pour more solution than in the wells shown in B ; this results in a better vizualization of the peaks.

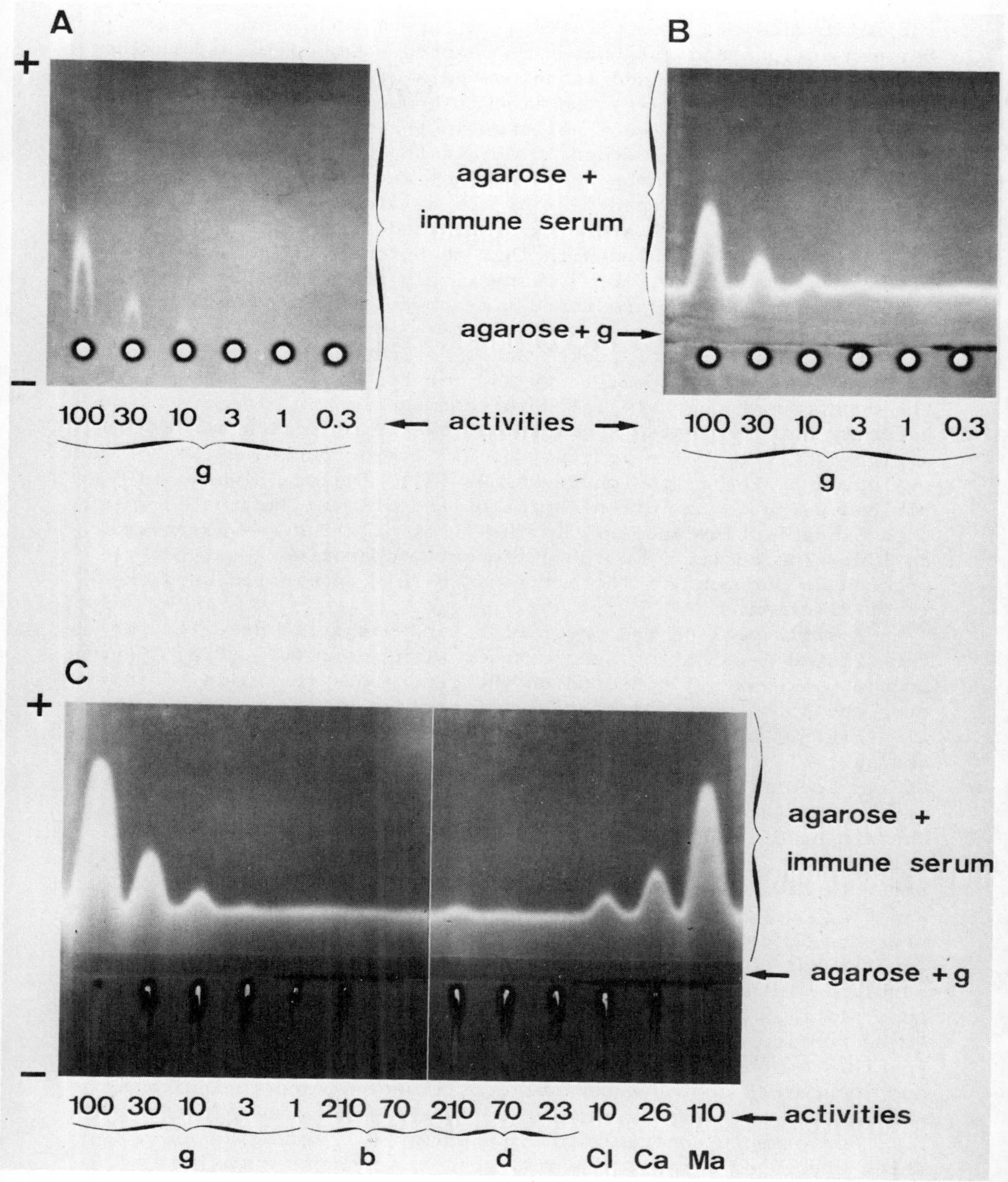
A
B
+
−
agarose +
immune serum
agarose + g
100 30 10 3 1 0.3
activities
100 30 10 3 1 0.3
g
g
C
+
−
agarose +
immune serum
agarose + g
100 30 10 3 1 210 70 210 70 23 10 26 110
activities
g
b
d
Cl
Ca
Ma

Figure 5

band in the gel containing antibodies, it forms complexes with the antibodies and precipitates as a line when a certain ratio between antigen and antibodies is reached. Nevertheless, in the zones where there are additional amounts of enzyme moved from the wells by electrophoresis, the line is broken and forms peaks, the size of which is in direct relation to the amount of additional antigen (Fig.5 B). In order to obtain the highest sensitivity, a maximum dilution of the wheat extract was used ; here the 5 mm broad gel strip contained 12 % of the diluted germinated seed extract (solution with activity designated as 100). The α - amylase G_2 was then detected with this technique for the solutions with activity 100, 30, 10, 3 respectively (Fig.5 B g). Thus, the latter technique appears three times more sensitive than the preceeding one.

The combination of rocket and line immunoelectrophoresis was then tested for its specificity and for its capacity to evaluate the constituent characteristic for germination in extracts of apparently not germinated seeds which,however,present a small α - amylase activity. The activity of all the preparations tested was evaluated by the diffusion technique (3). The scale was established using the dilution series of the diluted germinated wheat seed extract. The activity in the first dilution was designated as 100. The activity in the different preparations is then expressed as per cent of the activity in this germinated seed extract dilution.

A small peak on the precipitin band was still detected for the diluted germinating seed extract with activity 3 (Fig.5C, g,3). Such a peak was not detected on the line above the wells containing the solutions of bacterial α - amylase with activity 70 and 210 (Fig.5C, b). This was also true for much higher bacterial α - amylase activity so that the enzyme did not at all react with antibodies specific for the G_2 constituent. Concerning α - amylase of developing seeds no peak was detected on the line above the wells containing dilutions of the extract with activity 23 and 70 (Fig.5C, d, 23, d, 70). However for a less diluted extract, with activity 210, a small peak was detected (compare Fig.5C, d, 210 and Fig.5C, g, 3). It is not yet known whether this is due to the presence in developing seeds of very small amounts of antigenical constituent G_2 or to an effect which is due to the antigenic resemblance between G_1 and G_2 and which is not eliminated by the absorption as performed under our conditions ; this point remains to be further investigated. Nevertheless,the technique makes it possible to compare the total α - amylase activity of the seed and the corresponding amount of constituent G_2 and to indicate whether the activity is mainly due to this constituent.

For example, extracts of apparently not germinated seeds which presented a small α - amylase activity were submitted to the analysis (Fig. 5C, Cl, Ca, Ma). The activity in these extracts were estimated by the diffusion technique (3) and approximated respectively activities 10, 30 and 120 in our scale. In these extracts, the amounts of the antigenic constituent G_2 reflected by the size of the peaks were similar to the amounts of this constituent in the dilutions of the germinated seed extract displaying activity 10, 30 and 100 respectively.

Practically the results indicate that in these apparently ungerminated samples, the small α- amylase activity was mainly due to the constituent characteristic for germination.

E. Remarks. The immunochemical characterization of α - amylases from germinated seeds served to set up qualitative and quantitative techniques for distinguishing between α - amylases from germinated seeds and α - amylases exogenous to the seeds (fungi, bacteries), (6, 8).

We have described here a sensitive technique for identifying the antigenic α - amylase specific for germination in seeds showing a small α - amylase activity. The quantitative possibilities of the technique could be used at best by employing purified preparations of the α - amylase characteristic for germinating seeds (G_2), in order to establish a reference scale.

The same technique could be developed for detecting and quantitating the two other antigenic α - amylases evinced in germinating and developing wheat seeds. For this purpose, the immune serum specific for α - amylases from developing seeds must be used. An immune serum specific for the amylase constituent common to developing and germinating seeds (D_1 and G_1) could be obtained by absorbing the immune serum with the purified antigen specific for germinated seeds (G_2). As far as D_2 is concerned, since the enzyme migrates towards the cathode by electrophoresis at pH 8.6, the wells must be cut in the anodic side of the gel, the gel strip containing the extract of developing seeds being inserted between the gel with the wells and the gel containing the immune serum.

Antigenical relationships have been reported between α- amylases of certain cereals (32, 35). Thus, some of the immune sera could be used for detection in several cereals.

IV. Abstracts.

Two types of application of the immunochemical characterization of α- amylases in barley and wheat are presented.

The first type of application allows the elimination of α - amylase from seed extracts in a specific and easy way without any interference with β- amylase in electrophoretic or diffusion techniques in gel. Thus, electrophorograms of β - amylase from seed extracts also containing α - amylase can be obtained, and quantitation of β - amylase activity in extracts containing α - amylase can be carried out using a gel diffusion technique.

The second type of application aims at identifying in sound

seeds, the origin of the very low α - amylase activity present : whether it is due to the seed enzymes or to enzymes exogenous to the seeds (bacteria,fungi). It also tends at identifying whether the enzymes are characteristic of the developmental or of the germinating part of the life of the seed.

V. Literature cited.

1. SCRIBAN, R. - Brasserie (1965), 221, 3-15.
2. JACOBSEN, J.V., SCANDALIOS, J.C. and VARNER, J.E. - Plant. Physiol. (1970), 45, 367-371.
3. BRIGGS, D.E. - J. Inst. Brew. (1962), 68, 27-33.
4. DAUSSANT, J. - in "Immunological aspects of food", N. CATSIMPOOLAS (Editor) AVI pub. Co. (in press).
5. DAUSSANT, J., SKAKOUN, A. and NIKU-PAAVOLA, M.L. - J. Inst. Brew. (1974), 80, 55-58.
6. DAUSSANT, J. and SKAKOUN, A. - J. of Immunol. Methods (1974), 4, 127-133.
7. LOYTER, A. and SCHRAMM, M. - Biochem. Biophys. Acta (1962), 65, 200-206.
8. DAUSSANT, J. and CARFANTAN, N. - J. of Immunol. Methods (1975) 8, 373-382.
9. DAUSSANT, J. and CORVAZIER, P., FEBS letters (1970), 7, 191-194.
10. MANCINI, G., CARBONARA, A.O. and HEREMANS, J.F., in Immunochemistry, Pergamon Press (1965), 235-254.
11. LAURELL, G.B., Annal. Biochem. (1966), 15, 45-52.
12. LACROIX, L.J., WAIKAKUL, P. and YOUNG, G.M., Cereal. Res. Comm. (1976), 4, 139-146.
13. KRUGER, J.E., Cereal Res. Comm. (1976), 4, 187-194.
14. OLERED, R., Publication of the department of Plant Husbandry, Upsala, Almqvist and Wiksells (1967), 23, 1-106.
15. BELDEROCK, B., Field Crop Abstr. (1968), 21, 203-211.
16. GALE, M.D., Cereal Res. Comm. (1976), 4, 231-243.
17. MERCIER, C. and COLAS, A., Ann. Nutr. Alim. (1967), 21, 299-340.
18. KNEEN, E., Cereal Chem. (1944), 21, 304-314.
19. OLERED, R., Arkiv. Kemi. (1964), 22, 175-183.
20. SANDSTEDT, R.M. and BECKORD, O.C., Cereal Chem., (1946), 23, 548-559.
21. DAUSSANT, J. and RENARD, M., Cereal Res. Comm. (1976), 4, 201-212.
22. KRUGER, J.E., Cereal Chem. (1972 a), 49, 391-398.
23. OLERED, R. and JÖNSSON, G., J. Sci. Fd. Agric. (1970), 21, 385-392.
24. OLERED, R., Cereal Res. Comm., (1976), 4, 195-199.
25. DAUSSANT, J. and RENARD, M., FEBS letters, (1972), 22, 301-304.
26. DEDIO, W., SIMMONDS, D.H., HILL, R.D. and SHEALY, H., Can J. Plant. Sci., (1975), 55, 29-36.

27. KRUGER, J.E., Cereal Chem. (1972 b), 49, 379-390.
28. MacGREGOR, A.W., GORDON, A.G., MEREDITH, W.O.S. and LACROIX,L. J. Inst. Brew., (1972), 78, 174-179.
29. CHOJNACKI, G., BRYKCZYINSKI, J. and TYMIENIECKA, E., Cereal Res. Comm., (1976), 4, 111-114.
30. DAUSSANT, J., LAURIERE, C., CARFANTAN, N. and SKAKOUN, A., in Phytochemical Society Symposia, Academic Press (in Press).
31. GUILBOT, A. and DRAPRON, R., Ann. Physiol. Veg. (1963), 5, 5-18.
32. DAUSSANT, J. and GRABAR, P., Ann. Inst. Pasteur, (1966), 110, 79-83.
33. CLARKE, H.G.M. and FREEMAN, T., in "Prot. Biol. Fluids", H. Peeters (Editor), Elsevier Amsterdam (1967), 503-509.
34. KRØLL, J., Scand. J. of Immunology (1973), 2, (Suppl.1), 83-87.
35. ALEXANDRESCU, V. and MIHAILESCU, F., Revue Roumaine de Biochimie, (1973), 10, 89-94.

VI. Acknowledgements.

The author thanks Miss C. MAYER for skilful technical assistance, Mrs R. DAUSSANT for her help in correcting the english text and Dr R.L. ORY for having reviewed the article.

This study was supported in part by the "Centre Technique de la Brasserie et Malterie françaises", Nancy, France.

7

Immobilized Enzymes

ALFRED C. OLSON and ROGER A. KORUS

Western Regional Research Laboratory, Agricultural Research Service, U.S. Department of Agriculture, Berkeley, Calif. 94710

An immobilized enzyme has been defined as an enzyme that is constrained one way or another within the limited confines of a solid support. The subject of immobilized enzymes has received a significant amount of attention in recent years because of advantages that have been and might be achieved through successful applications in a number of fields including food processing. This chapter on immobilized enzymes has been divided into five parts beginning with an introduction in which we have tried to position immobilized enzymes within the larger context of the use of enzymes in food processing, pointing out advantages and limitations to the method. This is followed by a section on supports that have been used to immobilize enzymes particularly for food applications. A third section covers some examples of immobilized enzymes currently in use in food processing followed by a section on proposed applications of immobilized enzymes. Since much of the literature on the subject expresses results in terms that make comparison between different methods of immobilization difficult, the final section is a brief discussion of some of the common parameters used to describe immobilized enzyme systems. The chapter is a selected survey of the recent literature with some applications covered in more detail in other chapters in this book. The ingenuity, complexity and diversity of methods for immobilizing enzymes has been looked at from the point of view of their application to food processing, with some suggestions as to the future uses of immobilized enzymes in food processing.

As little as 20-30 years ago a great deal of art and tradition were still involved in the use of enzymes to process foods. Much of the technical knowledge we now have regarding enzymes and their application to food processing has been discovered and developed in the intervening years. Basic information on enzymes related to food can be found in books by Reed (1) and Whitaker (2, 3). More specialized books by Zaborsky (4), Pye and Wingard (5), Dunlap (6), Olson and Cooney (7) and Salmona, Saronio and Garattini (8) discuss the use of immobilized enzymes and related subjects. A number of reviews on immobilized enzymes have also appeared (9-15).

With the situation changing so rapidly the problem of describing new developments in applied food enzymology becomes very difficult. There is often a dichotomy of interest between basic scientific studies and commercial applications. Ideas, experiments, processes and products that look good in the laboratory may not meet marketing requirements, customer desires or the interest of a promoter. A large investment by a company in a new process involving enzymes will probably result in the company keeping that information proprietary as long as possible. To project what role enzymes, including immobilized enzymes, may have in food processing in the future is even more difficult. The exercise is not without merit, however, for an overview of the situation often clarifies objectives and brings into focus areas where progress can be made.

A study of recent developments in the use of enzymes in food processing can be divided into three areas of investigation: enzyme production, enzyme purification and enzyme application. All three areas are of importance and promise and are currently receiving a great deal of attention. For example, by applying knowledge from studies made in microbial genetics and induced enzyme synthesis it may be possible to prepare much larger amounts of intra- and extra-cellular food grade enzymes than was possible before. These enzymes can be used as heat stabilized whole cells or purified by a variety of procedures. A summary of the steps involved in the commercial production and purification of enzymes for food use is shown in Figure 1.

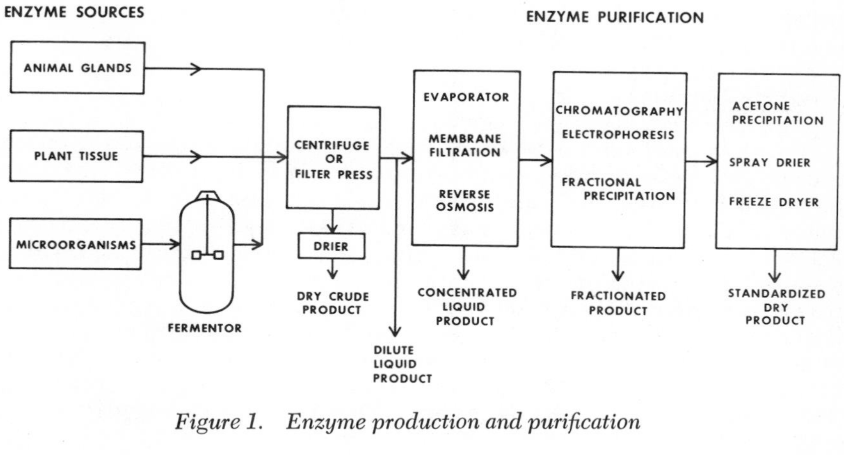

Figure 1. Enzyme production and purification

Advantages - 1) more rapid processing due to elimination of retarding effect of product accumulation e.g. 2 days → minute or hours

While some enzymes such as rennin and papain are obtained from animal and plant sources, the majority are from microorganisms. Purification involves some kind of collection followed by a concentration step. Enzyme preparation and purification that is done in batches makes reproducibility difficult and costs high. Efforts are underway in a number of laboratories to be able to continuously produce, purify, and use enzymes. In the commercial application of enzymes it should be advantageous to combine continuous production, purification and use of enzymes, including immobilized enzymes, at a single location.

The ability to immobilize enzymes, then, represents only one part of a larger research effort that is changing food processing. What are some of the advantages to immobilize an enzyme? Briefly, they include (1) the ability to use the enzyme many times more than the soluble enzyme can be used, (2) lower enzyme costs (because of 1), ~~(3) no enzyme left in the product,~~ (4) increased stability of the enzyme (immobilization has been shown to stabilize some enzymes and prolong their activity on exposure to substrate), (5) improved enzyme behavior (pH optima shifted to more advantageous pH on immobilization to certain supports), (6) continuous process possible, (7) better quality control of the product (because of 6), (8) less additional processing (because of 3), (9) lower labor costs, and (10) advantageous use of multiple enzyme systems. There are also potential disadvantages which include (1) the cost of the immobilization, (2) loss of enzyme activity on immobilization, (3) greater initial plant investment, (4) more technically complex process, (5) more skilled supervision of process required (because of 4), (6) higher labor costs (because of 5), (7) unique sanitation and toxicology problems, and (8) applicable mainly to soluble substrates. While these generalizations can provide some guidelines, final considerations that determine whether it is advantageous to immobilize an enzyme depend on the particular process one is considering.

Enzyme Supports

In this section we have covered information on enzyme immobilization in terms of supports that have been used for this purpose including polysaccharides, inorganic supports, fibrous proteins, synthetic polymers, hydrogels and hollow fibers. Enzymes can be constrained or held to these supports by adsorption, covalent attachment, crosslinking, entrapment, microencapsulation or by various combinations of the foregoing methods. Examples of these methods of constraint will be discussed where they have been used with particular support materials. The variety of support materials that have been proposed with some suitability for a food processing application is worth noting.

Polysaccharide Supports. A large number of polysaccharide and substituted polysaccharide materials have been used as supports for enzyme immobilization. These supports have been used for adsorption of enzymes in a number of food processing applications such as production of L-amino acids in Japan which uses DEAE-Sephadex (16, 17) and the Clinton Corn Processing Company process for high fructose corn syrup in which glucose isomerase adsorbed on DEAE-cellulose is used in reactors consisting of an assembly of shallow beds of the immobilized enzyme (18).

Cellulose is the major constituent of fibrous plants and the most abundant organic material in nature. The monomer units of cellulose are D-glucose. A large number of substituted celluloses have been prepared by replacing the hydroxyl groups by various substituents. Replacement by diethylaminoethyl (DEAE) substituents introduces an amino group which provides a polycationic, weakly basic anion exchange group. DEAE-cellulose has been one of the most widely used supports for enzyme immobilization. On the other hand, replacement of hydroxyl groups by carboxymethyl (CM) substituents introduces carboxyl groups which provide a weakly acidic cation exchange material.

Other popular supports for enzyme immobilization are Sephadex and substituted Sephadex. Sephadex is a beaded, crosslinked dextran. Dextrans are polysaccharides of glucose units that are produced by strains of *Leuconostoc mesenteroides*. Substituted Sephadex can be prepared in much the same way as substituted cellulose and shows a similar ability for enzyme adsorption.

Enzymes having a relatively high content of acidic amino acids remain firmly bound to DEAE-cellulose or DEAE-Sephadex even at high substrate concentration as long as certain physical conditions such as ionic strength and pH are maintained. The aminoacylase from *Aspergillus oryzae* adsorbed on DEAE-Sephadex loses 40% of its activity over a 32-day period when used at 50°C for continuous hydrolysis of acetylated L-methionine (19). Deteriorated columns could be reactivated by direct addition of aminoacylase. Higher enzyme activities were obtained with DEAE-Sephadex than with DEAE-cellulose. Upward flow in a packed column was used to avoid compression of the packed bed and excessive pressure build-up. The optimum conditions of pH, ionic strength and temperature must be determined to achieve and maintain good adsorption and good activity. If a strong adsorption does not occur, the enzyme is readily desorbed with subsequent loss of activity and contamination of the product. The selection of optimum conditions has mostly been a matter of trial and error.

Microcrystalline DEAE-cellulose powder is easily compressed. Only shallow packed beds can be used in a large scale commercial operation to avoid excessive pressure build-up.

This flow behavior can be improved by using porous DEAE-cellulose beads which show a high enzyme loading capacity for immobilization of glucose isomerase (20). Glucose isomerase immobilized by adsorption on DEAE-cellulose was found to be somewhat less stable than the free enzyme at 60°C. At 50°C the immobilized glucose isomerase has a half-life of 10-11 days under continuous operation (21).

Another promising polysaccharide support is chitin. Chitin forms the structural material of the skeletons of marine crustaceans and bears the same relation to invertebrates as cellulose does to plants. The monomer unit is glucosamine. On about five out of every six glucosamine units the amino nitrogen is blocked with an acetyl group. Native chitin contains approximately equal amounts of calcium carbonate and chitin. The calcium carbonate can be removed by acid giving a rigid porous material that is relatively inert to both chemical and microbial attack. Glucose isomerase (22) and glucoamylase (23) have been immobilized on chitin by simple adsorption. The following enzymes were immobilized on chitin with glutaraldehyde: lactase (acid tolerant), acid phosphatase, α-chymotrypsin (24) and glucoamylase (23).

When chitin is heated under pressure with strong alkali, it is deacetylated to produce chitosan. Typically 80% of the acetyl groups in chitin are removed leaving free amino groups in chitosan. Chitosan has also been used with glutaraldehyde to prepare immobilized acid-tolerant lactase (25).

These polysaccharides are relatively cheap, food grade materials and some have a high capacity for protein adsorption. As a group they are currently among the most widely used supports for enzyme immobilization in the food industry.

Inorganic Supports. Porous glass, alumina, hydroxyapatite, nickel oxide on nickel screen, silica alumina impregnated with nickel oxide, porous ceramics of various types, stainless steel and sand are some of the inorganic materials to which enzymes have been attached. Apparent advantages of inorganic supports for enzymes include: (1) structural stability over a wide range of pH, pressure, temperature, and solvent composition, (2) excellent flow properties in reactors, (3) inertness to microbial attack or attack by enzymes, (4) ease of adaptability to various particle shapes and sizes, and (5) good regeneration capability.

Attachment of enzyme to these supports has been by adsorption and by covalent attachment to the inorganic surface. Since Weetall (26, 27) first reported the use of porous glass to immobilize enzymes, many enzymes have been covalently attached to modified porous glass and related materials (4, 28, 29). Porous glass commonly used for this purpose has pore diameters in the range of 200-1500 A and particle size in the range of 20-80 mesh. It is 96% silica and can be reacted with an alkylamino silane to give an alkylamine glass as follows.

$$-O-\underset{\underset{O}{|}}{\overset{\overset{O}{\parallel}}{Si}}-OH \;+\; CH_3CH_2O-\underset{\underset{O\;CH_2CH_3}{|}}{\overset{\overset{O\;CH_2CH_3}{|}}{Si}}-CH_2CH_2CH_2NH_2 \longrightarrow$$

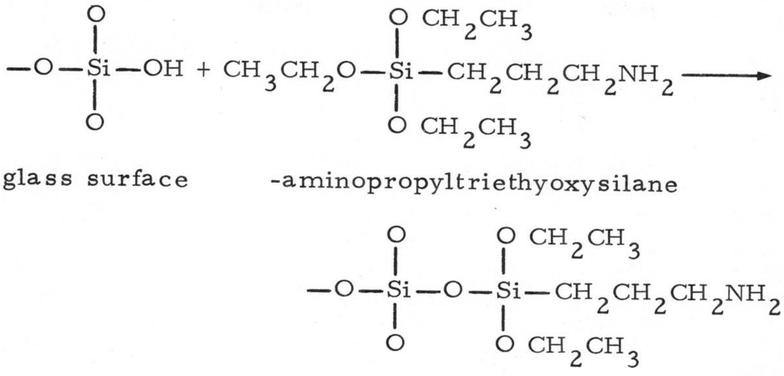

glass surface -aminopropyltriethyoxysilane

$$-O-\underset{\underset{O}{|}}{\overset{\overset{O}{\parallel}}{Si}}-O-\underset{\underset{O\;CH_2CH_3}{|}}{\overset{\overset{O\;CH_2CH_3}{|}}{Si}}-CH_2CH_2CH_2NH_2$$

alkylamine glass

Enzymes can then be coupled to the amine derivative by the car-bodiimide, thiourea or azo linkage methods which probably in-volve attachment to carboxyl, amine or aromatic residues, res-pectively, on the enzyme. An alternative procedure is to treat the alkylamine glass with glutaraldehyde followed by the enzyme to effect a stable coupling. Porous glass has been reported to be partly soluble and breaks down at pH values near 7 or above. This problem can be overcome by coating the glass with zirco-nium oxide, a very insoluble material, prior to use (30).

Sand is another silica support that has recently been used successfully to immobilize β-galactosidase (31), alcohol dehy-drogenase and urease (32). While sand has a lower effective surface area and enzyme loading capacity than porous glass, it is less expensive and thus may be suitable for large volume applications with low concentrations of substrate. Application of the enzyme to sand can be accomplished by preparing alkyl-amine sand, treating this with glutaraldehyde followed by the enzyme.

Porous ceramic carriers have also been developed and used successfully to bind enzymes. These supports are less expen-sive than porous glass and more insoluble. Aminoacylase has been immobilized on ceramic carriers composed of SiO_2, Al_2O_3, and TiO_2 (33). Coupling was again achieved by first preparing the silanized ceramics followed by treating with glutaraldehyde and finally adding enzyme.

Some applications of immobilized enzymes involve the sim-ple adsorption of enzymes. For example, a porous ceramic is used as an enzyme support in the production of 5'-mononucleo-tide flavor enhancers (34). Glucose isomerase has been suc-cessfully adsorbed on the internal surface of controlled pore alumina (35). The alumina was pretreated with a dilute solution of magnesium acetate and cobalt acetate prior to adding the enzyme. The resulting adsorbed glucose isomerase exhibited conversions in the range of 85% with a half life in column operation of 40 days.

Glucose oxidase has been covalently coupled to nickel oxide on nickel screens (36) and to nickel oxide impregnated silica alumina pellets (37). More recently Markey, Greenfield and Kittrell (38) studied the immobilization of glucose oxidase and catalase on a number of silanized inorganic supports using glutaraldehyde as a crosslinking agent. A decrease in activity of both enzymes was observed as particle size of the supports increased up to 1000 microns. This decrease was attributed to both external film and internal diffusional resistances. Particle composition, size, shape and pore volume are important factors in determining the effectiveness of these kinds of supports for enzymes (38, 39). Greenfield and Laurence (39) and Bouin et al. (40) point out that glucose oxidase immobilized simultaneously with catalase significantly extends the usefulness of the glucose oxidase activity by decomposing the hydrogen peroxide formed in the glucose oxidase reaction. Immobilized catalase and glucose oxidase undergo substrate and product inactivation, respectively, by hydrogen peroxide (41). As a consequence the industrial performance of these enzymes will probably be determined by how well this inactivation can be reduced and controlled.

Fibrous Proteins. The fibrous proteins offer a number of potential advantages as enzyme supports for food applications. They are natural materials, with good mechanical properties, offer a low resistance to substrate diffusion, and have an open internal structure with many potential binding sites for enzyme attachment and stabilization. They are abundant and inexpensive and can be processed into films, membranes and other forms that retain their structural identity.

Of the fibrous proteins the collagens have been the most widely studied as enzyme supports (42, 43). Collagen proteins in combination with elastin, mucopolysaccharides and mineral salts form the connective tissue responsible for the structural integrity of the animal body. The distinctive characteristic of the amino acid composition of a variety of vertebrate collagens is a high glycine content which is close to one-third of the total number of residues. Collagens are also rich in the imino acids proline and hydroxyproline which can amount to 20-25% of the total number of residues. The unique triple stranded helical structure of collagen is closely related to the high content of glycine and imino acid residues.

There is a complex regularity to the organization of the collagen molecules into microfibrils that results in the presence of holes spaced at repeating intervals along the length of the microfibril. It is believed that these regions are the primary sites for enzyme binding (44). Collagen swells in the presence of water, is hydrophilic, and at neutral pH takes up more than its own weight of water. Immobilization of enzymes on collagen involves the formation of a network of non-covalent bonds such as salt linkages (ionic interactions), hydrogen bonds and Van der Waals interactions acting together between collagen and the enzyme (43). While individually these bonds are weak, taken together they form a very stable network between enzyme and collagen.

Collagen-enzyme complexes have been prepared by three different procedures: (1) impregnating a preswollen membrane with enzyme, (2) macromolecular complexation in which an enzyme-collagen dispersion at the desired pH is cast and dried and then rehydrated for use, and (3) electrocodeposition in which enzyme is deposited in a collagen membrane under the influence of an electric field. The complexes can subsequently be treated with glutaraldehyde to induce crosslinks to strengthen the membrane and the bonding of the enzyme to the collagen. Collagen-enzyme membranes have been coiled and placed in a cylinder with suitable spacers to make efficient continuous reactors for enzyme conversion. A large number of enzymes, including whole cells, have been immobilized on collagen membranes. This technique seems applicable in general to enzyme immobilization.

Keratin is another fibrous protein that has been examined as an enzyme support. Keratin is found in the outer layers of vertebrates, is unreactive toward the environment and is mechanically strong and durable. Like the collagens, keratins vary depending on their source. They have a high content of the sulfur containing amino acids and crosslinking provided by the S-S bond of cystine which maintains structural integrity. Stanley and co-workers (45) have recently prepared a granular keratin from feather meal, a byproduct of the poultry industry, and have successfully bound lactase to the granules with glutaraldehyde. Feather keratin in reduced form has some unique attributes worth exploiting. The high concentration of free sulfhydryl groups could be used in purifying sulfur-containing enzymes through the technique of thiol-disulfide interchange (46) or simply by adsorption in some cases. Urease has been immobilized on feather keratin by adsorption (45).

Phenolic Resins. A resin made from phenol and formaldehyde has been found to be an excellent support for some enzymes by several investigators (47, 48, 49). A resin of this type is commercially available from Diamond Shamrock Chemical Company, Redwood City, California as Duolite ES-762. This porous resin is relatively stable to acids, bases and most organic solvents. The 10-50 mesh particles are rigid, irregular and granular, and resistant to attrition. They have a pore volume of about 0.6 cm^3 per cm^3 of resin and a surface area of about 100 m^2/cm^3. This type of resin is used as an adsorbant for proteins, aromatics and other compounds and as such it will bind enzymes without further treatment. However, the addition of glutaraldehyde to the resin either before or after application of the enzyme can increase the retention of enzyme activity significantly. Immobilization is rapid at room temperature and can be done in a batch process in an open vessel or with the resin packed in a column. In most cases the enzyme can be removed from the resin by treatment with dilute acid and/or base and new enzyme adsorbed to the support. This resin has been successfully used to immobilize lactase (47, 48), α-amylase (49), trypsin (50) and other enzymes. Modifications of the phenolformaldehyde resin are also possible. Resins substituted with various primary, secondary and tertiary methyl amino groups are also available. These anion exchange resins have been used to immobilize invertase (48), glucose isomerase and aspartase (51). The adsorption process with the unsubstituted resin most probably involves the phenolic rings interacting with aromatic amino acids in the protein. This type of interaction of proteins with phenolic compounds is also observed in their reaction with tannins and in particular tannic acid to form insoluble complexes. Such complexes with invertase and lactase are enzymatically active and remain insoluble if the complex is stabilized with glutaraldehyde (48). It is also necessary to deposit this complex on a support such as Celite for column operation. While these limitations on the use of tannins to precipitate enzymes appears great, the combinations of different tannins and supports that could be tried suggests that for some applications this approach could be useful.

Hydrogels. The method of gel entrapment involves the formation of a crosslinked polymer network in the presence of an enzyme or microbial cell. This technique seems to be particularly attractive for the immobilization of whole cells. With whole cell immobilization, the costly enzyme purification processes can be avoided. Sequential reactions can be carried out utilizing the various enzymes and cofactors which are left intact in the microbial cells. Within the microbial cells many enzymes are already immobilized and compartmentalized in order to carry out complex metabolic and synthetic reactions. By entrapping cells in a hydrophilic polymer network, a supporting structure can be provided that has little effect on cell properties.

The polymer most often used for entrapment is polyacrylamide. The crosslinking agent used to form a three dimensional network is usually N, N'-methylene bis(acrylamide). Polymerization is carried out in an aqueous solution containing enzyme or microbial cell, monomer, crosslinking agent, and free radical initiator. A three dimensional polymer network is formed entrapping enzyme or microbial cell.

Tanabe Seiyaku Company in Japan has developed two

processes for amino acid production based on microbial cell entrapment in polyacrylamide gels. In one process, L-aspartic acid is produced from fumaric acid using the aspartase activity of entrapped E. coli bacterial cells. In the other process, L-malic acid is produced from fumaric acid using the fumarase activity of Brevibacterium ammoniagenes (52). B. ammoniagenes cells entrapped in polyacrylamide gels have also been studied for the synthesis of coenzyme A from pantothenic acid, cysteine, and ATP. This synthesis uses five sequential enzymic steps and illustrates the advantage of using immobilized whole cells for multi-step systems (53).

Although it is usually assumed that enzymes entrapped in polyacrylamide gels remain unattached to the gel matrix and retain their native conformation, this may not always be true. Harrison has shown that the enzyme glucose 6-phosphate dehydrogenase is bound covalently to the gel matrix via free sulfhydryl or amino groups during gel entrapment (54).

Enzyme immobilization by gel entrapment has been recently reviewed (55). Although most enzyme entrapment has been in polyacrylamide, other hydrogels may be superior with regard to mechanical strength and elimination of enzyme leakage. A large number of enzymes have been entrapped in poly(2-hydroxyethylmethylacrylate) gels, (56). This gel may be preferred for some applications because of its superior mechanical strength. Enzymes have also been immobilized using a large number of synthetic monomers and polymers such as N-vinyl pyrrolidone, poly(vinyl alcohol), and poly(vinyl pyrrolidone) under radiant ray irradiation (57). Whole yeast cells have been entrapped in spherical agar pellets and used for sucrose hydrolysis (58).

Hollow Fibers. An immobilization technique which offers several advantages for food systems is the use of hollow fibers. Two types of hollow fiber devices that can be used for enzyme immobilization are the isotropic, cellulosic fibers made by Dow Chemical Company and the anisotropic, noncellulosic fibers made by Amicon Corporation and Romicon Corporation. The Dow semipermeable hollow fibers are made of cellulose acetate or other cellulosic materials. They are homogeneous (isotropic) and permit permeation of solute or solvent from either side of the fiber. Amicon and Romicon fibers are anisotropic and composed of an approximately 0.5 μm thick semipermeable membrane supported by a thick open-celled sponge (Figure 2). The X50 hollow fibers are copolymers of poly (vinyl chloride) and polyacrylonitrile and have a nominal molecular weight cut-off of 50,000. The P10 fibers are polysulfone polymers with a nominal molecular weight cut-off of 10,000.

The immobilization technique often used with the Amicon and Romicon hollow fibers consists of physically entraining an aqueous preparation of enzyme or whole cells within the porous sponge region of the fibers. Substrate solution is then passed

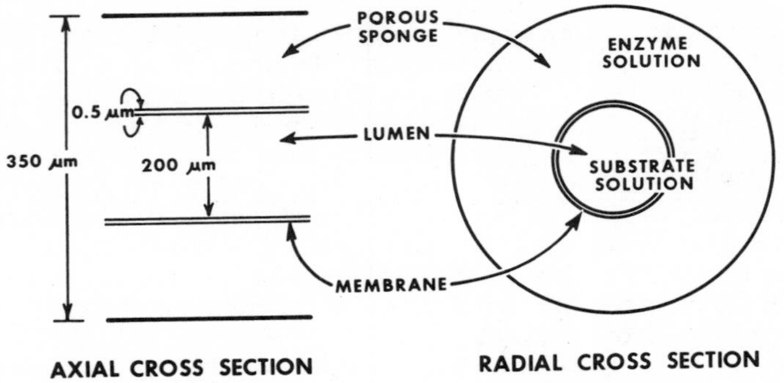

Figure 2. Structure of assymmetric hollow fiber

through the fiber lumen giving a continuous flow enzyme reactor. The enzyme-substrate system must be selected so that the enzyme is too large to pass through the fiber membrane while the substrate is sufficiently small so as to pass through the membrane. A theoretical model describing enzymic catalysis using hollow fiber membranes has been developed by Waterland et al. (59). Reactor conversion data for β-galactosidase agree well with predictions obtained from the theoretical model (60).

Hollow fiber enzyme reactors have also been used for immobilization of glucose isomerase (61), α-galactosidase and invertase (62). The stability of enzymes immobilized in hollow fiber reactors is approximately the same as for the corresponding enzymes in free solution. However, the Romicon and Amicon P10 hollow fibers require a preconditioning with an inert protein before use in order to avoid enzyme inactivation on contact with the fibers (61, 62).

Some of the advantages of using hollow fibers for enzyme immobilization are: high throughput with a low pressure drop can be achieved, no enzyme modification is required and there is no product contamination from chemicals used in the immobilization process, hollow fiber cartridges can be easily cleaned and reloaded with enzyme, and the fibers have a large surface-to-volume ratio and a thin membrane which offers a low resist ance to substrate transport.

Microbial Pellets. In a submerged culture of filamentous microorganisms, the mycelia are likely to aggregate to form pellets. Media composition and pH of the fermentation media can determine whether pulpy mycelia or pellets are formed (63). Microbial pellets can be viewed as immobilized whole cells where the intracellular enzymes are available for utilization. These untreated cells can be directly used as biocatalysts for industrial processes. This eliminates the need for enzyme purification and greatly simplifies the immobilization process.

Pellets produced by fermentation of the fungus Mortierella vinaceae have been successfully used in the beet sugar industry as an enzyme source of α-galactosidase. This enzyme hydrolyzes the indigestible sugar raffinose to sucrose. Raffinose reduces sugar yield by retarding the rate of sucrose precipitation from beet sugar molasses.

Kinetic studies of α-galactosidase containing pellets of M. vinaceae show that the behavior of pellets is similar to that of gel entrapped enzymes (64). There is a decrease in enzyme activity with increasing pellet size above 0.25 mm because of diffusional effects. For pellets smaller than about 0.25 mm in diameter, diffusional effects are absent. The effectiveness factors found experimentally for pellets of M. vinaceae with α-galactosidase activity were found to agree fairly well with a theoretical model describing intraparticle diffusion in spherical particles.

Commercial Application of Immobilized Enzymes

Detailed information on commercial processes using immobilized enzymes in food processing is generally proprietary. For some processes the only available information is the microorganism from which the enzyme used was obtained and a general description of the immobilization method. The commercial applications of immobilized enzymes presented in this section are well documented and represent significant advances in process development in the food industry.

Aminoacylase. The first large scale use of immobilized enzymes was the Tanabe Seiyaku Company process for L-amino acid production in Japan. This process uses aminoacylase adsorbed on DEAE-Sephadex for the continuous, automatically controlled preparation of L-methionine, L-phenylalanine, L-tryptophane and L-valine from racemic mixtures of those amino acids. The reduced costs of enzyme, substrate and labor have cut production costs of the immobilized enzyme process to 60% of that for the conventional batch process using soluble enzymes (65, 66).

A process flow sheet for the continuous production of L-amino acid from racemic acetyl-D, L-amino acid is shown in Figure 3. The acylated amino acid mixture is fed to a column containing aminoacylase immobilized by adsorption onto DEAE-Sephadex. This enzyme hydrolyzes only the acetyl-L-amino acid. The resulting mixture of acetyl-D-amino acid and L-amino acid can be separated by crystallization because of solubility differences. L-methionine has been produced with a reported output of 20 metric tons per month (67).

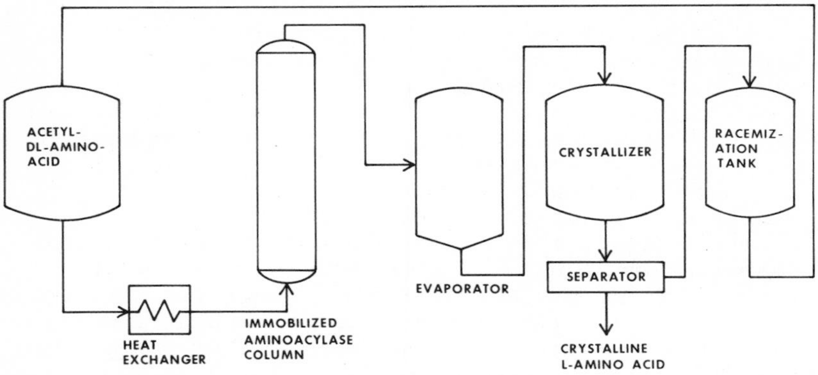

Figure 3. Process flow sheet for the continuous production of L-amino acid from racemic acetyl-D-L-amino acid

Glucose Isomerase. Until 1938 commercial corn syrups were produced by acid hydrolysis of corn starch. This process produced a syrup with a dextrose-equivalent (D. E.) of 56-58. A sweeter syrup was not produced because higher conversions by acid hydrolysis gave bitter flavors. With the introduction of saccharifying enzymes it became possible to increase the D. E. without producing bitter flavors. The glucoamylase currently used for starch saccharification produce corn syrups with a D. E. of 95-97. This high dextrose corn syrup, which is 70-75% as sweet as sucrose, can be transformed to a product which is equivalent in sweetness to sucrose by a partial isomerization of glucose (dextrose) to fructose. This isomerization has been carried out on a commercial scale since 1968. The commercial application of immobilized glucose isomerase to produce high fructose corn syrup is currently the largest volume use of an immobilized enzyme in the world. Approximately one billion pounds of high fructose syrup were produced in 1974 (9).

The commercial process for high fructose corn syrup production is shown in Figure 4. The process involves liquefying raw corn starch, saccharifying the starch to a high dextrose syrup using acid and enzymic hydrolysis, and refining the syrup. The carbon and ion exchange refined glucose syrup contains about 93% D-glucose on a dry basis. The remaining 7% is made up of other corn syrup saccharides. This glucose syrup is prepared for isomerization by the addition of required salts (Co and Mg) and adjustment of pH. The glucose syrup is pumped through a series of shallow bed reactors which contain glucose isomerase adsorbed onto DEAE-cellulose. The flow rate is controlled so that the product contains about 42% fructose. The flow rate can be reduced as isomerase activity

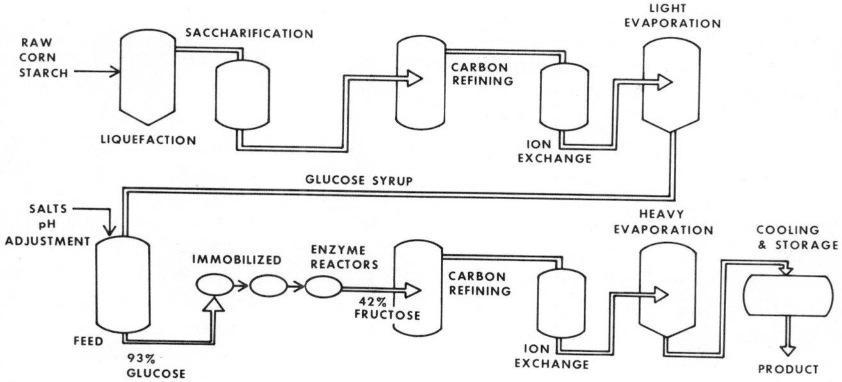

Figure 4. High-fructose corn syrup process flow sheet

diminishes in order to maintain this conversion. The fructose syrup is then refined and concentrated by evaporation. The refined high fructose corn syrup is substantially colorless, ash free and has little tendency to form color on storage (68). A higher fructose content can be obtained by separating glucose and fructose in ion exchange columns, thereby producing 90% fructose syrups on a large scale with high yields (69). These syrups are now commercially available.

α-Galactosidase. The beet sugar industry uses immobilized α-galactosidase to hydrolyze the raffinose in sugar beets. This enzyme can be held within pellets of the fungus M. vinaceae which are formed under specific fermentation conditions. Diluted sugar beet molasses is passed through a reactor containing the pellets. The pellets are very compressible and cannot be readily packed in a column without an excessive pressure build-up during continuous operation. An enzyme reactor that has been developed for the commercial process (70) consists of a horizontal vat divided into several chambers, each being agitated, and each separated by a screen to prevent the passage of pellets from one chamber to another. The molasses is fed into one end of the reactor and contacts the suspended pellets as it passes from chamber to chamber.

Typically, a sugar beet processing plant might process 3000 tons of beets a day and obtain water-soluble extracts representing 600 tons of sucrose and as much as 4 or 5 tons of raffinose (71). About 80% of the raffinose can be hydrolyzed as the molasses is passed through the agitated chambers. Since raffinose interferes with the crystallization of sucrose, sucrose recovery is increased by raffinose hydrolysis to a greater extent than the amount of raffinose hydrolyzed. Experience in Japan indicated that plant throughput was increased by at least 5% and recoveries of sucrose increased (71).

There is also interest in α-galactosidase for hydrolysis of the α-galactosides, raffinose, stachyose and verbascose, that are considered to be partially responsible for the flatulence associated with dry bean products such as soybean milk (62, 72). However, there have been no reports of a commercial application.

Ribonuclease. A recent application of immobilized enzymes in Japan is the production of 5'-mononucleotide flavor enhancers (34). Inosine-5'-monophosphate (IMP) and guanosine-5'-monophosphate (GMP) act synergistically with monosodium glutamate to enhance food flavors. The 5'-ribonucleotides, adenosine-5'-monophosphate (AMP), cytidine-5'-monophosphate (CMP), uridine-5'-monophosphate (UMP) and GMP, are produced from ribonucleic acids using the ribonuclease 5'-phosphodiesterase immobilized on a porous ceramic support. IMP is produced by the replacement of the amino group of AMP with a hydroxyl group. This reaction is catalyzed by 5'-adenylate deaminase immobilized on a porous ceramic support.

$$\text{RNA} \xrightarrow[\text{5'-phosphodiesterase}]{} \text{5'-ribonucleotides (AMP, GMP, CMP, UMP)}$$

$$\text{AMP} \xrightarrow[\text{5'-adenylate deaminase}]{} \text{IMP}$$

From these observations on what has succeeded in the field of immobilized enzymes applied to food processing we would suggest that for large volume, low unit price items, the more simple immobilization schemes on inexpensive food grade supports are necessary. Adsorption on DEAE cellulose and stabilized microbial pellets have succeeded. Of course, as the value of the product goes up a more complex scheme may be possible.

Potential New Applications of Immobilized Enzymes

Some examples of the many proposed uses of immobilized enzymes to food processing are listed in Table I. Each of these applications is a significant subject in and of itself. We have selected a few proposals to discuss in more detail as illustrative of what might be done with immobilized enzymes in food processing in the future.

Immobilized Starch Hydrolyzing Enzyme Systems. Sugar syrups prepared from enzymatically hydrolyzed starch are important ingredients in many modern food products. Syrups are available containing a wide range of sugars with varying sweet-

Table I. Proposed Uses of Immobilized Enzymes to Food Processing

New Application	Examples
Sugar hydrolysis	For fermentation processes lactase on whey, prior to wine fermentation
Clarification or haze removal	In beer and wine industries use papain, bromelain, and other proteases
Sterilization	Lysozyme to remove micoorganisms from air Catalase to remove excess H_2O_2
Food stabilization	Oxygen removal with glucose oxidase-catalase system
Prevent gelling	In frozen orange juice concentrates use pectic enzymes
Flavor modification	Debittering proteins with proteases Naringinase for grapefruit juice Limonate dehydrogenase for citrus juices
Nutritional improvement	Plastein reaction to add amino acid to protein Cellulases, hemicellulases, pectinases, etc., to improve food digestibility
Removing undesirable factors	Decaffeinating coffee Removing tannins from tea

ness and functional properties. The processes involved in preparing these syrups constitute one of the largest commercial uses of enzymes. Changes and improvements in enzyme sources and means of utilization can thus be of considerable importance to the industries involved, the quality and properties of the products produced and the consumer. Potential advantages for using immobilized enzymes in these processes include lower enzyme costs, continuous processes, lower labor costs, cleaner products, better quality control and less additional processing. With these potential advantages it is not surprising that well over 100 papers have been published in the last few years dealing with the immobilization of one or more of the enzymes involved in starch hydrolysis. Some of this work has been reviewed by Zaborsky (4) and Olson and Richardson (11).

The process of preparing sugar syrups from starch can be divided into several steps (Figure 5). The gelatinization and

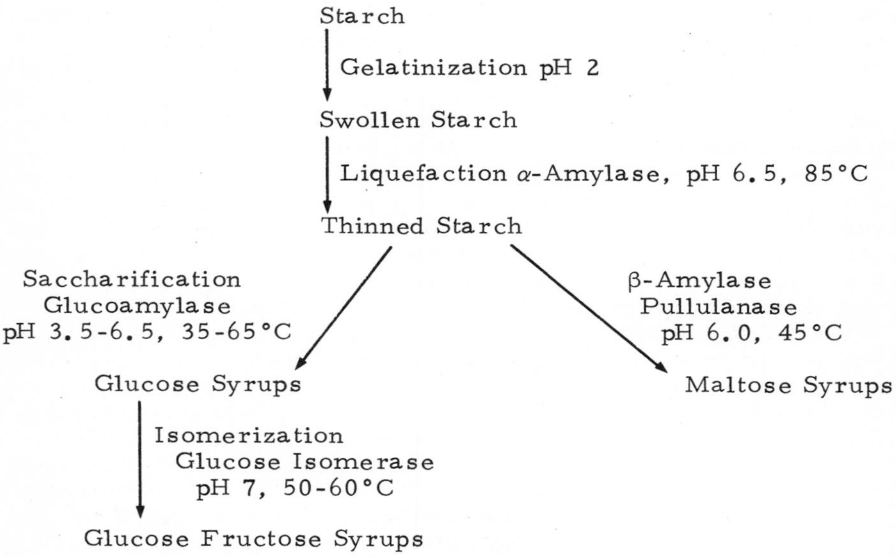

Figure 5. A summary of the steps involved in preparing sugar syrups from starch

liquefaction steps can be combined in a single acid-enzyme process in which an α-amylase is added to a 30-40% (w/w) starch slurry at pH 6.5. The mixture is subsequently heated to 105-110°C for 5-15 minutes for gelatinization and then cooled and held at 95°C for 1-20 hours for liquefaction by the α-amylase. While α-amylase preparations from Bacillus subtilis have been used to thin starch, a recently commercially introduced, more heat stable α-amylase from Bacillus licheniformis works even better in a single step process with lower total Ca^{++} requirements for enzyme temperature stability (73).

Immobilized α-amylase might also be used industrially. The enzyme has been entrapped in gels (74), covalently bound to crosslinked polyacrylamide derivatives (75), and immobilized on collagen membranes (76). α-Amylase attached to polystyrene particles in a continuous stirred reactor α-amylase showed a greater extent of multiple attack on starch than the soluble enzyme with an increase in the production of glucose (77). When α-amylase was immobilized in an ultrafiltration reactor some of the enzyme passed through the membrane (78). To overcome this difficulty Wykes et al. (79) increased the size of the soluble enzyme by coupling it to soluble s-triazine derivatives of dextran, DEAE-dextran and carboxymethylcellulose. The larger soluble enzyme derivatives did not pass through the pores of the ultrafiltration membranes as readily as the non-derivatized enzyme but were effective in hydrolyzing starch solutions.

α-Amylase hydrolyzes bonds in the interior of the starch molecule to produce oligosaccharides and various amount of glucose, maltose and maltotriose depending on the source of the enzyme. It is used commercially for the liquefaction of gelatinized starch before saccharification by adding the enzyme to a slurry and heating to 80-90°C. Under these conditions the enzyme hydrolyzes enough of the starch to effect liquefaction but in the process it can be inactivated by the high temperature. Immobilized α-amylase should probably be used for the liquefaction procedure under such conditions that it is not inactivated. Wheat starch has been successfully gelatinized between 62-72°C and partially hydrolyzed at 72°C by B. subtilis α-amylase immobilized on cyanogen bromide activated carboxymethylcellulose in a stirred tank reactor (80). The immobilized α-amylase produced relatively more glucose and maltose than the soluble enzyme.

β-Amylase hydrolyses α-1,4-glycosidic linkages from the non-reducing ends of starch chains to give maltose. The enzyme is being used commercially in the brewing, distilling and baking industries to convert starch to fermentable sugar. High maltose syrups, produced with a mixture of α-amylase and β-amylase, are used in the confectionary industry because they are nonhydroscopic and do not crystallize as readily as high glucose syrups. β-Amylase has been immobilized to crosslinked polyacrylamides (75), entrapped in gel lattice (81), attached to alkylamine porous glass with glutaraldehyde (82), and chemically attached to a crosslinked copolymer of acrylamide-acrylic acid using a water soluble carbodiimide (83). β-Amylase acting alone on starch gives maltose and a large portion of a branched undegraded amylopectin β-limit dextrin. In order to increase the yield of maltose from starch Martensson (84, 85) has immobilized β-amylase together with pullulanase, a debranching enzyme. β-Amylase was attached to the support first because it was more stable to the carbodiimide coupling reagent. Coupling yields of 40% β-amylase protein and 38% pullulanase protein were obtained with residual enzyme activities of 22% and 32% respectively. This two enzyme system was considered advantageous as a steric opening of the substrate is performed by one enzyme facilitating the action of the other enzyme. Marked increased operational stability was observed for the immobilized two enzyme system compared to the free enzymes in solution. Pullulanase could also be used alone to produce low D. E. syrups of commercial value.

Glucoamylase hydrolyzes α-1,4-glucan links removing

glucose units consecutively from the nonreducing ends of starch molecules. The soluble enzyme is used in the enzymatic conversion of starch to glucose. The potential advantages of using immobilized glucoamylase have been well recognized. The enzyme ionically bound to DEAE-cellulose (86) will remain so as long as certain physical conditions such as low ionic strength and pH (about 4) are maintained. Columns of DEAE-cellulose bound enzyme must be long to allow for sufficient residence time for practical conversions. As a result the cellulose-enzyme complex packs very tight restricting the flow. It is, however, possible to use glucoamylase on DEAE-cellulose in a continuous stirred tank reactor (CSTR) at 55°C for 3-4 weeks without loss of enzyme (87).

Glucoamylase can be covalently bound to DEAE-cellulose by first activating the support with 2-amino-4,6-dichloro-s-triazine and then coupling the enzyme to the activated support (88, 89, 90). As with most immobilized enzymes the observed catalytic behavior of the bound enzyme was different from that of the free enzyme. On immobilization the pH optimum shifted from 5 to 4.3, K_m decreased from 6.1 mM to 1.4 mM, and temperature stability at 50°C increased.

Part of the glucoamylase molecule is a carbohydrate which offers an unusual means for attaching the enzyme to a cellulose support. Christison (91) has oxidized the carbohydrate with sodium metaperiodate and coupled the resulting oxyglucoamylase to carboxymethylcellulose hydrazide.

Amberlite IR-45 (OH), a strong anion exchange resin, will bind significant amounts of glucoamylase (92). The bead form of this support could make it superior to cellulose supports for long column operations since compaction of the support should be less. Glucoamylase has also been covalently bound to various derivatives of macroreticular polystyrene beads (93). A derivative prepared by nitration of the beads followed by reduction to a poly(p-aminostyrene) was a good support to which glucoamylase could be bound with glutaraldehyde.

Stable gels can be prepared containing glucoamylase which can hydrolyze dextrins to glucose and from which there is no leakage of enzyme. For example, glucoamylase has been immobilized in thin polyacrylamide gels by photopolymerizing aqueous solutions of acrylamide, a crosslinking monomer (alkylglycidyl ether), an enzyme reactive monomer (glycidyl acrylate) and glucoamylase (94). Maeda and co-workers (95) have prepared immobilized glucoamylase in polyacrylamide gels by γ-ray irradiation but without the use of any crosslinking agent. The enzyme has also been entrapped in poly(vinyl alcohol) (95) and poly(vinyl pyrrolidone) gels (96) by initiating polymerization of solutions of monomers and the enzyme with γ-rays. Packed beds of matrix bound glucoamylase in polyacrylamide gels have been used to convert 28 D.E. starch hydrolyzate syrups to syrups containing 90% or more dextrose (94).

Columns of glucoamylase immoblized on chitin with glutaraldehyde in bench scale operations gave remarkably good operational characteristics. Using a feed of α-amylase treated starch containing 30% solids a product was obtained in which the dextrose content was 95.6%. Analyses by high pressure liquid chromatography indicated that unhydrolyzed residue was essentially restricted to disaccharides (23).

Collagen membranes containing glucoamylase have been prepared by adding enzyme to collagen homogenates prior to casting the membrane (codispersion) and by soaking a preformed membrane in a solution of the enzyme (impregnation) (97). In both cases the enzyme was subsequently fixed to the membrane by crosslinking with glutaraldehyde. The impregnated prepared membrane bound less enzyme than the codispersed membrane but retained higher activity for a longer period of time. The duration of exposure and level of glutaraldehyde used to fix the enzyme to the membranes were critical for the successful immobilization of active enzyme.

Glucoamylase has been immobilized to a number of inorganic carriers including controlled pore glass (30, 98, 99), acid-activated molecular sieve and alumina (100). Enzyme loadings of 100 mg/g carrier with full retention of enzyme activity could be obtained on a molecular sieve that had been previously activated with 2N HCl. The gel-like nature of the preparation, however, makes it unsuitable for column work. Alumina supported glucoamylase prepared by several different procedures showed lower activity than the molecular sieve preparations. They did have good mechanical and flow characteristics suitable for column operation.

When glucoamylase coupled to porous glass was used in column operation, some decline in enzyme activity was observed that was attributed to dissolution of the glass carrier. This problem was solved by coating the glass with ZrO_2 (30). When glucoamylase was coupled to arylamine-glass by the azo-linkage procedure, a number of covalent links are formed between glass and protein molecules and the bond energies of these links are sufficient to prevent shearing of the enzyme from the support under stresses that could be expected in column or packed bed reactors (101). Losses in enzyme activity from reactors using such bound glucoamylase would have to be due to some other causes such as enzyme denaturation or poisoning of the enzyme by an inhibitor.

Several studies have been made on the pilot plant production of glucose with glucoamylase immobilized to porous silica (102, 103, 104). In one such study (102) a 1 ft^3 column was prepared by filling the column with alkylamine silica beads in water. After the column was fully packed a glutaraldehyde solution was pumped into it to replace the water. Following this, the column was washed with pH 7 phosphate buffer and a 20% glucoamylase solution in the same buffer was recirculated through the column at room temperature for 2 hours. The column was sterilized with a saturated aqueous solution of chloroform. Bound in this way glucoamylase was found to be very stable in pilot plant operation. Activity losses at 40°C were probably due to plugging of carrier pores with incompletely processed substrate. Microbial contamination was minimal and could be kept under control by washing the column with a saturated solution of chloroform in water. The reactor was capable of producing 1000 lbs/day of glucose at 40°C and pH 4.5. Residence time of the feed in the column was important in obtaining high D.E. and glucose concentration in the product. At long residence times conversion of glucose to isomaltose became significant. α-Amylase treated dextrin as feed gave the highest value of D.E. and glucose concentration.

Manufacture of Cheese. In the commercial production of cheese the controlled curdling, or coagulation, of milk is accomplished with proteolytic enzymes called rennets. In order to develop a continuous, automated process for cheese production it would be advantageous to use immobilized rennets. In this process cooled milk (about 15°C) would pass over the immobilized enzyme column where proteolysis of the milk protein, casein, would occur. No coagulation actually occurs until the milk is heated. Heating of the enzyme-treated milk would take place in a separate secondary step to induce curdling. The major difficulties encountered in using immobilized rennet are low activity and poor stability of the immobilized enzymes, release of enzyme during continuous operation, column plugging, and microbial contamination (105).

Cofactor Regeneration. Commercial enzymic processes in the food industry use degradative hydrolytic enzymes. Immobilized enzymes may be applied to synthesize or assemble relatively complex molecules such as expensive pharmaceutical chemicals. A big difficulty in such an application is the requirement for expensive cofactors that will probably have to be regenerated by means of an immobilized enzyme system. It is unlikely that such an application will soon be used by the food industry. Preliminary cost estimates indicate that, even with efficient cofactor recycling, only production of a very high cost product is economically feasible (9, 106).

Analysis. Enzymes including immobilized enzymes (107, 108) are being used more routinely in analytical systems. Glucose analyzers are already being marketed by Yellow Springs Instrument Company (Yellow Springs, Ohio) and Leeds & Northrup Company (North Wales, Pa.). Using different systems, these analyzers employ amperometric detectors and immobilized glucose oxidase. In the Yellow Springs instrument an enzyme-containing membrane is fixed directly to an electrode sensor which responds to the H_2O_2 produced during glucose oxidation. The Leeds & Northrup instrument consists of a column of immobilized enzyme followed by a detector.

Cellulose, Hemicellulose, and Lignin Degradation. One of the biggest potential applications of enzymes to food systems is the enzymic conversion of cellulose to glucose. While not involving immobilized enzymes as we have been considering them the problems involved in this system are related.

In woody plants cellulose is physically protected from enzyme hydrolysis by the interpenetration of lignin into a crystalline cellulosic network. This structure must be disrupted by chemical or enzymic treatments or by milling before enzymes can attack the cellulose. To help determine the problems involved and the economic viability of cellulose degradation a highly instrumented prepilot plant containing fermentors, enzyme reactors, and auxilliary equipment has been set up at the U. S. Army Natick Development Center (109).

One of the processing schemes proposed for cellulose hydrolysis is shown in Figure 6 (110). Cellulase is produced by

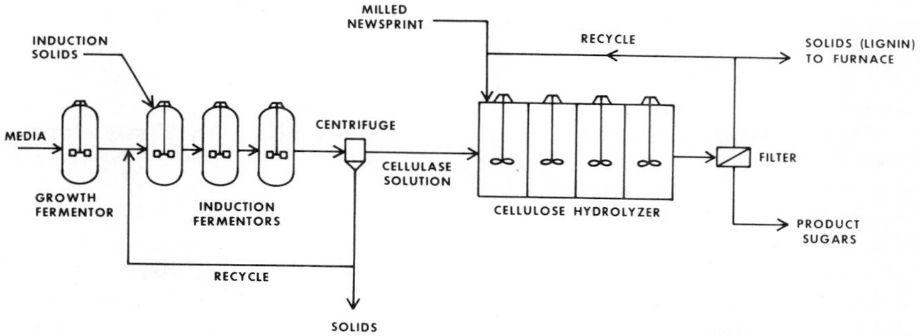

Applied Polymer Symposium No. 2

Figure 6. Berkeley process for the enzymatic conversion of cellulose to glucose (110)

Trichoderma viride in a two-stage fermentation system with microbial cell growth with soluble sugars in a growth fermentor followed by cellulase induction on solid cellulose in induction fermentors. While this process was designed to handle waste newsprint, once the feasibility of cellulose hydrolysis is established, it may also be advantageous to convert hemicellulose and lignin (111) into simple, soluble substances. It would then be possible to treat crop residues, garbage and food processing plant wastes and produce important raw materials and energy. The sugars produced in this process could be recovered or converted into ethyl alcohol, methane or single cell protein by continuous fermentation.

The form of hemicellulose that is found in greatest abundance in food products such as corn is xylan. Xylan can be hydrolyzed by xylanase and β-xylosidase enzymes to give a high yield of the monosaccharide xylose (82). Although xylose has little value as a foodstuff, it can be converted to xylitol by catalytic hydrogenation or used as a carbon source for fermentation. Xylitol has a sweetness equivalent to sucrose and has been suggested as a replacement for sucrose in food formulations (112).

Appendix: Terminology and Theoretical Aspects of Immobilized Enzymes

Most immobilized enzymes exhibit altered kinetic behavior when compared with the corresponding soluble enzymes (113). Activity of enzymes almost always decreases as a result of immobilization. This can occur because of a conformational change in the enzyme structure during immobilization or because the enzyme is in a different environment after immobilization. A conformational change is likely to occur if the enzyme is covalently bound to a support. Environmental effects are greatest when the enzyme environment is most unlike a free aqueous enviroment. For example, acidic functional groups in the vicinity of an enzyme will cause a shift in maximum activity to a higher pH.

Reduction in immobilized enzyme activity can also occur as a result of diffusional effects. In the immobilization process the enzyme can be bound to a surface or held within a matrix Figure 7). In order for an enzyme catalyzed reaction to occur the substrate must first diffuse from solution to the support surface and, in the case of an entrapped enzyme, diffuse within the support to reach the enzyme. After reaction the products must then diffuse back into the solution. The removal of product is especially important if the reaction is reversible or inhibited by product. In both cases enzyme activity is reduced by increased diffusional resistance.

In order to compare different immobilized enzyme preparations, it is important to have a standardized nomenclature. Recommendations for nomenclature standardization (114) suggest that the activity of an immobilized enzyme should be expressed as an initial reaction rate in terms of micromoles of substrate converted per second per mg of dry immobilized enzyme preparation.

A very useful concept in discussing mass transfer effects is the effectiveness factor, η, which is defined as the ratio of the rate of reaction for an immobilized enzyme to the rate of reaction for the same amount of enzyme in solution (i.e. in the absence of diffusional effects). The effectiveness factor can be readily determined experimentally and gives a good indication

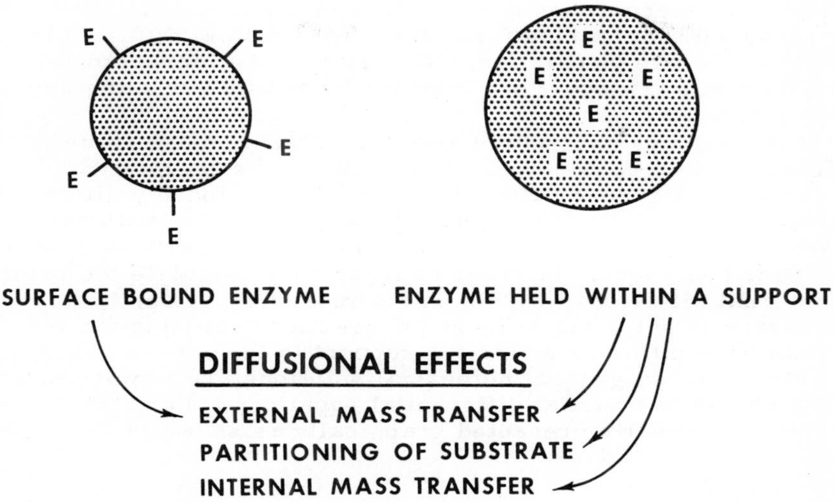

Figure 7. Schematic of surface-bound and entrapped enzymes illustrating the important diffusional effects for each

of the extent of diffusional effects. These effects arise from mass transfer considerations either external or internal to the enzyme support surface. For enzyme attached to a surface only external mass transfer effects are involved and the rate of reaction will be determined by the rate at which substrate can be brought to the surface. This flux of substrate to the surface can be characterized by a mass transfer coefficient, k_c where

$$\text{Flux} = k_c\,(c - c_s)$$

where c = concentration of substrate in solution and c_s = concentration of substrate at the support surface.

This flux of substrate is greatly affected by the movement of fluid past the support surface. The value of k_c increases as the flow increases. An estimate of the flux for spherical particles can be made by setting $c_s = 0$ and letting $k_c = 2D/d_p$ where D is the substrate diffusivity and d_p is the particle diameter. This is the value of k_c for a stagnant fluid and represents a lower limit for k_c (115). For a more exact estimate of k_c, correlations are available for flow through a packed bed (116). There will be no diffusional reduction in the reaction rate if the calculated overall flux of substrate is greater than the total free enzyme activity.

External diffusional rate limitation will be apparent experimentally when activity is measured as a function of flow rate. In the absence of external diffusional effects no change in activity will occur when the flow rate is changed. Activity will increase with increasing flow rate when external diffusion is limiting the rate of reaction. This is because increased flow will facilitate transport of substrate from solution to the support surface.

For enzymes held within a matrix internal mass transfer effects must be considered. Partitioning of substrate will occur at the support surface because of the solubility difference between free solution and within the support. External diffusional effects can be estimated and experimentally examined as described above. In order to rigorously calculate the extent of internal diffusional effects on the rate of reaction, it is necessary to solve the differential equations equating the rate of substrate diffusion within the support to the rate of reaction of substrate. In general numerical solutions are required by these non-linear partial differential equations (117, 118). The results are usually presented graphically as shown in Figure 8.

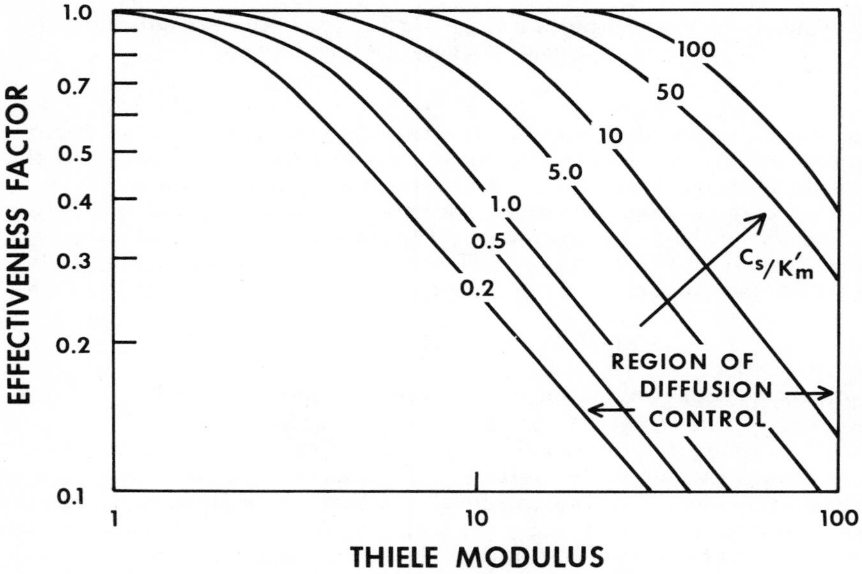

Figure 8. Double logarithmic plot of effectiveness factor against the Thiele modulus for various values of c_s/K_m'

The Thiele modulus is a measure of the ratio of the rate of reaction to the rate of substrate diffusion within the matrix. It can be expressed in a number of different forms, but in general

$$\text{Thiele modulus} = (\text{support dimension}) \left(\frac{\text{immobilized enzyme activity}}{K'_m \times \text{substrate diffusivity}}\right)^{1/2}$$

The support dimension is the radius for spherical particles or the thickness for membranes and K'_m is the Michaelis constant for the immobilized enzyme.

The effectiveness factor has a value of one when there is no diffusional limitation, assuming that enzyme activity has not been altered as a result of immobilization. As can be seen from Figure 8 the effectiveness factor approaches one for small values of the Thiele modulus. Also, the effectiveness factor is increased as c_s becomes larger than K'_m since high substrate concentrations saturate the immobilized enzyme support. This increases substrate concentration in the interior of the support making more substrate available to the immobilized enzyme. The effectiveness factor is increased by small support dimensions, low immobilized enzyme activity, high substrate diffusivity, and a high substrate concentration. All of these factors reduce diffusional effects.

As immobilized enzyme activity is increased by adding more enzyme per unit volume of support, the effectiveness factor decreases since high enzymic activity causes the reaction to become diffusion controlled. In the region of diffusion control in Figure 8 the rate of the immobilized enzyme reaction is proportional to the square root of the enzyme concentration.

A primary consideration in immobilized enzyme use is the ability to retain activity under conditions of storage and usage. When the rate of reaction is limited by diffusional effects, there is an apparent increase in the half-life of the immobilized enzyme (119). Figure 9 shows the apparent increase in stability that is possible for an enzyme uniformly distributed within a supporting matrix. In the absence of diffusional effects this immobilized enzyme has a true half-life of one time unit. The apparent half-life is increased at low substrate concentrations, high enzyme activity, large support dimensions, and low substrate diffusivity within the support or, in other words, by those factors that cause a reaction to be diffusion controlled. When the reaction is completely controlled by diffusion, the rate will be proportional to the square root of the enzyme concentration and the apparent half-life will be twice the true half-life of the enzyme. A true measure of immobilized enzyme stability is obtained only under conditions where diffusional effects are absent.

In reviewing the literature on the kinetics of immobilized enzymes, many theoretical articles can be found which make such restrictive assumptions that they are of virtually no use in most practical systems. For instance, in order to obtain closed form analytical solutions for immobilized enzyme

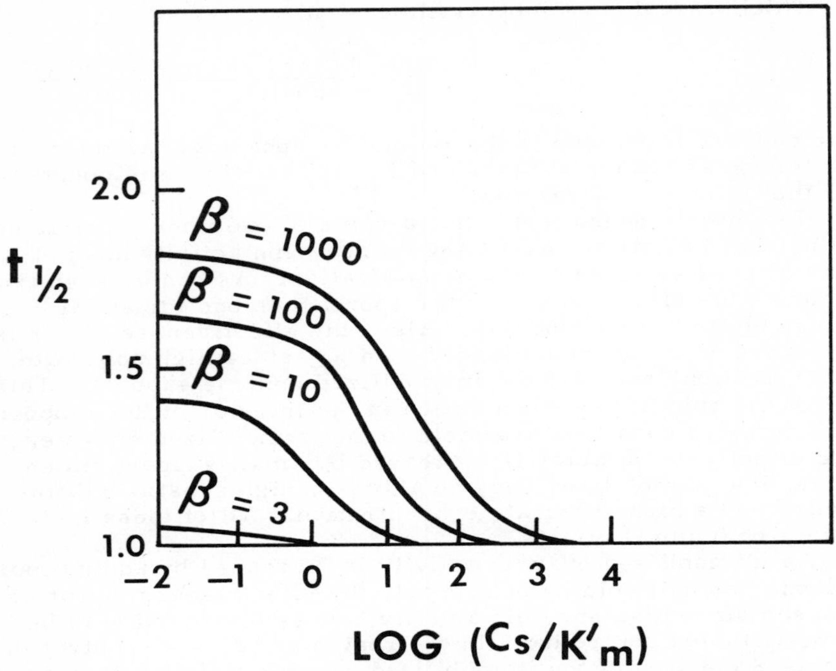

Figure 9. Apparent half-life of an enzyme which is uniformly distributed within a spherical particle (119)

reactor performance, many authors assume that the enzyme reaction is first order in substrate concentration ($c \ll K'_m$). This is certainly not the case for most reactions of interest in food systems.

Literature Cited

1. Reed, Gerald, "Enzymes in Food Processing," second edition, 573 pages, Academic Press, New York, 1975.
2. Whitaker, J. R., "Principles of Enzymology for the Food Sciences," 636 pages, Marcel Dekker, Inc., New York, 1972.
3. Whitaker, J. R., Editor, "Food Related Enzymes," 365 pages, American Chemical Society Advances in Chemistry Series No. 136, 1976.
4. Zaborsky, O. A., "Immobilized Enzymes," 175 pages, CRC Press, Cleveland, Ohio, 1973.
5. Pye, E. Kendall and Wingard, Lemuel B. Jr., Editors, "Enzyme Engineering," Volume 2, 470 pages, Plenum Press, New York, 1974.

6. Dunlap, R. B., Editor, "Immobilized Biochemicals and Affinity Chromatography," 377 pages, Plenum Press, New York, 1974.
7. Olson, A. C. and Cooney, C. L., Editors, "Immobilized Enzymes in Food and Microbial Processes," 268 pages, Plenum Press, New York, 1974.
8. Salmona, M., Saronio, C. and Garattini, S., Editors, "Insolubilized Enzymes," 226 pages, Raven Press, New York, 1974.
9. Skinner, K. J., Chem. and Eng. News (1975) 53(33), 22-41.
10. Richardson, T., J. Food Sci. (1974) 39, 645-646.
11. Olson, N. F. and Richardson, T., J. Food Sci. (1974) 39, 653-659.
12. Stanley, W. L. and Olson, A. C., J. Food Sci. (1974) 39, 660-666.
13. Hultin, H. O., J. Food Sci. (1974) 39, 647-652.
14. Lee, Y. Y. and Tsao, G. T., J. Food Sci. (1974) 39, 667-672.
15. Smiley, K. L. and Strandberg, G. W., Advances in Applied Microbiology (1972) 15, 13-38.
16. Tosa, T., Mori, T., Fuse, N., and Chibata, I., Biotechnol. Bioeng. (1967) 9, 603-615.
17. Tosa, T., Mori, T., and Chibata, I., Agr. Biol. Chem. (1969) 33, 1053-1059.
18. Schnyder, B. J., Die Starke (1974) 26, 409-412.
19. Tosa, T., Mori, T. and Chibata, I., Agr. Biol. Chem. (1969) 33, 1047-1052.
20. Chen, L. F. and Tsao, G. T. (1976), submitted for publication.
21. Park, Y. K. and Toma, M., J. Food Sci. (1975) 40, 1112-1114.
22. Stanley, W. L., Watters, G. G., Kelly, S. H., Chan, B. G., Garibaldi, J. A., and Schade, J. E., Biotechnol. Bioeng. (1976) 18, 439-443.
23. Stanley, W. L., Watters, G. G., Kelly, S. H., and Olson, A. C. (1976), submitted for publication.
24. Stanley, W. L., Watters, G. G., Chan, B., and Mercer, J. M., Biotechnol. Bioeng. (1975) 17, 315-326.
25. Masri, M. S., Randall, V. G. and Stanley, W. L., Polymer Preprints (1975) 16(2), 70-75.
26. Weetall, H. H., Nature (1969) 223, 959-960.
27. Weetall, H. H., Science (1969) 166, 615-617.
28. Messing, R. A., Process Biochemistry (1974) 9 (9), 26-28.
29. Weetall, H. H., Food Product Development (1973) 7 (3), 46-52.
30. Weetall, H. H. and Havewala, N. B., Biotechnol. Bioeng. Symp. (1972) 3, 241-266.

31. Byrne, M. J. and Johnson, D. B., Biochem. Soc. Trans. (1974) 2, 496-497.
32. Brotherton, J. E., Emery, A. and Rodwell, V. W., Biotechnol. Bioeng. (1976) 18, 527-543.
33. Weetall, H. H. and Detar, C. C., Biotechnol. Bioeng. (1974) 16, 1537-1544.
34. Samejima, H., "Enzyme Engineering," Volume 3, Plenum Press, to be published.
35. Messing, R. A. and Filbert, A. M., J. Agr. Food Chem. (1975) 23. 920-923.
36. Weetall, H. H. and Hersh, L. S., Biochim. Biophys. Acta (1970) 206, 54-60.
37. Herring, W. M., Laurence, R. L. and Kittrell, J. R., Biotechnol. Bioeng. (1972) 14, 975-984.
38. Markey, P. E., Greenfield, P. F. and Kittrell, J. R., Biotechnol. Bioeng. (1975) 17, 285-289.
39. Greenfield, P. F. and Laurence, R. L., J. Food Sci. (1975) 40, 906-910.
40. Bouin, J. C., Dudgeon, P. H. and Hultin, H. O., J. Food Sci. (1976) 41, 886-890.
41. Altomare, R., Greenfield, P. F. and Kittrell, J. R., Biotechnol. Bioeng. (1974) 16, 1675-1680.
42. Barndt, R. L., Wang, S. S. and Leeder, J. G., J. Food Sci. (1976) 41, 494-497.
43. Vieth, W. R. and Venkatasubramanian, K., Chem. Tech. (1974) 4, 47-55.
44. Bernath, F. R. and Vieth, W. R., "Immobilized Enzymes in Food and Microbial Processes," A. C. Olson and C. L. Cooney, Editors, p. 157-185, Plenum Press, New York, 1974.
45. Stanley, W. L., Watters, G. G., Chan, B. G. and Kelly, S. H., submitted for publication.
46. Carlsson, J., Axen, R., Broklehurst, K. and Crook, E. M., Eur. J. Biochem. (1974) 44, 189-194.
47. Okos, M. R., Ph.D. Dissertation, Ohio State University, 1975.
48. Olson, Alfred C. and Stanley, William L., J. Agr. Food Chem. (1973) 21, 440-445.
49. Boundy, J. A., Smiley, K. L., Swanson, C. L. and Hofreiter, (1976), submitted for publication.
50. Grulke, Eric A. and Okos, Martin R., unpublished results.
51. Samejima, H. and Kimura, K., "Enzyme Engineering," Volume 2, E. K. Pye and L. B. Wingard, Editors, p. 131-135, Plenum Press, New York, 1974.
52. Chibata, I., "Enzyme Engineering," Volume 3, Plenum Press, to be published.
53. Shimizu, S., Morioka, H., Tani, Y. and Ogata, K., J. Ferment. Technol. (1975) 53, 77-83.
54. Harrison, R. A. P., Anal. Biochem. (1974) 61, 500-507.

55. O'Driscoll, K. F., Advances in Biochem. Eng., to be published.
56. O'Driscoll, K. F., Hinberg, I., Korus, R. and Kapoulas, A., J. Polymer Sci.: Symposium No. 46 (1974) 227-235.
57. Maeda, H., Biotechnol. Bioeng. (1975) 17, 1571-1589.
58. Tosa, K. and Shoda, M., Biotechnol. Bioeng. (1975) 17, 481-497.
59. Waterland, L. R., Michaels, A. S., and Robertson, C. R., Amer. Inst. Chem. Eng. Journal (1974) 20, 50-59.
60. Waterland, L. R., Robertson, C. R. and Michaels, A. S., Chem. Eng. Commun. (1975) 2, 37-47.
61. Korus, R. A. and Olson, A. C., J. Food Sci., accepted for publication, 1976.
62. Korus, R. A. and Olson, A. C., Biotechnol. Bioeng., accepted for publication, 1976.
63. Kobayashi, H. and Suzuki, H., J. Ferment. Technol. (1972) 50, 625-632.
64. Kobayashi, H. and Suzuki, H., Biotechnol. Bioeng. (1976) 18, 37-51.
65. Tosa, T., Mori, T. and Chibata, I., Agr. Biol. Chem. (1969) 33, 1053-1059.
66. Tosa, T., Mori, T., Fuse, N. and Chibata, I., Biotechnol. Bioeng. (1967) 9, 603-615.
67. Anonymous, Chem. Eng. (1970) 77, 44-46.
68. Mermelstein, N. H., Food Technol. (1975) 29 (6), 20-25.
69. Anonymous, Food Processing (1975) 36 (11), 26-28.
70. Shimizu, J. and Kaga, T. (1972) U.S. Patent 3664927.
71. Scott, D., "Enzymes in Food Processing," second edition, G. Reed, Editor, 496-498, Academic Press, New York, 1975.
72. Rackis, J. J., Sessa, D. J., Steggerda, F. F., Shimizu, T., Anderson, J. and Pearl, S. L., J. Food Sci. (1970) 35, 634-639.
73. Slott, S., Madsen, G. and Norman, B. E., "Enzyme Engineering 2," E. K. Pye and L. B. Wingard, Editors, 343-350, Plenum Press, New York, 1974.
74. Bernfeld, P. and Wan, J., Science (1973) 142, 678-679.
75. Barker, S. A., Somers, P. J., Epton, R. and McLaren, J. V., Carbohyd. Res. (1970) 14, 287-296.
76. Strumeyer, D. H., Constantinides, A. and Freudenberger, J., J. Food Sci. (1974) 39, 498-502.
77. Ledingham, W. M. and Hornby, W. E., FEBS Lett. (1969) 5, 118.
78. Butterworth, T. A., Wang, D. I. C. and Sinskey, A. J., Biotechnol. Bioeng. (1970) 12, 615-631.
79. Wykes, J. R., Dunnill, P. and Lilly, M. D., Biochim. Biophys. Acta (1971) 250, 522-529.
80. Linko, Y., Saarinen, P. and Linko, M., Biotechnol. Bioeng. (1975) 17, 153-165.

81. Marshall, J. J. and Whelan, W. J., Chem. Ind. (1971) 1971 (25), 701-702.
82. Reilly, Peter, J., 1976, personal communication.
83. Martensson, Kaj, Biotechnol. Bioeng. (1974) 16, 567-577.
84. Martensson, Kaj, Biotechnol. Bioeng. (1974) 16, 579-591.
85. Martensson, Kaj, Biotechnol. Bioeng. (1974) 16, 1567-1587.
86. Bachler, M. J., Strandberg, G. W. and Smiley, K. L., Biotechnol. Bioeng. (1970) 12, 85-92.
87. Smiley, K. L., Biotechnol. Bioeng. (1971) 13, 309-317.
88. Wilson, R. J. H. and Lilly, M. D., Biotechnol. Bioeng. (1969) 11, 349-362.
89. Kay, G. and Lilly, M. D., Biochim. Biophys. Acta (1970) 198, 276-285.
90. O'Neill, S. P., Dunnill, P. and Lilly, M. D., Biotechnol. Bioeng. (1971) 13, 337-352.
91. Christison, J., Chem. Ind. (1972) 1972 (5), 215-216.
92. Park, Y. K. and Lima, D. C., J. Food Sci. (1973) 38, 358-359.
93. Baum, G., Biotechnol. Bioeng. (1975) 17, 253-270.
94. Walton, H. M. and Eastman, J. E., Biotechnol. Bioeng. (1973) 15, 951-962.
95. Maeda, H., Yamauchi, A. and Suzuki, H., Biochim. Biophys. Acta (1973) 315, 18-21.
96. Maeda, H., Suzuki, H., Yamauchi, A. and Sakimae, A., Biotechnol. Bioeng. (1974) 16,11517-1528.
97. Strumeyer, D. H., Constantinides, A. and Freudenberger, J., J. Food Sci. (1974) 39, 498-502.
98. Marsh, D. R., Lee, Y. Y. and Tsao, G. T., Biotechnol. Bioeng. (1973) 15, 483-492.
99. Marsh, D. R., Ph.D. Dissertation, Iowa State University, Ames, Iowa, 1973.
100. Solomon, B. and Levin, Y., Biotechnol. Bioeng. (1975) 17, 1323-1333.
101. Weetall, H. H., Havewala, N. B., Garfinkel, H. M., Buehl, W. M. and Baum, G., Biotechnol. Bioeng. (1974) 16, 169-179.
102. Lee, D. D., Lee, Y. Y., Reilly, P. J., Collins, E. V., Jr. and Tsao, G. T., Biotechnol. Bioeng. (1976) 18, 253-267.
103. Lee, D. D., Lee, Y. Y. and Tsao, G. T., Die Starke (1975) 27, 384-387.
104. Marsh, D. R. and Tsao, G. T., Biotechnol. Bioeng. (1976) 18, 349-362.
105. Sardinas, J. L., Process Biochem. (1976) 11, 10-17.
106. Baricos, W. H., Chambers, R. P. and Cohen, W., Enzyme Technology Digest (1975) 4, 39-53.

107. Bowers, Larry D. and Carr, Peter W., Anal. Chem. (1976) 48, 544A-559A.
108. Weetall, H. H., Anal. Chem. (1974) 46, 602A-615A.
109. Mandels, M. and Sternberg, D., J. Ferment. Technol. (1976) 54, 267-286.
110. Wilke, C. R. and Yang, R. D., Applied Polymer Symposium No. 28 (1975) 175-188.
111. Kirk, T. K., Biotechnol. Bioeng. Symp. (1975) 5, 139-150.
112. Andres, C., Food Processing (1976) 37 (4), 100.
113. Pitcher, W. H., Jr., Catal. Rev.-Sci. Eng. (1975) 12, 37-69.
114. Sundaram, P. V. and Pye, E. K, "Enzyme Engineering," Volume 2, E. K. Pye and L. B. Wingard, Jr., Editors, 449-452, Plenum Press, New York, 1974.
115. O'Neill, S. P., Biotechnol. Bioeng. (1972) 14, 675-678.
116. Lee, Y. Y., Fratzke, A. R., Wun, K. and Tsao, G. T., Biotechnol. Bioeng. (1976) 18, 389-413.
117. Kobayashi, T. and Laidler, K. J., Biochim. Biophys. Acta (1973) 302, 1-12.
118. Korus, R. A. and O'Driscoll, K. F., Can. J. Chem. Eng. (1974) 52, 775-780.
119. Korus, R. A. and O'Driscoll, K. F., Biotechnol. Bioeng. (1975) 17, 441-444.

One of us, RAK, was an NRC-ARS Postdoctoral Research Associate during the preparation of this paper.

8

Conversion of Aldehydes to Alcohols in Liquid Foods by Alcohol Dehydrogenase

C. ERIKSSON, I. QVIST, and K. VALLENTIN

SIK, The Swedish Food Institute, Fack, S-40021 Göteborg, Sweden

Enzymic aldehyde-alcohol conversion

Aliphatic aldehydes and alcohols are frequently found in the volatile fraction of most kinds of foods, e.g., dairy products, meat, poultry, fish, edible oils, vegetables, potato, fruits, and berries. Aldehydes and alcohols can be formed in a food in the normal or abnormal metabolism of tissues and microorganisms, in Strecker degradation during food processing or in lipid oxidation during storage of fresh, dried or frozen foods or food components. The formation of certain odor-potent aldehydes and alcohols from unsaturated fatty acids under the influence of enzymes and hemoproteins, particularly in vegetables, has been summarized earlier (1).

Aliphatic aldehydes, alcohols and esters can be transformed into each other as scheduled by the reaction route.

$$R\text{-}CHO \overset{1}{\rightleftharpoons} R\text{-}CH_2OH \overset{2}{\rightleftharpoons} R\text{-}COO\text{-}R'$$

Step 1 is catalyzed by the enzyme alcohol dehydrogenase (ADH) in the presence of reduced or oxidized nicotinamide-adenine dinucleotides (NADH, NAD^+), while in step 2 formation of an ester can be achieved by ester synthetase systems, in the presence of a carboxylic acid (R'-COOH), a reaction that is less well known for these types of esters. In step 2 the hydrolysis of aliphatic esters involves carboxylic-ester hydrolases. Green pea skins contain such a hydrolase, which was purified by ammonium sulfate fractionation, gel chromatography and two consecutive steps of ion exchange chromatography. This enzyme was found to hydrolyze ethyl-, n-propyl-, n-butyl-, n-pentyl, n-hexyl-, and n-hex-*trans*-2-enylacetate in increasing rate order, while etylpropionate, -buturate, -valerate, and -hexanonate were not hydrolyzed by this enzyme (Yamashita, I., and Eriksson, C., unpublished results).

Earlier investigations have revealed that plant alcohol dehydrogenases have a wide specificity over aliphatic straight chain saturated and unsaturated alcohols and aldehydes. This is true for the enzyme from pea (2), orange (3), potato (4), and tea (5). The equilibrium constant of the reaction

$$\text{Alcohol} + \text{NAD}^+ \rightleftharpoons \text{Aldehyde} + \text{NADH} + \text{H}^+$$

was found to favor alcohol formation at the pH and coenzyme relations that are normally found in plant material. The equilibrium constant also varies due to the nature of the alcohol - aldehyde pair (2).

This variation is exemplified in Table I, which shows the calculated mol percentage of n-hexanal and n-hex-*trans*-2-enal, each in equilibrium with the corresponding alcohol for two different pH and NAD/NADH values.

Table I. Composition of aldehyde-alcohol equilibrium mixtures. Theoretical figures.

NAD/NADH		10	10	100	100
pH		6.0	7.0	6.0	7.0
n-Hexanal	(%)	0.014	0.14	0.14	1.4
n-Hexanol	(%)	99.986	99.86	99.86	98.6
n-Hex-*trans*-2-enal	(%)	1.4	12.5	12.5	58
n-Hex-*trans*-2-enol	(%)	98.6	87.5	87.5	42

Other aldehyde-alcohol pairs will produce still other figures according to chain length, presence of double bonds (number and position), branching etc.

Odor properties of aldehydes and alcohols

The importance of the above knowledge for the flavor of a food was further studied in experiments, where the odor detection concentration in water of n-hexanal, n-hex-*trans*-2-enal, n-hex-2,4-dienal, n-hept-*trans*-2-enal, n-oct-*trans*-2-enal, and the corresponding alcohols was determined (6). It was found that the odor detection concentration of both the aldehydes and the alcohols decreased with the chain length of C_6-C_8 saturated compounds, while it increased with the number of double bonds in C_6 aldehydes and alcohols. This can be seen in Table II, which also shows that the odor detection concentration of alcohols is in all cases significantly higher than that of the aldehydes.

Table II. Odor detection concentration (ODC) of some aldehydes and alcohols (Figures from reference 6)

Compound	ODC (M)	$\frac{ODC_{alc.}}{ODC_{ald.}}$
n-Hexan-1-ol	4.8×10^{-5}	259
n-Hexanal	1.9×10^{-7}	
n-Hex-*trans*-2-en-1-ol	6.7×10^{-5}	21
n-Hex-*trans*-2-enal	3.2×10^{-6}	
n-Hex-*trans*-2,*trans*-4-dien-1-ol	2.4×10^{-4}	49
n-Hex-*trans*-2,*trans*-4-dienal	5.0×10^{-6}	
n-Hept-*trans*-2-en-1-ol	3.7×10^{-5}	83
n-Hept-*trans*-2-enal	3.4×10^{-7}	
n-Oct-*trans*-2-en-1-ol	6.6×10^{-6}	231
n-Oct-*trans*-2-enal	2.9×10^{-8}	

The results concerning the equilibrium and odor properties provided the background of further experiments to reduce the amount of aldehydes in foods by converting them to alcohols by the addition of ADH. This reaction can for instance be applied to liquid foods, whose polyunsaturated fatty acids, although present in small amounts, have oxidized sufficiently to give rise to off-flavor due to aldehyde formation. Such application studies are described later in this article.

Due to the contribution of several aldehydes to this kind of off-flavor and to differences in their reaction rate and equilibrium concentration as well as in odor detection concentration of both aldehydes and alcohols, such a reaction will, or should, not always merely quench the aldehyde odors. The result might therefore be a changed odor sensation from a complex mixture of different aldehydes and alcohols. This sensation might thus not only depend on presence of odor but also an possible quality changes, that occur when aldehydes are converted to alcohols or *vice versa*. For this reason an attempt was made to find out whether corresponding aldehydes and alcohols also differ in odor quality.

Water solutions of n-hex-*trans*-2-enol and n-hex-*trans*-2-enal of varied strength, as well as four different mixtures of these compounds were subjected to descriptive odor analysis. In performing this analysis a specially trained panel with 12 members was used.

The panel was the same as the one used earlier for descriptive odor analysis of soy and rape seed protein (7). The odor notes used in the analysis were selected from the literature and from other investigations in our laboratory on odor properties of pure chemicals. These odor notes were pre-tested with some of the samples to be analyzed. Finally the intensity of **thir-teen** odor notes (Table III) were estimated, using a 0 - 9 point scale. The maximum score, 9, represents the strongest intensity one would expect in normal, everyday life, while the minimum score, 0, meant absence of that particular odor note.

The samples, containing varying concentrations of either n-hex-trans-2-enol or n-hex-trans-2-enal or mixtures of them, were presented as 20 ml portions at room temperature in 50 ml Erlenmeyer flasks provided with clear-fit stoppers. Detailed procedures for the purification of the compounds and sample preparation have been published elsewhere (6).

Table III. Selected odor notes for water solutions of n-hex-trans-2-enol and n-hex-trans-2-enal.

Odor strength	Sweet
Sharp pungent	Musty
Green, like grass	Floral
Green, leafy	Nut-like
Sour, acid	Oily, fatty
Fruity, citrus	Rubber
Fruity, other	Candle-like

In the analysis altogether 10 different samples were evaluated (Table IV) in random order. In each session three different samples were analyzed, until each sample had been evaluated four times.

Table IV. Concentration in mg per litre (ppm) of n-hex-trans-2-enol and n-hex-trans-2-enal in ten samples used for odor description analysis.

Sample No.	n-Hex-trans-2-enol	n-Hex-trans-2-enal
1	10	-
2	40	-
3	100	-
4	-	0.5
5	-	5
6	-	20
7	10	0.5
8	10	5
9	40	0.5
10	40	5

A computer program was used to calculate the mean and standard deviation of the intensity of each odor note for each sample and also to test significant differences between the samples.

It was found that the most typical odor notes of n-hex-trans-2-enol were "Rubber" and "Candle-like" while those of n-hex-trans-2-enal were "Fruity, other" and "Green, leafy" and to a lesser extent "Green, like grass" and "Floral". Odor notes which were perceived equally for both compounds were "Sour, acid", "Fruity, citrus" and "Oily, fatty" (cf. Table III).

The results are exemplified in Figure 1 where the changes in six of the initially selected thirteen odor notes as influenced by concentration are shown. It is obvious that an increase in concentration of the single compounds expectedly is followed by an increase in perceived odor intensity of each individual of the six notes (samples 1-3 and 4-6). This is also true for the mixtures (samples 7-10) except for the odor note "Rubber" which decreased in intensity on increasing the aldehyde concentration alone (cf. sample 7 and 8, 9 and 10 respectively). Most probably, the odor note "Rubber", produced primarily by the alcohol, was masked by the increased intensity of other odor notes introduced by the increased aldehyde concentration.

In conclusion, a conversion between this particular aldehyde and alcohol will result in an overall change in odor intensity with little change in odor quality.

Application studies.

Beer, milk and apple juice are well-known examples of liquid foods which have been reported to develop oxidative flavor due to lipid oxidation, although both beer and apple juice contain only minor amounts of lipids. At this stage these products contain aldehydes like n-hexanal, n-hex-trans-2-enal, n-hept-trans-2-enal, etc. Initial studies revealed that the products mentioned contained no active ADH and no or very small amounts of NAD^+ and NADH. It was also found that n-hexanal added to beer, milk, and apple juice could rapidly and effectively be converted into n-hexanol by the addition of ADH and NADH.

Milk was chosen as the final object of investigation since the pH of apple juice, beer and milk is around 3.5, 4.1, and 6.2, respectively, and on the fact that particularly ADH, but also NADH, is less stable at a pH lower than 6. The milk used was produced by a fistulated cow that had been fed sunflower oil in

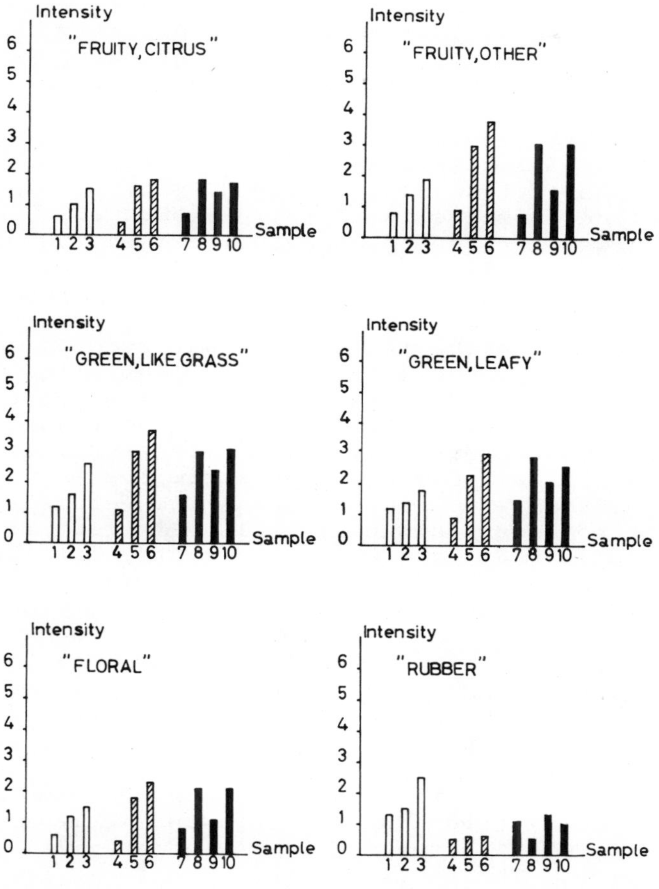

Figure 1. Mean panel intensities of selected characteristic odor notes used in descriptive analysis of 10 water solutions containing n-*hex*-trans-2-*enol (samples 1–3),* n-*hex*-trans-2-*enal (samples 4–6), and mixtures of the two compounds (7–10). The concentration of the odor substances are listed in Table IV.*

order to increase the proportion of polyunsaturated fatty acids in the milk fat. Such milk undergoes rapid lipid oxidation immediately after milking, much larger amounts of n-hexanal and similar aldehydes being formed than in ordinary milk. The sunflower oil was administrated to the cow through a tubing directly into the abomasum in order to prevent the unsaturated fatty acids from being hydrogenated in the rumen, which is the case after oral intake of unsaturated fat. The fatty acid fraction of the milk produced contained 24% oleic, 19% linoleic and 1% linolenic acid. The milk was pasteurized at 63°C for 30 min, 2.5 mg of tetracycline (Dumocyclin, Dumex, Copenhagen, Denmark) being added per litre of milk to increase the microbiological stability. The stabilized milk was then divided into two batches, one which was allowed to contain the original 3.5 per cent of fat (whole milk), and another where the fat content was reduced to approximately 1% (reduced fat milk). The following two experiments were made. In the first experiment one hundred ml portions of full fat milk and reduced fat milk were stored for 4 days at 4°C in sterilized stoppered flasks. Every day four samples each of either whole milk or reduced fat milk were withdrawn for gas chromatographic n-hexanal analysis. In two of these samples from each batch, 3 mg of yeast ADH (300 U/mg, Boehringer, Mannheim, W. Germany) and 50 mg of NADH (78% β-NADH, Boehringer, Mannheim, W.Germany) per 100 ml were added 30 min prior to the n-hexanal analysis. The remaining two samples of each batch were not treated and used as controls. The reaction took place at 20°C. The n-hexanal analyses were performed on days 1 and 3 with the whole milk and on days 2 and 4 with the reduced fat milk.

In the second experiment 3 mg of ADH and 50 mg of NADH per 100 ml were added to both the whole milk and reduced fat milk on the second day after milking and pasteurizing. The samples were then stored for 9 days at 4°C. Controls with no enzyme and coenzyme added were run parallel. The n-hexanal analyses were made on days 3, 5, 7, and 9 with both the enzyme-coenzyme treated and the untreated milk.

For n-hexanal determination two 100 ml samples of milk and 200 ml of distilled water for dilution were transferred on each occasion to a perviously described head space sampling device connected to a Perkin-Elmer 900 gas chromatograph (8). Three hundred ml of head-space gas was forced through a precolumn concentration accessory by air pressure for 30 min whereby the volatile organic compounds in the head space gas were condensed in a cool trap. The condensed material was allowed to enter the injector of the gas chromatograph, kept at 110°C by sudden heating of the trap with an oil bath at 140°C. The separation was performed on an tubular stainless steel column, 0.76 mm I.D. x 181 mm, coated with SF 96/Igepal CO880 (95/5). The oven temperature was programmed 20-140°C at 2°/min after an initial 3 min isothermal period, and the nitrogen gas flow was 12 ml/min. The FID signal was fed into an Infratronics CRS-101 electronic integrator, with digital print-out equipment. The integrator values of the hexanal peak (previously identified by mass spectrometry) were used to represent the n-hexanal content of the head space.

The results given in Tables V and VI show that both discontinuous (first experiment) and continuous (second experiment) treatment of the two kinds of milk either reduced the concentration of the initially formed n-hexanal instantaneously or kept it reduced throughout the experiment. In the gas chromatograms one could also see that the n-hexanol peak rose along with the decline of the n-hexanal peak and that other aldehydes and alcohols behaved similarly. The initial n-hexanal concentration was in the range of 0.5 - 1 ppm, as compared to earlier studies on milk with added n-hexanal.

Table V. Concentration of n-hexanal in the headspace over milk stored at 4°C. Immediate effect of added alcohol dehydrogenase (ADH) and coenzyme (NADH). (Integrator values).

	Days after milking			
	1	2	3	4
Skim milk no addition	860		910	
Skim milk ADH+NADH, 30 min	130		125	
Whole milk no addition		810		850
Whole milk ADH+NADH, 30 min		100		120

Table VI. Concentration of n-hexanal in the headspace over milk stored at 4°C for 9 days. Long term effect of alcohol dehydrogenase (ADH) and coenzyme (NADH) added on day 2. (Integrator values).

	Days after milking			
	3	5	7	9
Skim milk no addition	800	910	1030	1035
Skim milk ADH+NADH added day 2	125	125	125	125
Whole milk no addition	760	820	875	930
Whole milk ADH+NADH added day 2	100	100	100	130

The further application of enzymic conversion of aldehydes and alcohols by the ADH-NADH system must probably await the development of suitable reactors. In material containing enough NADH or NAD^+ in correct concentration ratios this coenzyme content can be utilized in the reaction. When, however, the coenzyme is insufficient or lacking, coenzyme from an external source, such as spent brewer's yeast, must be used. By a standard procedure (9) we found that brewer's yeast contains about 2.5 mg NAD^++ NADH/g dry yeast in a NAD^+/NADH ratio from approximately 2/1 to 10/1 depending on the way of preparation. A high proportion of NADH can be obtained by reduction of NAD^+ either chemically, electrolytically (10) or enzymatically.

Since the number of molecules of coenzyme needed will be the same as the number of convertible aldehyde or alcohol molecules, it will in most cases be both impractical and uneconomical to apply the ADH-NAD(H) technique without disposal of re-usable enzyme-coenzyme systems. Immobilization of enzymes and coenzymes on suitable carriers seems to be a fruitful evolution to enable future conversions of this type. In the case of ADH and NADH both the enzyme and a coenzyme analogue have been immobilized on Sepharose 4 B, which allowed the formation of a binary ADH-NADH complex. The immobilized binary complex was also shown to be functional in the same way as the soluble one (11). The oxidized and reduced coenzymes can easily be transferred into each other _in situ_ by applying alternative substrates. Other suggestions to retain and regenerate coenzyme are microincapsulation (12) and membrane technique (13).

Abstract.

Enzymic interconversion of aliphatic aldehydes, alcohols, and carboxylic esters is briefly surveyed. Odor properties are presented, both quantitative ones, as odor detection concentration and qualitative ones, as odor descriptions of certain aldehydes and alcohols, which are normally found in foods after lipid oxidation. Experiments were performed to reduce the amount of preformed aldehydes, particularly n-hexanal in milk containing polyunsaturated fat, by addition of alcohol dehydrogenase and NADH.

Acknowledgment.

The authors are indebted to Mrs. M. Knutsson, M.Sc., Astra-Ewos AB, Södertälje, Sweden, for delivering polyunsaturated cow's milk for this study.

Literature cited.

(1) Eriksson, C., in "Industrial Aspects of Biochemistry", ed., Spencer, B., Vol. 30, part II, p. 865, North Holland/American Elsevier, Amsterdam, 1974.

(2) Eriksson, C.E., J. Food Sci. (1968) 33, 525.

(3) Bruemmer, J.H. and Roe, B., J. Agr. Food Chem. (1971) 19, 266.

(4) Davies, D.D., Patil, K.D., Ugochukwu, E.N. and Towers, G.H.N., Phytochemistry (1973) 12, 523.

(5) Hatamaka, A. and Oghi, T., Agr. Biol. Chem. (1972) 36, 2033.

(6) Eriksson, C.E., Lundgren, B and Vallentin, K., Chemical Senses and Flavor (1976) 2, 3.

(7) Qvist, I. and von Sydow, E., Lebensm.-Wiss. u. Technol. (1976) in press.

(8) von Sydow, E., Andersson, J., Anjou, K., Karlsson, G., Land, D. and Griffiths, N., Lebensm.-Wiss. u. Technol. (1970) 3, 11.

(9) Klingenberg, M., in "Methoden der enzymatischen Analyse", ed., Bergmeyer, H.U., Vol. 2, p. 1975, Verlag Chemie, Weinheim, 1970.

(10) Alizawa, M., Coughlin, R.W. and Charles, M., Biotechnology and Bioengineering (1976) 18, 209.

(11) Gestrelius, S., Månsson, M-O. and Mosbach, K., Eur. J. Biochem. (1975) 57, 529.

(12) Campbell, J. and Ming Swi Chang, T., Biochim. Biophys. Acta (1975) 397, 101.

(13) Chambers, R.P., Ford. J.R., Allender, J.H., Baricos, W.H. and Cohen, W., in "Enzyme Engineering", ed., Kendall Pye, E., and Wingard, Jr, L.B., Vol. 2, p. 195, Plenum Press, New York, 1974.

9

Physiological Roles of Peroxidase in Postharvest Fruits and Vegetables

NORMAN F. HAARD

Department of Biochemistry, Memorial University of Newfoundland, St. John's, Newfoundland, Canada, A1C5S7

Peroxidase (E.C. 1.11.1.7, hydrogen donor oxidoreductase) was initially reported as a constituent of mushroom extracts in 1855 (1) and was later named by Linossier in 1898 (2). Since this time it has been the subject of continued interest, experimentation and speculation (3). Peroxidase appears to be ubiquitous to the living state having been identified in animal, plant, microbial and viral systems. Peroxidase is a common constituent of higher plants and may occur at concentrations up to several percent on a fresh weight basis. There is extensive literature describing changes in peroxidase activity and isoenzyme spectra as a function of ontogenic change, conditions of stress (e.g. water deficit, freeze injury, chill injury, hypersensitivity, pathogen interaction, hyperoxygenicity, etc.) and as a specific characteristic of plant varieties and cultivars. Although thousands of experimental studies with peroxidase have been published, we have a relatively poor grasp of the function(s) and metabolic control(s) of this enzyme *in vivo*. Today, there are many who believe that peroxidase is a vestige of the past, having no essential function in higher plants. Alternatively, other individuals find it tempting to link peroxidase with a myriad of important events which include respiratory control, gene control, hormone metabolism and the biosynthesis and biodegradation of a wide range of secondary plant metabolites. There have been extensive studies describing the action of peroxidase on substances which yield bright colors on oxidation, but which have no apparent physiological role. Unfortunately, there has been far too little done to understand the physiologically relevant substrates for this enzyme.

Irrespective of their physiological role, it has been well established that peroxidase can contribute to deteriorative changes in flavor, texture, color and nutrition in properly processed fruits and vegetables (4-6). It has generally been observed that peroxidase is quite stable to adverse conditions encountered during food processing - such as elevated temperature, freezing, ionizing radiation, dehydration and even when inactivated by such treatments is capable of re-naturation (7). Regeneration of peroxidase activity can pose a serious limitation to a food processing operation. For example, fruits and vegetables subjected to high temperature - short time treatments (HTST) are particularly prone to peroxidase regeneration and associated quality change during storage. Regeneration of peroxidase occurs within hours or days following thermal processing and may occur even after several months in frozen fruits and vegetables. Peroxidase regeneration appears to involve refolding of the polypeptide chain (8) and may relate to the fact that peroxidase contains a conjugated carbohydrate moiety. The resistance of peroxidase to thermal inactivation, together with its ubiquity and direct involvement in deterioration of food quality, has led to the wide use of peroxidase acitivity as an index of processing efficacy by blanching and other heat treatments.

Peroxidase is also of utility to the food technologist. For example, it is used extensively in conjunction with glucose oxidase for the detection and determination of glucose. In this reaction the hydrogen peroxide generated by the glucose oxidase catalyzed conversion of glucose to gluconic acid is used in the peroxidatic reaction to produce a readily measurable chromophore from a hydrogen donor such as O-dianisidine. The coupled reaction of glucose oxidase and peroxidase may also find use in food processes designed to rid a system of residual glucose or oxygen (9). Also, the finding that isoenzyme profiles of peroxidase may be highly specific for different plant tissues also leads to the suggestion that peroxidase be used as a means of checking adulteration or contamination in plant extractives such as flours and protein isolates.

In this treatise, I will examine several lines of experimentation designed to understand the physiological role of peroxidase in postharvest fruits and vegetables. In most case, harvested fruits and vegetables undergo ontogenic change similar to that which occurs on the parent plant. Accordingly, the reactions to be discussed are somewhat different from those occurring in processed foods where the structure and functional capability of the tissue has been grossly disrupted. It is clear that at least part of the mystery surrounding peroxidases is due to the presence of unusually large numbers of isoenzyme species and to the observation that peroxidase can catalyze a variety of reactions, in some cases with apparently low substrate specificity. For this reason it is important to review some general information on this enzyme.

Classification

Peroxidases have been classified as iron containing peroxidases and flavoprotein peroxidases (3). The iron enzymes are further subgrouped into ferriprotoporphyrin peroxidases and verdoperoxidases. The first group all contain ferriprotoporphyrin III as a prosthetic group and exhibit a red-brown color when highly purified. The ferriprotoporphyrin type peroxidases are common in higher plants and have also been identified in animals (e.g. tryptophan pyrrolase, thyroid iodine peroxidase) and microorganisms (e.g. yeast cytochrome C peroxidase). The verdoperoxidases contain an iron porphyrin prosthetic group which differs from ferriprotoporphyrin III and which is not removed on treatment with acidic acetone as occurs with the former group of peroxidases. Examples of this enzyme are lactoperoxidase found in milk and myeloperoxidase from myelocytes. Flavoprotein peroxidases, contain FAD as prosthetic group, and have been purified from microorganisms (e.g. Streptococcus faecalis)and animal tissues. At this time we know relatively little about the latter two groups of peroxidase in higher plants. However, the ferriprotoporphyrin III peroxidases from horseradish, Japanese radish and turnip have been fairly well characterized. The horseradish enzyme has a molecular weight of approximately 40,000 and contains one ferriprotoporphyrin III group per molecule. Four of the six coordination bonds are taken up in interaction with the pyrrole ring nitrogens. One of the remaining coordination bonds appears to be associated with a carboxyl group of the protein and the other is coordinated to an amino group or to a water molecule. The horseradish enzyme and apparently other sources of peroxidase (10) contain conjugated carbohydrate which appears to impart unusual stability to the molecule. The horseradish enzyme is stable in solution from pH 4 to 12, retains 50% of its activity after heating at 100 C for 12 minutes and is stable to low water activity and freezing.

Because of their characteristic absorption maxima in the U.V. and in the Soret region the peroxidases are often characterized by the ratio of absorption at 403 and 275 nm or the R.Z. value (Reinheitszahl). A highly purified isoenzyme of horseradish peroxidase may have an R.Z. value greater than 3.0.

Catalytic Properties

Four general types of catalytic activity have been found in association with peroxidases. These are the peroxidatic, oxidatic, catalatic and hydroxylation reactions. The peroxidatic reaction, more generally thought to be of most physiological significance, has been studied more extensively than the other three reactions.

Peroxidatic Reaction. In general, peroxidatic reactions occur in the presence of a wide variety of hydrogen donors, including p-cresol, quaicol, resorcinol, benzidine and O-dianisidine. Certain peroxidases appear to have a greater affinity for specific hydrogen donors such as NADH, glutathione or cytochrome C (11). One approach to elucidate hydrogen donors of physiological importance is the use of affinity chromatography (12). The preferred oxidant for the peroxidatic reaction is hydrogen peroxide although other peroxides are effective substrates. In postharvest tissues the peroxidatic reaction of most obvious importance at this time is lignification which can profoundly influence the toughening of vegetables such as beans and asparagus. It has also been suggested that the peroxidatic reaction functions to protect the cellular milieu from peroxides which may cause an imbalance in redox potential and damage membranes, enzymes, etc. (13).

Oxidatic Reaction. The oxidatic reaction requires the presence of molecular oxygen and a suitable hydrogen donor. Examples of so-called redogenic hydrogen donors are indole-3-acetic acid (IAA), oxalacetic acid, dihydroxyfumaric acid, ascorbic acid and hydroquinone. The IAA oxidase function of peroxidase appears to be extremely important in postharvest fruits and vegetables. There is also recent evidence that cytokinins are oxidized by peroxidase. It may also be that the oxidatic function may contribute to deteriorative reactions such as membrane lipid oxidation and oxidation of essential thiol groups. Treatment of peroxidases which are efficient in peroxidatic reaction with certain sulfhydryl reagents will convert them to forms capable of oxidatic catalyzed oxidation of hydrogen donors such as membrane lipids (14). It is tempting to speculate that peroxidase functions in such a deteriorative way in senescing or stressed fruits and vegetables since increases in peroxidase activity invariably precede or accompany the hypersensitive reaction.

Catalatic and Hydroxylation Reactions. In the absence of hydrogen donor, peroxidase can convert hydrogen peroxide to water and oxygen although this reaction is some 1000 times slower than the peroxidatic and oxidatic reactions. Finally in the presence of certain hydrogen donors, such as dihydroxyfumaric acid, and molecular oxygen, peroxidase can catalyze hydroxylation of a variety of aromatic compounds notably tyrosine, phenylalanine, p-cresol and benzoic acid. The metabolism of phenolic substances is of particular importance to the quality of postharvest fruits and vegetables in that they may act as effectors of hormone metabolism, intermediates in lignin biosynthesis and may result in discoloration resulting from their enzymic or non-enzymic oxidation.

Isoenzymes

The occurrence of multiple forms of peroxidase was first noted by Theorell (15) working with horseradish roots. This tissue contains seven major isoenzyme forms, which are similar in molecular weight and amino acid composition but may be separated by ion exchange chromatography or electrophoresis (16). The cationic species contain greater numbers of arginine residues while the anionic species are rich in glutamic acid, phenylalanine and tyrosine. Other sources of isoperoxidases are similar to the horseradish species in their constancy of molecular weight and differences in electrophoretic behavior. Peroxidase from ripening banana fruit is composed of at least twelve isoenzymes which have isoelectric points ranging from approximately 3.3 up to 9.5 (17). This wide range in P_I makes it impossible to isolate and purify all isoenzymes by one technique since certain species are invariably lost by precipitation or adsorption phenomena. There are many studies showing changes in isoperoxidase with respect to growth, differentiation, age and reaction to environmental stresses. We will come back to some specific examples of such change in postharvest systems.

While isoenzyme forms often differ in catalytic efficiency (18), far too little work has been done to elucidate the physiological significance of these multiple forms. One must certainly test the possibility that isoperoxidases are an artifact of the isolation and purification procedures employed (19). It has also been shown that certain isoenzymes may be formed from preexisting species during plant development *in vivo* or *in vitro* during extraction (20). There have been, however, numerous experiments employing inhibitors of transcription and translation, deuterium oxide label and incorporation of labeled-amino acids which demonstrate the *de nova* synthesis of new isoperoxidases during the plant's life cycle (21,22).

Cellular Localization

There are several reports shwoing that peroxidase is localized in various sectors of the cell including ribosomes (23), nucleus (24), nucleolus (24), mitochondria (25), cell walls (26) and in the intercellular spaces (27). Peroxidase may be bound to given cell particulates by ionic interaction or by covalent bonding (26). In most cases, only certain isoenzyme species are found in association with a cellular organelle or fraction. Unfortunately it is often difficult to distinguish localization representative of the native cell and that which arises as a result of cellular disruption. It is, however, intriguing to speculate that nature's rationale for multiple molecule forms of peroxidase lies in their affinity for specific microenvironments within and between the cells. Such microcompartmentation may provide an added dimension of specificity to the peroxidases. In addition, it is known that association of peroxidase with cell surfaces can profoundly affect their catalytic properties. This phenomenon, called allotopic control, will be illustrated in the following case studies.

Control of Lignin Biosynthesis

Lignification may occur in postharvest fruits, vegetables, cereals and pulses and have a profound influence on the edible quality of the tissue. The presence of lignin and associated

TABLE I. Comparison of browning and grit cell formation in different pear cultivars.

Cultivar	Total polyphenolics	Browning	Sclereids
1	40.0	+ + +	+ + +
2	52.5	+ + +	+ + +
3	97.5	+ + + + + +	+
4	22.5	+	+ + + + + +

TABLE II. Total recoverable peroxidase from developing pear fruit. Peroxidase activity, assayed with hydrogen peroxide as oxidant and O-dianisidine as H-donor, has the units A 460 nm/min/mg N at 25 C. 'Yuzuhada' fruit exhibit excessive lignification relative to 'Bartlett' fruit. Data from (34).

days before harvest	peroxidase	
	'Bartlett'	'Yuzuhada'
35	1.19	0.62
28	1.28	1.18
21	0.84	0.82
14	0.89	0.52
7	1.10	0.63
0	1.18	0.82

TABLE III. Soluble peroxidase from developing pear fruit. Peroxidase activity, assayed with hydrogen peroxide as oxidant and O-dianisidine as H-donor, has the units A 460 nm/min/mg N at 25 C. 'Yuzuhada' fruit have excessive lignification compared to 'Bartlett' fruit. Data from (34).

days before harvest	peroxidase	
	'Bartlett'	'Yuzuhada'
21	0.65	0.39
14	0.81	0.29
7	1.08	0.37
0	1.18	0.57

fibrous materials can reduce the digestability coefficient of proteins by as much as eighty percent. Toughening of vegetables, such as asparagus and beans, is due to lignification of fibrovascular tissue which occurs shortly after harvest (28). Similarly, certain fruits contain aggregates of highly lignified cells (sclereids) which impart a gritty texture to the pulp (29). The extent of lignification in such cases is markedly influenced by variety or cultivar type, cultivation practices and postharvest handling.

Peroxidase is involved with conjunction of phenylpropanoid units during the deposition of lignin (30). While other enzymes participate in lignin biosynthesis, evidence points to the peroxidase catalyzed reaction as an important level of control (31). Tissue culture studies have shown that lignin formation occurs only when peroxidase is associated with cell surfaces and that calcium ions elicit solublization of the wall bound enzyme (32, 33). The concept that peroxidase involvement in lignification is subject to allotopic control provided a clue to understanding and solving a problem encountered by fruit growers.

"Yuzuhada Disorder". Pear fruit may exhibit a physiological disorder called "Yuzuhada" which is symptomized by excessive sclereid development (29). "Yuzuhada" is transmitted as a hereditary characteristic but may develop in any cultivar which is subjected to adverse growing conditions. Our survey of some forty selections of pear fruit indicated an inverse relationship between free phenolic substances, noteably chlorogenic acid, and the degree of sclereid formation (Table 1)(34). This observation suggested any given pear selection will invariably either be highly subject to enzymic browning or to lignification and the most acceptable selections were intermediate in levels of free phenolics and in lignin content. From these findings we hypothesized that mineral nutrition and localization in the developing fruit determine the extent of peroxidase association with the lignin template(s) which, in turn, delimits the conjugation of phenylpropanoid units into lignin. The thesis was supported by our study of peroxidase distribution in soluble and particulate pools. Comparison of peroxidase in a pear selection extremely prone to sclereid development with a less susceptible variety at various stages of fruit development revealed that peroxidase localization and not its net activity were positively related to sclereid formation. Although total peroxidase was consistently higher in the pear selection exhibiting minimal lignification (Table 2), it was primarily localized in the soluble phase (Table 3). Alternatively, a relatively high fraction of peroxidase was associated with cell particulates in the pear selection exhibiting "Yuzuhada" disorder (Table 4). Histochemical examination of these fruit for peroxidase confirmed the observation that wall-bound peroxidase was more extensive in the fruit exhibiting excessive lignification. The calcium content of the pulp was also consistent with

TABLE IV. Cell wall and membrane-bound peroxidase from developing pear fruit. Peroxidase activity assayed with hydrogen peroxide as oxidant and O-dianisidine as H-donor has the units A 460 nm/min/mg N at 25°C. "Yuzuhada" fruit exhibit excessive lignification relative to 'Bartlett' fruit. Data from (34).

days before harvest	peroxidase 'Bartlett'	'Yuzuhada'
21	0.19	0.43
14	0.09	0.23
7	0.02	0.26
0	0.00	0.23

TABLE V. Calcium content of developing pear fruit. Calcium was determined by atomic absorption spectrophotometry on pulp. Note lower levels of calcium during early stages of development where lignification is initiated. Data from (34).

days before harvest	calcium (ppm) 'Bartlett'	'Yuzuhada'
41	46	16
34	56	30
27	68	32
21	56	40
14	54	36
7	48	36
0	52	52

our thesis (Table 5). The lower calcium content of "Yuzuhada" fruit throughout early development may have provided an environment conducive to peroxidase association with lignin template. The influence of calcium on peroxidase release from cell walls is shown in Figure 1. Although additional study is needed to confirm this model it provides an attractive explanation relating mineral nutrition of the tree to important quality parameters of the fruit.

Fiber Formation in Postharvest Asparagus. Asparagus spears undergo texture changes that occur during maturation, either in the field or in storage, and these changes progress from the butt end of the stem to the tip. Toughening of asparagus is due to lignification of fibrovascular tissue and occurs within hours after harvest in spears stored at ambient conditions (35). In view of the apparent involvement of peroxidase-wall associations in developing pear fruit, we thought it would be of interest to see whether similar controls are operative in the asparagus stem. If so, brief treatment of spears with solutions containing calcium salts would be expected to prevent lignification after harvest. Attempts by us to achieve such beneficial effects were unsuccessful apparently because of the greater tenacity of asparagus peroxidase for the fibrovascular bundles. Levels of calcium effective in arresting lignin deposition (0.5M $CaCl_2$) invariably caused extensive tissue damage which was followed by pathogen invasion. The solublization of peroxidase from asparagus tissue required approximately five-fold higher levels of Ca^{++} than was observed for pear fruit (Figure 2)(28). While we observed no differences in the degree of wall binding of peroxidase and relatively little difference in extractable activity during aging of the spears (Figure 3), the initiation of rapid lignification was closely paralleled by the emergence of new isoperoxidase species (Figure 4). The induction of new isoperoxidase species and lignin deposition occurred within a few hours after cutting and were dependent on protein synthesis. The emergence of new isoperoxidase species progresses from the butt end of the spear to the tip. A similar progression from the butt to the tip is observed in the toughening process. The emergence of the predominent new isoperoxidase species correlated closely with lignin deposition in the three segments of the spear (Figure 5). These findings suggest that de nova synthesis of new isoenzyme species is causally related to the rapid onset of fiber formation in asparagus.

Why is it that the simple process of cutting triggers the rapid biosynthesis of lignin? It has also been observed that the height of cut during harvest dramatically influences the rate of toughening of spears (Figure 6). Spears cut at increasing heights above the ground level toughen much faster than those cut at below ground level as was the practice prior to the introduction of mechanical harvesting. Our data indicate that the cutting operation leads to formation of wound ethylene. Spears cut above

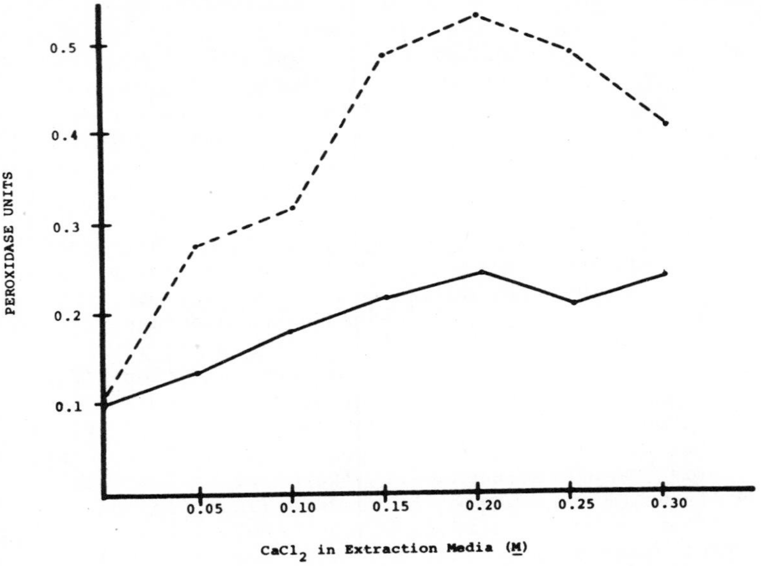

Journal of Food Science

Figure 1. Effect of calcium chloride on peroxidase release from pulp tissue of Bartlett (——) and Yuzuhada (– – –) pear fruit (29)

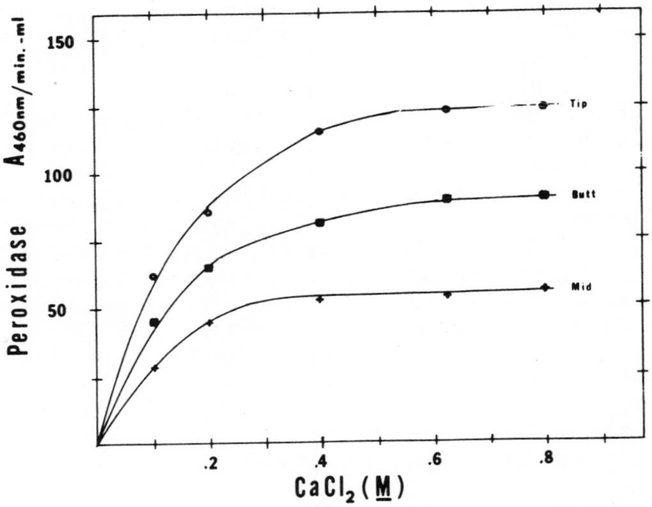

Journal of Food Science

Figure 2. Solubilization of peroxidase from freshly harvested asparagus spear tissue by calcium chloride. Spears were divided into tip (○ – ○), mid (+ – +), and butt (■ – ■) sections prior to extraction (29).

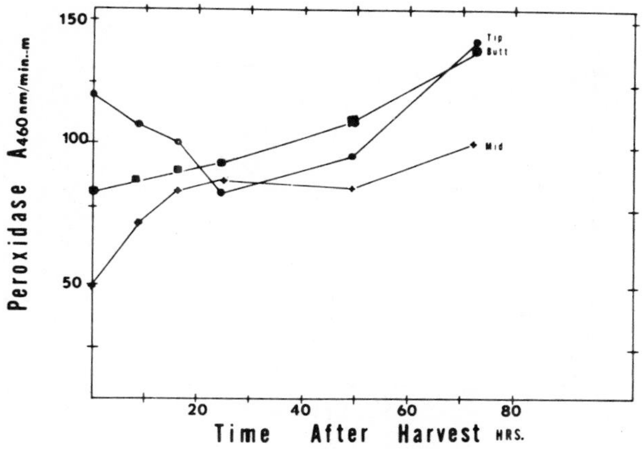

Journal of Food Science

Figure 3. Recovery of total peroxidase from spear tissue stored at 20°C for up to 72 hr (29)

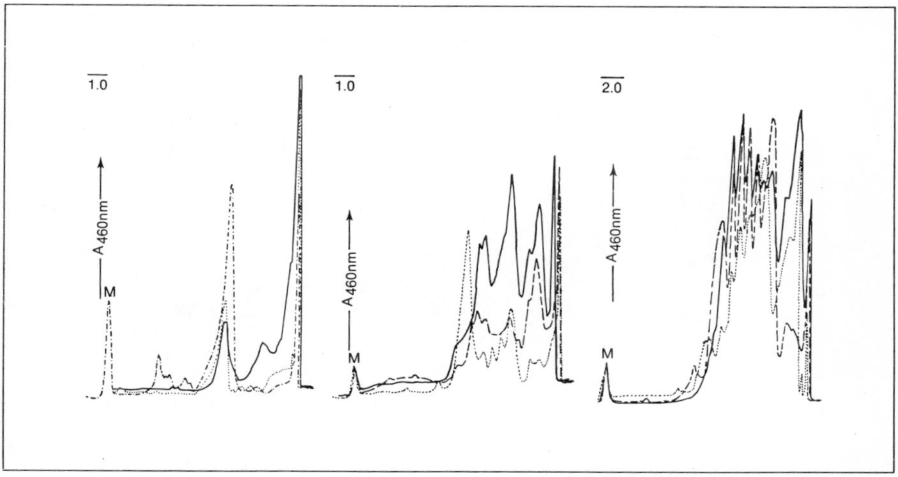

Journal of Food Science

Figure 4. Densitometric scans (460 nm) of polyacrylamide gels of 10 μL peroxidase extracts from: (A) spears frozen in liquid nitrogen immediately after harvest; (B) spears stored at 20°C for 24 hr after harvest; and (C) spears stored at 20°C for 48 hr after harvest. Spears were divided into tip (– · –), mid (– – –), and butt (——) sections (29)

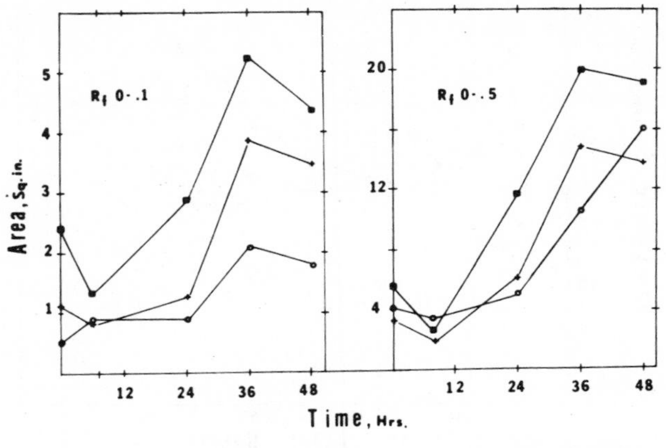

Figure 5. Area in square inches of densitometric scans with R_f *0.0–0.1 and* F_f *0.0–0.5 as a function of storage time at 20°C. One inch represents* A_{460nm} *of 0.1 and approximately 1-cm distance along the gel. Tip (○ – ○), mid (+ – +), and butt (■ – ■) sections, respectively, showed increased lag in emergence of isoenzyme species with the indicated* R_f*(29).*

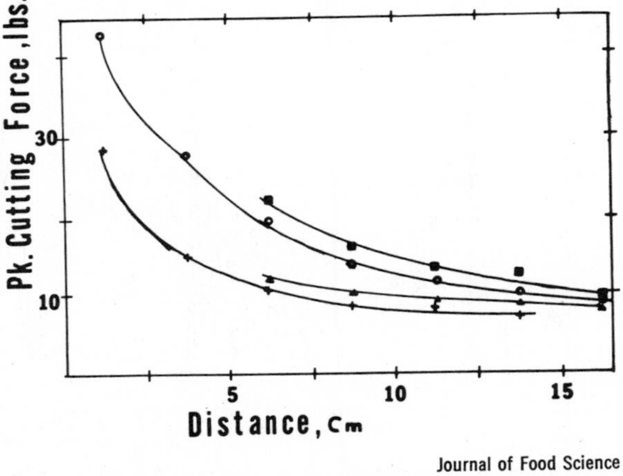

Figure 6. Effect of height of cut toughening of spears stored at 10°C for 2 hr (▲ – ▲; + – +) and 72 hr (■ – ■; ○ – ○). Spears were cut at 5 cm above ground level (▲–▲; ■–■) or at ground level (○–○; +–+). Each data point represents an average of readings from four spears (29).

ground level produce ethylene at a greater rate than those cut below ground level (Table 6). Moreover, treatment of spears with exogenous ethylene accelerates the emergence of new isoperoxidase species and, as well, the rate of spear toughening (28). Our suggested role of ethylene in lignification also lends nicely to previous observations that controlled atmospheres employing elevated carbon dioxide depress the onset of fiber development (36). Carbon dioxide is a competitive inhibitor of all known ethylene mediated events in plant tissue (37).

Biosynthesis of Stress Metabolites

Sweet potato roots accumulate a variety of metabolites in response to stress (38). These include phenolic substances, coumarin derivatives and a family of fifteen and nine carbon furanoterpenoids (Figure 7). These furanoterpenoid stress metabolites are of interest to us because they accumulate at low levels in marketable sweet potatoes (39) and are toxic to mammals (40). Ipomeamorone and ipomeamaronal are hepatotoxins and the nine carbon compounds 4-ipomeanol, 1-ipomeanol, 1.4-ipomeadiol and ipomeanine are lung and kidney toxins. Evidence has been provided that these terpenes are fungistatic agents which contribute to disease resistance of the sweet potato root (38). Earlier studies showed that on infection with the black rot organism (*Ceratocystis fimbriata*) roots accumulate peroxidase in the healthy layer of cells adjacent to the zone of incipient infection (41). These findings led to the notion that a specific isoperoxidase species acts as an effector of protein synthesis by virtue of its ability to modify lysine residues in histone proteins, and thereby acts to elicit the formation of furanoterpenoids (42). If this concept is correct it may help us understand why root tissue accumulates low levels of furanoterpenoids as a result of stresses imposed during storage and handling. We also had the idea that a unique isoperoxidase may be used as a simple indicator for the presence of stress metabolites in root tissue.

Ethylene can increase the resistance of sweet potatoes to infection by *C. fimbriata* (43). Such resistance is also associated with increased peroxidase and polyphenoloxidase activity (38).

Earlier studies indicated that ethylene released from sweet potato tissue by invasion of *C. fimbriata* took part in increasing peroxidase activity (44). Matsuno and Uriani (45) reported that sweet potato roots contain 4 major and 3 minor isoperoxidase species. Black rot infected species exhibited no change in isoperoxidase numbers while cut-injured and ethylene treated root slices contained a new cationic species called component H. Induction of component H was accompanied by the formation of a lignin-like substance on the cut surface. These workers provided additional evidence that component H was efficient in the conjugation of phenylpropanoid monomers into lignin. Thus, while cut injury and ethylene led to increases of constitutive enzymes and induction of a new isoperoxidase, infection only resulted in increases in the constitutive-enzymes. While these data appear to negate any hypothesis suggesting the involvement of a unique isoperoxidase in the elicitor response we continued with this work because previous studies did not account for isoperoxidases associated with the cellular particulate fractions.

Our study showed that sweet potato roots contain ionically bound peroxidase which increased slightly as a result of cut injury and quite dramatically as a result of ethylene treatment and infection (Figure 8). The rate at which peroxidase activity increased after inoculation was dependent on the virulence of the mold. That is, a more virulent culture of C. fimbriata gave a response like that shown in Figure 8, while a less virulent mold showed a response similar to that observed for ethylene treatment also shown in this figure. These results indicate that changes in peroxidase activity do occur in the ionically bound fraction and that factors associated with infection, other than ethylene, appear to effect the speed at which this change occurs. The resulting accumulation of furanoterpenoids parallels peroxidase induction in virulent and non-virulent interactions. Sweet potato roots also contained a "covalently bound" peroxidase; however, the total activity of this fraction was not influenced by cut injury, infection or ethylene treatment (52).

Comparison of isoperoxidases in cut-injured, ethylene treated and black rot infected roots revealed differences in the soluble fraction of peroxidase (Figure 9), but no evidence of unique isoperoxidase or isoperoxidase patterns in the ionic or covalent bound fractions. While isoperoxidase changes were clear and reproduceable in roots stressed in the described manner, we found no evidence for a unique isoenzyme associated with infection which was not also associated with either ethylene treatment or simple cut injury. The control root tissue contained four major isoenzymes (6-9) and eleven additional components. Cut-injured tissue contained the same 15 isoperoxidases and, in addition, changes indicative of 3 new zones (4a, 5a, 10a). The most evident change resulting from cut injury was the emergence of zones 12 and 14 which were barely detectable in the untreated tissue. A similar change in isoperoxidase pattern was also observed in the ethylene treated root slices, although zones 12 and 14 increased to a much greater extent in the latter sample. Infection with the black rot organism led to somewhat different isoperoxidase patterns. Zones 12 and 14 increased as occurred in the other samples and zone 13 increased to a greater extent than cut injured tissue but to a lesser degree than ethylene treated tissue. In addition, there was a more striking increase in zones 2, 4, 4a and 5 relative to the other treatments, although we could find no evidence for a unique isoperoxidase. The broad range of electrophoretic mobilities and the complexity of changes which occur makes evaluation of the data difficult and

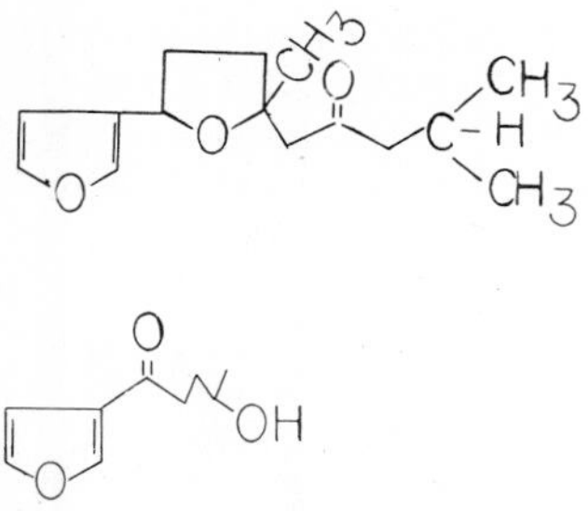

Figure 7. Stress metabolites from sweet potato root ipomeamorone, top—a hepatotoxin and 4-ipomeanol, bottom—a lung toxin

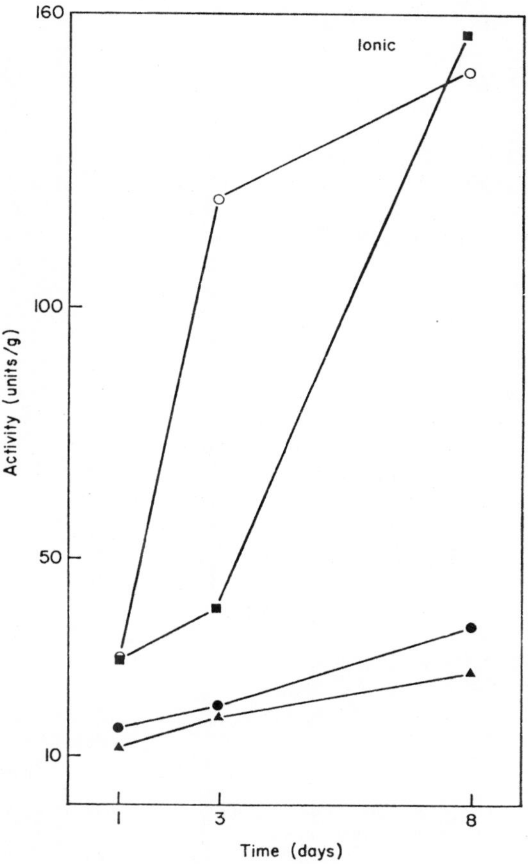

Physiological Plant Pathology

Figure 8. Ionically bound peroxidase from control (▲), cut-injured (●), ethylene-treated (+), and infected (○) sweet potato root sections (66)

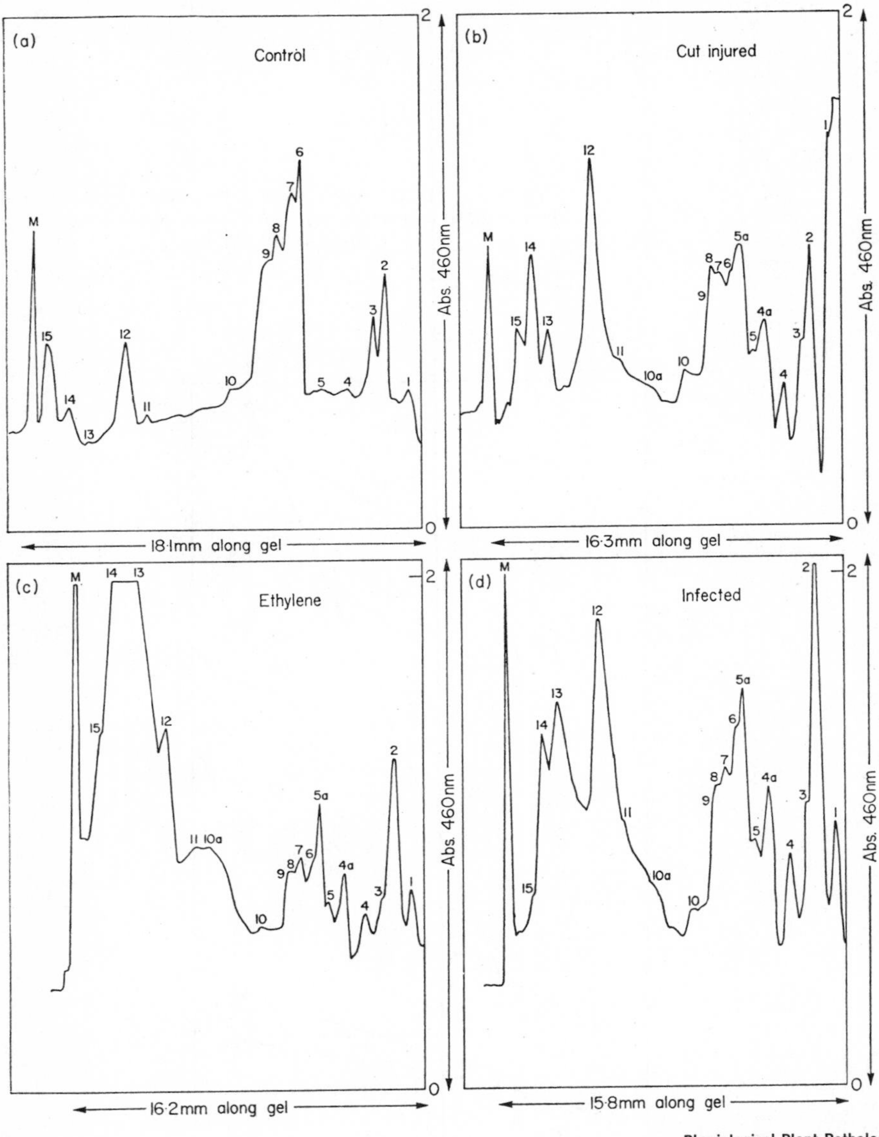

Physiological Plant Pathology

Figure 9. Densitometric scans of soluble isoperoxidases isolated from control, cut-injured, ethylene-treated, and infected sweet potato root sections after eight days' treatment. Data are representative of three additional trials. A 15 μL sample was applied to each respective gel. M *in densitometric scans indicated position of marker dye. Isoenzyme changes were similar for extracts obtained after a 1- and 3-day treatment. No peroxidase migrated into the gel when the electrodes were reversed (66).*

it is possible that an elicitor isoenzyme is present. Some studies on the oxidatic function I'll describe later indicate we may have measured the wrong function in this system.

Control roots contain two readily detectable isoperoxidases in the ionically bound fraction (Figure 10). Cut injury resulted in striking increases in zone 1 and the appearance of 3 new isoenzymes (1a, 3, 4). Changes occurring as a result of C. fimbriata infection and ethylene treatment were essentially identical to those during cut injury. Again, the relative increases were different. Zone 1a from ethylene treated tissue increased more dramatically than in the control while zone 1 was similar in all samples. These changes are similar to those reported by Matsuno and Uritani (45) and they identified these cationic zones as component H. As previously discussed, they provided evidence that component H functions in lignin deposition. The covalent bound peroxidase contained nine readily identifiable isoperoxidase zones in control tissue and 3 new species (3a, 7a, 9) appeared as a result of treatments (32). Again, there was no isoperoxidase change clearly unique to infection. From these data we conclude that changes in isoperoxidase, which closely parallel stress metabolite formation in the infected tissue, are not causally related to terpene induction, Figure 11.

More recent studies in our laboratory indicate that the IAA oxidase function of peroxidase shows dramatic increases which occur differently in ethylene treated or cut injured tissue (Figure 12). The rise in IAA oxidase again parallels stress metabolite formation. These data lead us to suggest that it is the oxidatic function of peroxidase and not the peroxidatic function which may relate to the elicitor response. It is tempting to speculate that a unique IAA oxidase catalyzes oxidation of IAA to specific metabolites which, in turn, de-repress gene expression or otherwise affects biosynthesis of furanoterpenoids. Our preliminary data has shown that the oxidation products of IAA resulting from catalysis by the enzyme isolated from infected tissue differ from those reported for other sources of IAA oxidase.

Role of Peroxidase in Fruit Ripening

Peroxidase activity appears to increase during ripening and senescence in fruits, e.g. mango (46), grape (47), apple (48), pear (49) and banana (50). As peroxidase has been implicated in numerous events of clear importance in plant senescence it is possible that it plays a regulatory role. These events include lignin synthesis in certain systems, as described previously, ethylene biogenesis (51), membrane integrity (52), respiration control (53) and oxidative metabolism of auxin (54) and cytokinin. We have taken an interest in the involvement of auxin metabolism in the process of fruit ripening and have characterized the indole-3-acetic acid oxidase function of peroxidase from normal ripening and mutant cultivars of tomato fruit. Two views have

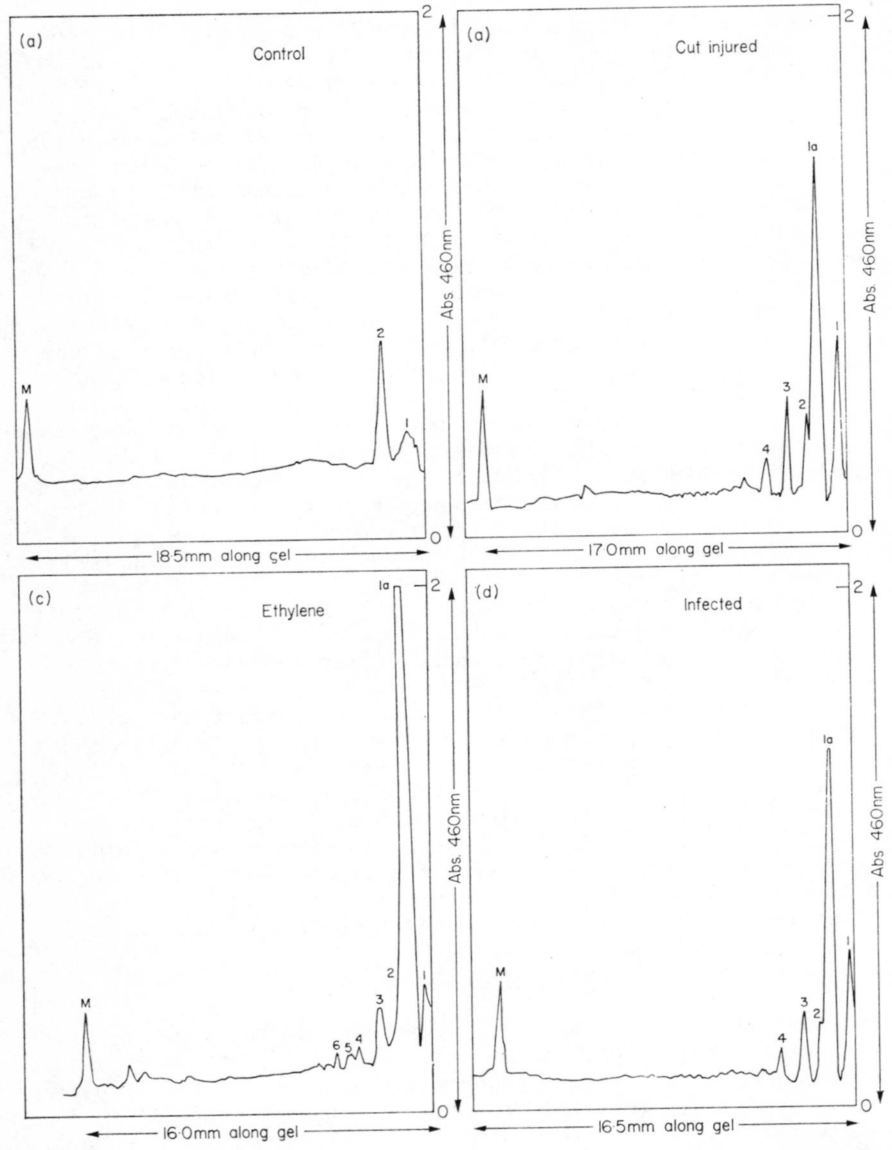

Physiological Plant Pathology

Figure 10. Densitometric scans of ionically bound isoperoxidases isolated from control, cut-injured, ethylene-treated, and infected sweet potato root sections after eight days' treatment. Data are representative of three additional trials. A 60 μL sample was applied to each respective gel. M *in densitometric scans indicates position of marker dye. Isoenzyme changes were similar for extracts obtained after a 1- and 3-day treatment. No peroxidase migrated into the gel when the electrodes were reversed* (66).

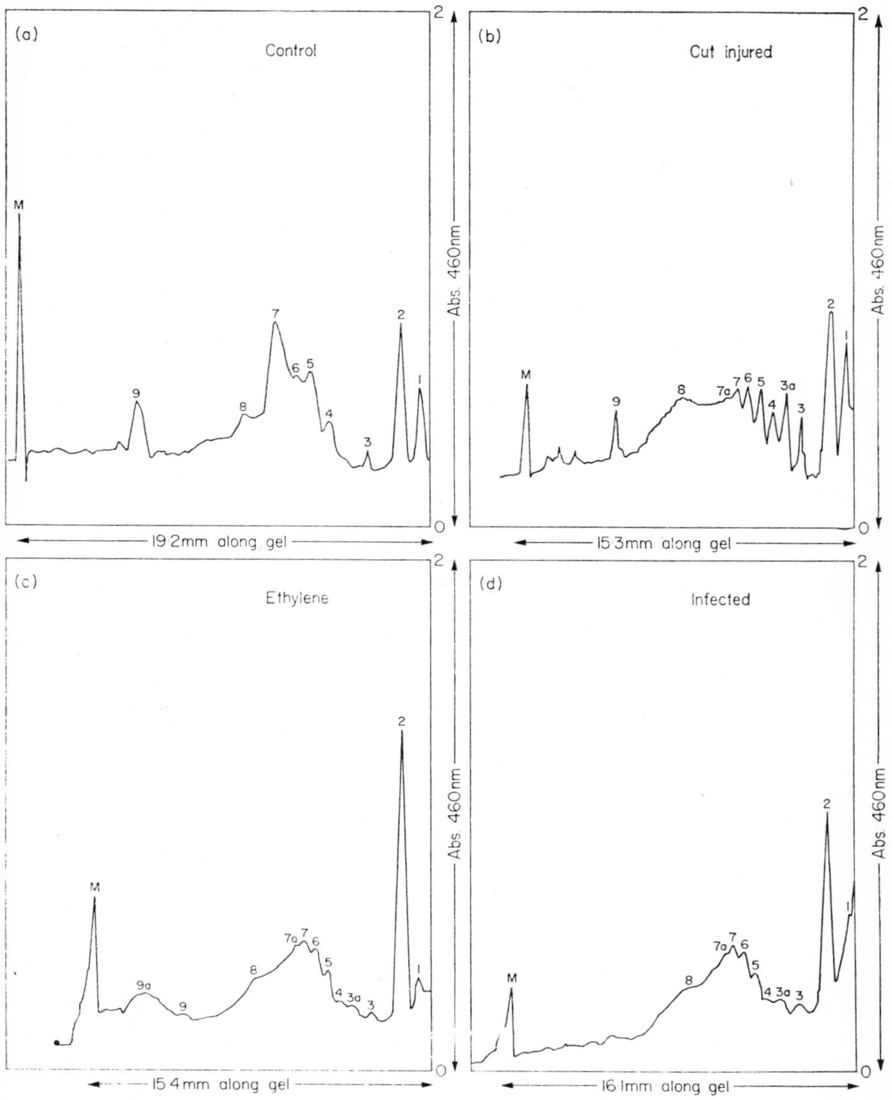

Physiological Plant Pathology

Figure 11. Densitometric scans of covalently bound isoperoxidases isolated from control, cut-injured, ethylene-treated, and infected sweet potato root sections after eight days' treatment. Data are representative of three additional trials. A 100 μL sample was applied to each respective gel. M *in densitometric scans indicates position of marker dye. Isoenzyme changes were similar for extracts obtained after a 1- and 3-day treatment. No peroxidase migrated into the gel when the electrodes were reversed* (66).

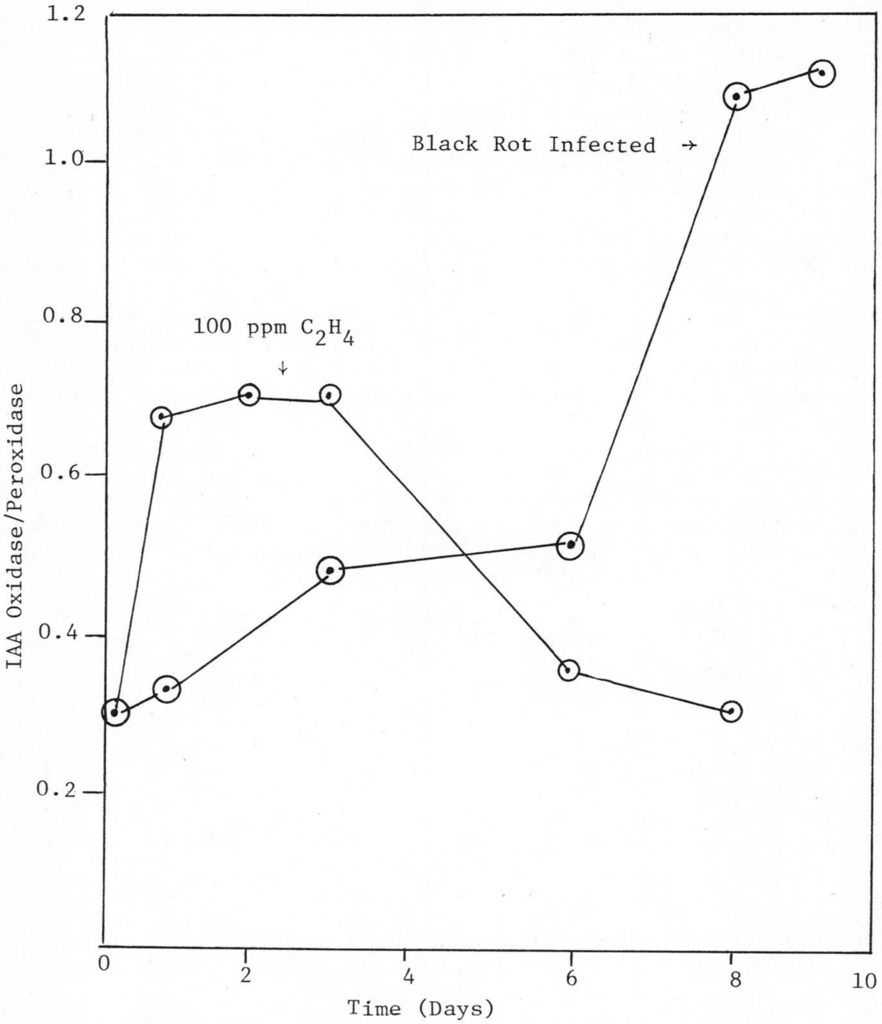

Figure 12. Change in ratio of oxidatic to peroxidatic function of peroxidase from sweet potato root after treatment with exogenous ethylene or infection with C. fimbriata

emerged regarding the involvement of IAA in fruit ripening. On the one hand it has been argued that IAA serves to repress genes which code for enzymes involved with fruit ripening and subsequent senescing reactions. These reactions involve both anabolic (e.g. pigment synthesis) and catabolic events (e.g. protein hydrolysis). The suggestion that the presence of auxin acts to repress ripening is supported by various findings that application of exogenous IAA to fruit causes a transient delay in the process (55,56). A difficulty with this view is that levels of exogenous auxin effective in delaying ripening range from .01 to 1mM (Figure 13), while concentrations of IAA in mature fruit are several orders of magnitude lower. One can, however, argue that the applied hormone must be introduced at a concentration gradient sufficient to reach a particular cellular compartment at effective levels. However, recently it was shown that μM levels of IAA delay ripening of avocado fruit (57). These studies are also complicated by the wound reaction resulting from working with tissue slices or infiltration procedures. Alternatively, it has been suggested that, rather than IAA being a negative effector, oxidation product(s) of IAA serves as a positive effector of ripening (58). This view is supported by findings that IAA oxidase activity increases during ripening of fruit (59) and by reports that methylene oxidole, an oxidation product of IAA, as catalyzed by the horseradish enzyme, stimulates fruit ripening (60). Also, it is known that an analogue of auxin, 1-(2,4 dichlorophenoxy) isobutyric acid (CPIBA) may serve to modulate the activity of IAA oxidase from fruit in different ways. In banana, where CPIBA represses fruit ripening (61) the analogue is known to inhibit the IAA oxidase from this tissue. Alternatively, CPIBA was shown to accelerate ripening of pear fruit where it also activated IAA oxidase (62)(Figure 14). Similarly, it has been reported that the activity of IAA oxidase at low IAA concentration is decreased dramatically at chilling temperatures which coincidently result in impaired or repressed ripening (63)(Figure 15). Recently, we have attempted to determine factors responsible for control of IAA oxidase in ripening tomato fruit.

Our examination of IAA oxidase from normal tomato cultivars revealed no large differences in kinetic properties of the purified enzyme as a function of ripening. Similarly, the enzyme from the RIN mutant, a cultivar which fails to ripen, had similar kinetic properties and recoverable activity as was observed for the normal ripening cultivars. However, it was observed that components in the phenolic fraction from ripening tomato fruit served to lower the apparent K_m of the IAA oxidase for auxin (Table 7). These data indicate that IAA oxidase may be modulated by changes in the phenolic pool. The influence of phenolics on IAA oxidase is complicated since dihydroxy phenols such as chlorogenic acid may inhibit the enzyme in a way manifested by a lag in the initiation of IAA oxidation. Alternatively, monohydroxyl phenols may activate the reaction by decreasing the lag time, ser-

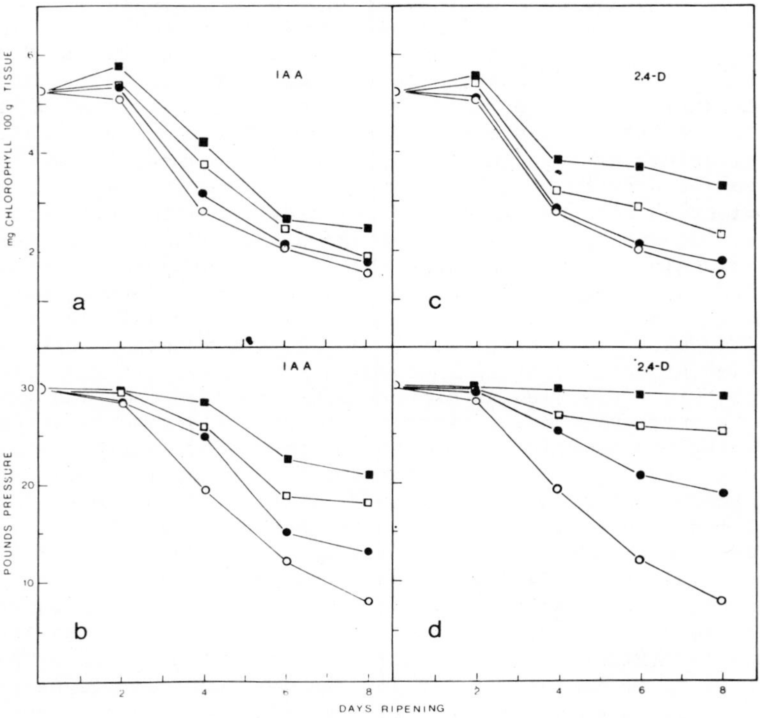

Postharvest Biology of Fruits and Vegetables

Figure 13. The effect of IAA on chlorophyll breakdown is shown in a *and* c. *The effect of IAA and 2,4-D on softening is presented in* b *and* d. *The auxin concentrations in the infiltration solutions were: zero (mannitol control) (○); 0.01mM (●); 0.1mM (□); and 1.0mM (■) (56).*

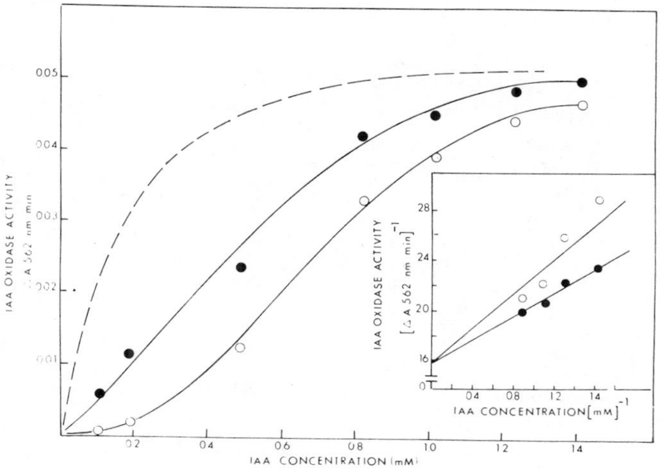

Plant Physiology

Figure 14. Substrate–velocity plot of IAA oxidase from pear in the presence of CPIBA. The sigmoidal kinetics of the enzyme (○) become less pronounced in the presence of 0.5mM CPIBA (●). The normal or expected hyperbolic kinetics are shown for comparison (– – –) (62).

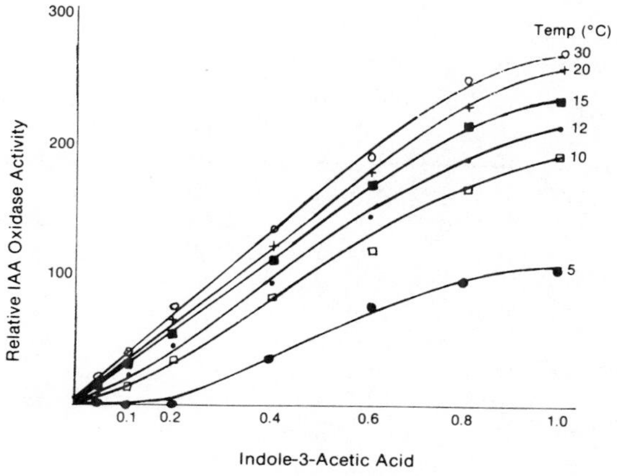

Plant Physiology

Figure 15. Influence of temperature on the activity of IAA oxidase isolated from preclimacteric banana fruit (63)

TABLE VI. Ethylene evolution from asparagus spears as a function of height of cut. Data from (28).

Time after cutting (min)	Height of cut (cm)	Ethylene evolution	
		µl/Kg-Hr	µl/Hr-spear
45	–5	2.05	0.0486
	0	2.13	0.0400
	+5	3.71	0.0532
90	–5	2.61	0.0620
	0	3.13	0.0590
	+5	4.91	0.0702
165	–5	2.61	0.0020
	0	3.12	0.0688
	+5	5.51	0.0788

TABLE VII. Influence of polyphenolic substances from tomato on kinetic parameters of tomato IAA oxidase.

Source phenolics	K_M (mM)	Vmax (µg O_2/min-ml)
None	0.75	21.4
Vendor, Mature green	0.71	17.7
Initial climacteric	0.57	17.0
Climacteric	0.31	10.2
RIN Mutant, aged	0.78	13.3

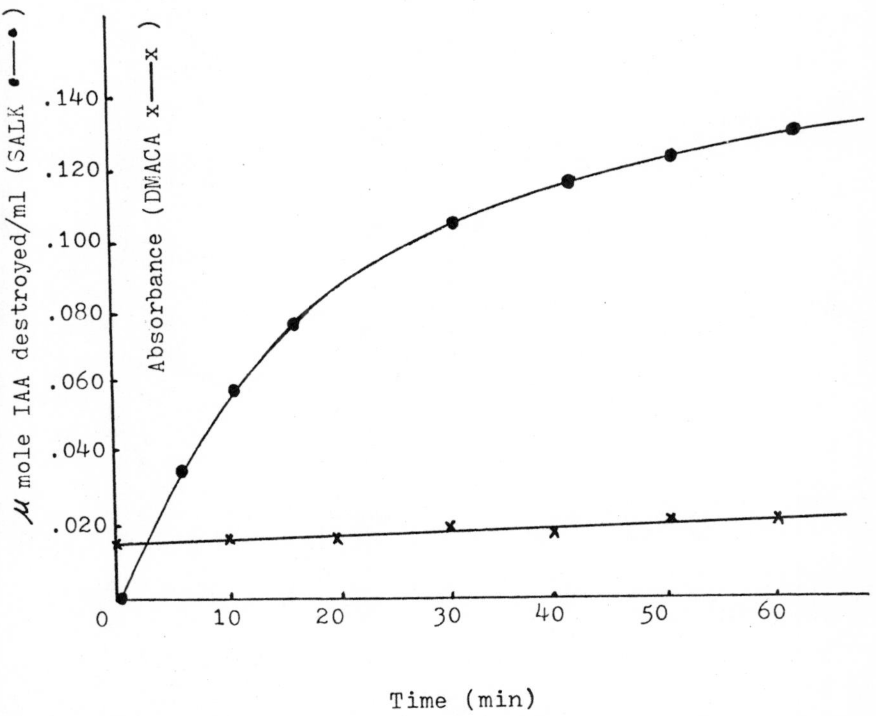

Figure 16. Comparison of DMACA test and Salkowski test of tomato IAA oxidase activity. Reaction mixtures, final volume is 2.0 mL, contained 0.20mM IAA, 0.10mM DCP, 0.10mM Mn^{+2}, 50mM sodium phosphate (pH 6.1), and 0.10mL enzyme extract from RIN mutant tomatoes.

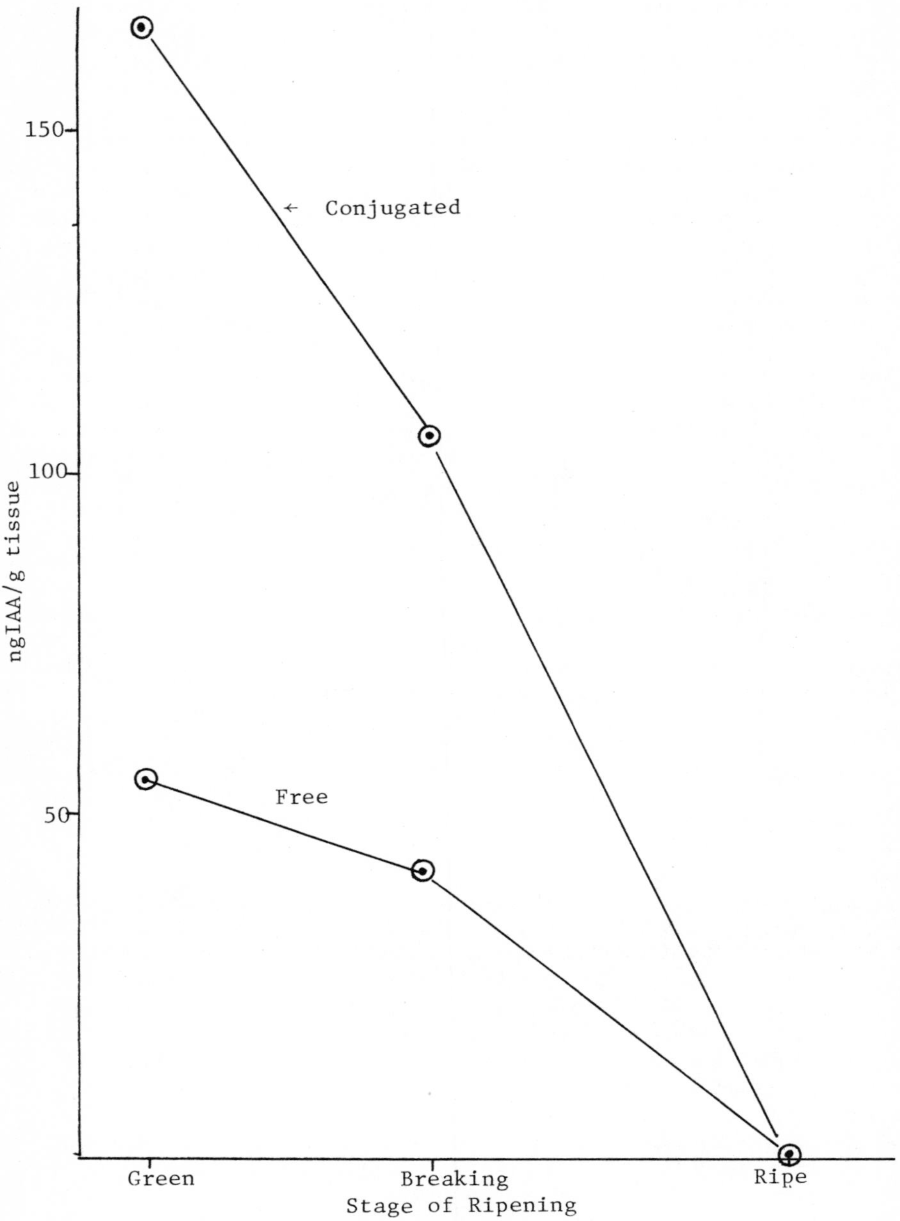

Figure 17. Change in IAA recovered in conjugated and free form from ripening tomato fruit. Data for detached mature green fruit (Vendor).

ving as a cofactor for the reaction or, as shown in these data, by otherwise affecting the kinetic properties of the enzyme. A most interesting feature of the tomato enzyme is that it yields different products than horse radish peroxidase and other sources of the enzyme that have been studied. Assay of tomato IAA oxidase by the dimethylaminocinnamaldhyde procedure (64) or by a fluorometric method (65) reveal no apparent activity although it is clear that IAA is being oxidized (Figure 16). We have ruled out the formation of methyloxindole, methyleneoxindole and indole-3-aldehyde although the product exhibits a similar U.V. spectra as the aldehyde. A component with identical chromatographic properties as the principle product of this reaction also accumulates as a result of the oxidation of radiolabeled IAA in ripening tomato fruit. Accordingly we are intriqued by the possibility that a specific oxidation product of IAA serves to modulate the "ripening genome".

We have also recently shown that IAA declines in tomato concommitently with ripening and that the rate of change is affected by environmental factors which coincidentally also affect fruit ripening (Figure 17). It was also of considerable interest to find the vine ripened tomato fruit accumulate more oxidation products of IAA during ripening apparently because IAA is continually supplied to the fruit after the initiation of the process. This may possibly indicate that the messenger RNA's and their respective enzymes associated with ripening have a relatively short half life thus requiring continuous gene expression for optimal development of ripening indices.

Alternatively, the RIN mutant contained no detectable free IAA, and relatively low levels of conjugated IAA. This may mean that factors which control the formation and storage of IAA and the conversion of IAA from bound to free forms are an abberant control site in these fruit.

Literature Cited

1. Schoenbein, C., Verh. Naturf. Ges. Basel (1885) 1: 339.
2. Linossier, M., Soc. Biol., Paris (1898) 50: 373.
3. Saunders, B., Holmes-Siedle, A., Stark, M., "Peroxidase". Butterworths, London, 1964.
4. Blundstone, H., Woodman, J., Adams, J., "Biochemistry of Fruits and their Products". A.C. Hulme, Editor. vol. 2, p. 561. Academic Press, New York, 1971.
5. Ben Aziz, A., Grossman, S., Ascarelli, I., Budowski, P., Phytochem. (1971) 10: 1445.
6. Markakis, P., "Postharvest Biology and Handling of Fruits and Vegetables". N. Haard and D. Salunkhe Editors. p. 62. AVI, Westport, Conn., 1975.
7. Schwimmer, S., J. Food Sci. (1972). 37: 350.

8. Wang, S. and DiMarco, G., J. Food Sci. (1972), 37: 574.
9. Reed, G., "Enzymes in Food Processing". p. 243. Academic Press, New York, 1975.
10. Morita, Y. and Kameda, K., Mem. Res. Inst. Food Sci., Kyoto Univ. (1958) 14: 49.
11. Altshchul, A., Abrams, R., Hogness, T., J. Biol. Chem. (1940) 136: 777.
12. Huang, A., Chism, G., Haard, N., J. Food Sci. (1975) 40: 1327.
13. Levitt, J., "Responses of Plants to Environmental Stresses". p. 147. Academic Press, New York, 1972.
14. Yamauchi, T., Yamamoto, S., Hayaishi, O., J. Biol. Chem. (1975) 250: 7127.
15. Theorell, H., Ark. Kemi. Min. Geol. (1943) 16A: No. 2.
16. Shannon, L., Kay, E., Lew, J., J. Biol. Chem. (1966) 241: 383.
17. Nagle, N. and Haard, N., J. Food Sci. (1975) 40: 576.
18. Shinshi, H. and Noguchi, M., Phytochem. (1975) 14: 1255.
19. Keilin, D. and Hartree, E., Biochem. J. (1951) 49: 88.
20. Liu, E. and Lamport, D., Arch. Biochem Biophys. (1973) 158: 822.
21. Shannon, L., Uritani, I., and Imaseki, H., Plant Physiol. (1971) 47: 493.
22. Anstine, W., Jacobsen, J., Scandalio, J. and Varner, J., Plant Physiol. (1970) 45: 148.
23. Penon, P., Cecchini, J., Miassod, R., Ricard, J., Teissere, M. and Pinna, M., Phytochem. (1970) 9: 73.
24. Raa, J., Physiol. Plant (1973) 38: 132.
25. Darimont, E. and Baxter, R., Planta (1973) 110: 205.
26. Ridge, I. and Osborne, D., J. Exper. Bot. (1970) 21: 843.
27. Haard, N., Phytochem. (1973) 12: 555.
28. Haard, N., Sharma, S., Wolfe, R. and Frenkel, C., J. Food Sci. (1974) 39: 450.
29. Ranadive, A. and Haard, N., J. Food Sci. (1973) 38: 331.
30. Brown, S., Bio. Sci. (1969) 19: 115.
31. Freundenburg, K., "Lignin Structure and Reactions". Martin, Editor. American Chemical Society, Advances in Chemistry Series 59, Washington, D.C., 1966.
32. Helper, P., Fosket, D. and Newcomb, E., Am. J. Botan. (1970) 57: 85.
33. Lipetz, J. and Garro, A., J. Cell Biol. (1965) 25: 109.
34. Ranadive, A. and Haard, N., J. Sci. Fd. Agr. (1973) 38: 331.
35. Sharma, S., Wolfe, R. and Haard, N., J. Food Sci. (1975) 40: 1021.
36. Lipton, W., Amer. Soc. Hort-Sci. Proc. (1965) 86: 347.
37. Burg, S. and Burg, E., Plant Physiol. (1967) 42: 144.
38. Uritani, I., Ann. Rev. Phytopathology (1971) 9: 211.
39. Coxon, D., Curtis, R., and Howard, B., Fd. Cosmet. Toxicol. (1975) 13: 87.
40. Wilson, B., Boyd, M., Harris, T. and Yang, D., Nature (1970) 231: 52.
41. Kawashima, K. and Uritani, I., Agr. Biol. Chem. (1963) 27:409.

42. Stahmann, M. and Demorest, D., "Fungal Pathogenicity and the plant's Response". p. 405. Academic Press, New York, 1973.
43. Stahmann, M., Clare, B., Woodbury, W., Plant Physiol. (1966) 41: 1505.
44. Imaseki, I., Plant Physiol. (1970) 46: 172.
45. Matsuno, H. and Uritani, I., Plant and Cell Physiol. (1972) 13: 1091.
46. Mattoo, A. and Modi, A., Plant Physiol. (1969) 44: 308.
47. Ivanova, T., and Ivanova, A., Ann. Tech. Agric. (Paris) (1968) 17: 333.
48. Kuhive, A., Dokl. Akod. Nank. Azerb. SSR (1969) 25: 109.
49. Ranadive, A. and Haard, N., J. Food Sci. (1972) 37: 381.
50. Haard, N., Phytochem. (1973) 12: 555.
51. Yang, S., Arch. Biochem. Biophys. (1967) 122: 481.
52. Dilley, D., "The Biochemistry of Fruits and Their Products". p. 195. Academic Press, New York, 1970.
53. Aylward, F. and Haisman, D., Adv. Food Res. (1969) 17: 1.
54. Gortner, W. and Kent, M., J. Biol. Chem. (1958) 233: 731.
55. Vendrell, M., Australian J. Biol. Sci. (1969) 22: 601.
56. Frenkel, C., Dyck, R. and Haard, N., "Postharvest Biology of Fruits and Vegetables". p. 19. AVI, Westport, Conn., 1975.
57. Tingwa, P., and Young R., Plant Physiol. (1975) 55: 937.
58. Hoyle, M., "Mechanisms of Regulation of Plant Growth". p. 659. Royal Society of New Zealand, Wellington, 1974.
59. Frenkel, C., Plant Physiol. (1972) 49: 757.
60. Frenkel, C., Plant Physiol. (1975) 56: 647.
61. Haard, N., J. Food Sci. (1973) 38: 639.
62. Frenkel, C. and Haard, N., Plant Physiol. (1973) 52: 380.
63. Haard, N., J. Food Sci. (1973) 38: 907.
64. Meudt, W. and Gaines, T., Plant Physiol. (1967) 42: 1395.
65. Haard, N., Chism, G. and Nagle, N., Anal. Biochem. (1975) 69: 627.
66. Haard, N. and Marshall, M., Physiological Plant Pathology (1976) 8: 195.

Acknowledgements

Part of the work reported here was supported by a grant from the United States Public Health Service (USPHS 1226001086A1). The data presented here represents the collaborative efforts of several individuals from our laboratory including Dr. A.S. Ranadive, Dr. N. Nagle, Dr. G. Chism, Dr. A. Huang, Dr. S. Sharma, Mr. B. Wasserman, Mr. M. Marshall and Mr. D. Timbie.

10

Enzymes Involved in Fruit Softening

RUSSELL PRESSEY

R. B. Russell Agricultural Research Center, Agricultural Research Service, U.S. Department of Agriculture, Athens, Ga. 30604

Softening of the flesh is one of the most dramatic changes accompanying the ripening of many fruits. Although other parameters of quality are important, the peak of fruit ripeness is usually associated with a fairly narrow range of firmness. Furthermore, texture that is considered optimal for fruit consumed fresh may not be best for fruit that is processed. Softening undoubtedly reflects changes in cell walls of fruit tissues as the fruit progresses through ripening into senescence. Once the process is initiated in mature fruit, the period of acceptable texture may be short, even with refrigeration and controlled atmosphere storage. A prerequisite to improved methods of controlling softening, whether to initiate and accelerate it or to prevent softening beyond the optimum stage, is an understanding of cell wall structure, its changes, and the enzymes involved in its degradation.

I. Histology of Fleshy Fruits.

Appreciation of the biochemical aspects of fruit softening requires some knowledge of the development and histology of fruits. Fruit growth may be considered to begin in the floral primordium (1). The ovary wall becomes the pericarp of the fruit. During development, the ground tissue of the pericarp remains relatively homogeneous and parenchymatic. Initial development occurs mainly through cell multiplication. The postfertilization period is marked by cell enlargement, although cell division also continues in ovaries of large-fruited plants (2). The pericarp may become differentiated into three distinct parts: the exocarp, the mesocarp, and the endocarp.

Fruits are highly diversified in their morphology, and fruit development is not restricted to the ovary but often involves noncarpellary parts of the flower (2). If the entire ground tissue develops into fleshy tissue, the fruit is a berry. All the fleshy tissue of a berry may originate from the ovary wall, as in the grape. In contrast, a considerable part of the tomato fruit consists of placenta. The body of the tomato is pericarp developed from the ovary wall and consists of outer, radial, and inner walls. During development, the placenta shows much more active cell division than the ovary wall. The result is locular cavities in the pericarp that contain the seeds imbedded in a jelly-like tissue (2).

Mohr and Stein (3) have described the fine structural changes in the outer pericarp of tomato fruit during development and ripening. Shortly after fertilization, the vacuoles of each cell of the pericarp enlarge to form one large vacuole, so that a layer of protoplasm lines the inner surface of the cell walls. Intercellular spaces develop subsequently, and cell walls partially separate along the middle lamella region, starting at the intercellular spaces. As the fruit nears maturity, the cells become very large (100-500 μ diameter) and separation of cell walls increases. Degeneration of protoplasm begins, with the disappearance of membraneous structures. In the ripe, red fruit, separation of adjacent cells is common, and degeneration of protoplasmic components is pronounced.

The development of many fruits involves maturation of the ovary wall into a pericarp with a conspicuous stony endocarp surrounded by a fleshy mesocarp. Important stone fruits include the olive, peach, plum, apricot, and cherry. The fleshy tissue of the peach consists of loosely packed parenchyma cells that increase in size from the periphery toward the interior. The mesocarp grows by rapid cell enlargement. Intercellular spaces are common in peach tissue. Cell wall thickness increases during peach development from 0.5μ, after cessation of cell division, to about 1μ, during the pit hardening phase (4). Increases in cell wall thickness are more marked in the late phases of cell growth, and reach a maximum of about 2μ in hard ripe fruit. During subsequent ripening, the cell walls decrease in thickness. There are two distinct types of peaches, the so-called "freestone" and "clingstone" varieties. Early reports (5) described differences in cell wall thickness between the two types, but Reeve (4) could not confirm this and observed that the cell walls decrease in thickenss to the same degree in clingstone as in freestone fruit. Cell walls do not appear to rupture during ripening (4).

The apple is an example of another variation in fruit development. The flesh of the apple is derived from the floral tube as well as from the carpellary tissue. The ovary region (the core) consists of five carpels imbedded in fleshy parenchymatic exocarp. The floral tube region of the apple, which forms the bulk of the edible part, also consists of parenchyma cells. During growth, cell division ceases at about 3 weeks after full bloom (6). Subsequent increase in fruit size is due mainly to cell enlargement and increase in the volume of intercellular spaces, which are abundant in apple tissue (7).

According to Nelmes and Preston (7), rapid wall synthesis during cell extension is followed by reassimilation of the pectic materials just before maturity. The cell walls decrease in thickness at this stage.

Despite the diversities in fruits in size, shape, structure, and development, patterns of ripening tend to be quite similar. Chlorophyll in the chloroplasts of the outermost cells decreases and eventually disappears. Carotenoids and anthocyanins develop and give fruits their characteristic colors. Other changes are the development of specific flavors, increases in sweetness, and decreases in acid content. The loss of firmness associated with ripening can be attributed to changes in the structure of cell walls which will be discussed next.

II. Cell Wall Structure.

The flesh of succulent fruits consists largely or entirely of parenchyma cells (2). These cells are relatively undifferentiated, but are highly complex physiologically because they possess living protoplasts. They are generally polyhedral or elongated with thin primary walls. The walls of two contiguous cells are separated by the intercellular substance or middle lamella which is rich in pectic substances (8). The distinction often is not obvious between the middle lamella and the cell wall which appear as a unit. Mature parenchyma tissue has abundant intercellular spaces (9).

Cell wall composition of plants is usually determined on the ethanol-insoluble fraction, which consists primarily of the cell wall and middle lamella. Jermyn and Isherwood (10) have found that this fraction of ripe pears contains 21.4% glucosan, 3.5% galactan, 1.1% mannan, 21% xylan, 10% araban, 11.5% polygalacturonic acid, and 16.1% lignin. Tomato cell walls contain 17% cellulose, 22% pectin, 17% protein, 21% araban-galactan, and 13-23% xylose plus glucose (11). Knee (12) has reported the composition of acetone-insoluble residues of apples in mg/g tissue are: arabinose, 2.58; xylose, 0.52; galactose, 3.81; glucose 32.8; and galacturonic acid, 3.53. The composition of dates (percent, dry weight basis of the whole fruit) is: 0.8% cellulose, 1.5% hemicellulose a, 0.8% hemicellulose b, 3.7% pectin, and 0.3% lignin (13).

Cellulose is the skeletal substance of the plant cell wall and, as the above data show, it is a major wall component in most fruits. It is a linear β-1,4-glucosan. There have been occasional suggestions that other monosaccharides may be present (14), but cellulose is by far the most homogeneous polysaccharide in the cell wall. The lengths of individual cellulose chains have been estimated at 6000-8000 monomers (15). These chains are combined into bundles, called microfibrils, that are up to 250°A wide and contain 2000 molecules in a transection (16). The arrangement of molecules apparently is orderly because X-ray studies have demonstrated regions of crystallinity in the microfibrils. The fribrillar system is interpenetrated by capillaries of various sizes, which are occupied by water, paracrystalline cellulose, and noncellulosic components of the cell wall. In the newly formed primary wall, the orientation of microfibrils is predominately transverse, but becomes more disperse during cell enlargement.

The bulk of the primary wall is noncellulosic polysaccharides, forming a dense, amorphous gel in which the cellulose microfibrils are imbedded. Meristematic and parenchymous tissues are particularly rich in polysaccharides called the pectic substances (pectin). Pectin is the main component of the middle lamella, as was indicated earlier, but it is also present in the primary cell wall. The structure of pectin may vary with the source. For example, a pure homogalacturonan has been isolated from sunflower heads (17) and Jackfruit (18). But the basic structure of pectin from most sources consists of long blocks of linear ∝-1,4-galacturonan interspersed by rhamnose through C1 and C2 (19, 20, 21, 22). Talmadge et al. (23) suggested that the pectin from suspension cultured sycamore cells consists of blocks of 8 units of galacturonic acid interrupted by units of rhamnose-galacturonic acid-rhamnose.

Neutral sugars other than rhamnose may be present in fruit pectins. For example, purified pectin from apples contained, in addition to 1.2% rhamnose, 9.3% arabinose, 1.4% galactose, 0.80% xylose, and traces of fucose, 2-O-methylxylose and 2-O-methylfucose (21). Galacturonosylgalactose and galacturonosylxylose have been tentatively identified in the hydrolyzates of apple pectin (21), but the linkages remain unknown. It is not clear whether the neutral sugars occur in the galacturonan chain or as branches on the chain. Rees and Wight (24) estimated that about 60% of the pectins from mustard cotyledons carry side chains. Branching of galacturonic acid is thought to occur mainly through C3. Pectins from fruit tissues are considered to be much less branched, however (20, 21, 22).

The only galacturonosyl-galacturonic acid linkage identified in pectin is ∝-1,4. On this basis, it is unlikely that branches of galacturonan, if they exist, are attached directly to the galacturonan backbone. It is possible, however, that such branches are attached through neutral sugars on the chain.

The carboxyl group in galacturonic acid introduces another variable in the structure of pectin. These groups are partly esterified with methanol, but the distribution of ester groups is not known. It has been suggested that the free carboxyl groups may be involved in intermolecular linkages (25). Calcium interacts with free carboxyl groups to form insoluble salts. By forming intermolecular bridges, calcium could contribute to the structure and binding properties of pectin in the cell wall.

Doesburg (26) has proposed that movement of calcium in cell walls may cause solubilization of pectin during fruit ripening.

Because arabinan and galactan often accompany the galacturonan in preparation of fruit pectins, many early workers considered the pectin substances in terms of the three polysaccharides. A possible reason for this apparent association may be simply that their solubilities are similar, although recent studies on pectin from suspension-cultured sycamore cells suggested covalent interactions (23). Hirst and Jones (27) separated the arabinans from apple and citrus pectins and found that they consist mainly of arabinose in a highly branched structure. Later workers (28) questioned the existence of a homoarabinan. Barrett and Northcote (21) isolated, from apple pectin, an arabinan-galactan complex containing nearly equal quantities of the monomers, but could not separate the two components. In contrast, Talmadge et al. (23) showed that a branched arabinan and a linear galactan are released from sycamore cell pectin by endopolygalacturonase treatment. The galactan in fruit pectins has not been characterized, but by analogy with Lupinus albus galactan (29), is assumed to be a linear β-1,4-polymer. In addition to the essentially homogeneous arabinans and galactans, highly branched polysaccharides containing both arabinose and galactose are common in plants (21).

The remaining noncellulosic polysaccharides in the cell wall can be solubilized with increasing concentrations of KOH. The hemicelluloses are characterized by diversity in composition, linkages, and branching. Mannose, galactose, arabinose, xylose, glucose, and other monomers may be present. These alkali-soluble polysaccharides are complex mixtures that vary with the source and method of extraction. The stumbling block in characterizing them has been difficulty in separating the components. There have been a few reports of homogeneous polysaccharides obtained by differential extraction and precipitation. A β-1,4-xylan with glucuronic acid side chains has been isolated from pear cell walls (30). A complex glucan has been extracted from mango fruit (31). Callose, a β-1,3-glucan, may be present in fruit tissues (32). Bauer et al. (33) obtained from sycamore cells, a neutral fraction consisting almost exclusively of xyloglucan. The structure of this polymer is a repeating unit containing 4 residues of β-1,4-linked glucose and 3 residues of xylose, with single xylose branches on 3 of the glucosyl residues.

Plant cell walls contain a small amount of protein that has relatively high proportions of seryl and hydroxyprolyl residues (34, 35). It has been proposed that this protein serves a structural role on the basis that cell wall polysaccharides form glycosidic linkages with the hydroxy-acid residues (36). Knee (37) confirmed that cell walls from apples contain a low level of glycoprotein rich in hydroxyproline. This glycoprotein is relatively soluble and can be separated from the galacturonan by ion exchange chromatography. Knee concluded that the glycoprotein and galacturonan are not linked covalently, but he suggested that the two components may be physically associated in the cell wall.

The current view on cell wall structure is that the polysaccharides and protein form a polymeric network, involving covalent linkages and hydrogen bonding (38). The model proposed for the structure of suspension-cultured sycamore cell walls is based on the premise that xyloglucan is hydrogen-bonded to cellulose. The reducing ends of the xyloglucan side chains are connected to the rhamnogalacturonan through the linear galactan. The rhamnogalacturonan, in turn, is connected to the hydroxyproline-rich protein through a branched arabinogalactan. A second chain of rhamnogalacturonan chain is attached, through arabinogalactan, to the same protein molecule and the order of components is reversed to another cellulose chain. The structural component of the cell wall can thus be considered as a macromolecule.

III. Enzymes Involved in Cell Wall Changes of Ripening Fruits.

1. Cellulase. Because of its abundance in fruit tissue and prominent role in cell wall structure, cellulose should be an important factor in fruit texture. Working with many varieties of apples, Kertesz et al. (39) found that the level of cellulose correlates with initial firmness of freshly harvested fruit, but they concluded that subsequent softening of apples is not due to changes in cellulose. Bartley (40) confirmed that the cellulosic glucose content of apple cell walls did not change in the ripening fruit. In peaches, Nightingale et al. (41) observed a decrease in cellulose content during ripening, but they used a method that is now considered unreliable. Jermyn and Isherwood (10) found a similar small decrease in cellulose in ripening pears. In a highly organized component like cellulose, changes in molecular orientation in the microfibrils could be more important than changes in cellulose levels. Using this approach, Sterling (42) examined the physical state of cellulose in ripening peaches with X-ray techniques. He found that the crystalline micelles enlarge in diameter during peach ripening. Sterling attributed the micellar enlargement to cellulose degradation rather than to formation of new microfibrils. He conceded, however, that the limited degradation of cellulose cannot contribute greatly to peach softening.

The insolubility of cellulose has hindered detection of cellulolytic enzymes in fruits. Many workers have circumvented the problem by using carboxymethylcellulose (CMC), a soluble derivative of cellulose, as the substrate. The high viscosity of aqueous solutions of CMC allows the use of the very sensitive viscometric method for measurement of cellulase activity which is extremely low in many plant tissues. The first report of cellulase in higher plants was by Tracy (43) who found activity in tobacco and other plants. Dickinson and McCollum (44) and Hall (45) detected similar activity in extracts of ripe tomatoes. The long incubation periods (20 hr) required to measurably reduce the viscosity of CMC indicate the low activity in tomatoes.

The cellulase in the locular material of tomatoes is soluble in water, whereas a high concentration of NaCl is required to solubilize the enzyme from pericarp tissue (45). Hall (46) found that cellulase is present in the outer pericarp of young, green tomatoes. The activity decreases during tomato maturation, and then increases at the onset of ripening (46). In contrast, the cellulase in the inner pericarp and the placental tissue is low in young fruit and remains low until the turning stage, when the activity increases sharply, especially in the placental portion. Hall concluded that cellulase may be involved not only in tomato softening during ripening but also in cell enlargement during fruit development.

Sobotka and Watada (47) confirmed that cellulase is present in mature green tomatoes and that it increases sharply with ripening. They measured fruit firmness and cellulase viscometrically in two commercial varieties and several breeding lines. The varieties with firm fruit exhibited lower cellulase activity than those with soft fruit. The rate of softening of each variety was closely associated with an increase in cellulase, but a sharp decrease in tomato firmness early in the ripening process was not accompanied by an increase in cellulase.

Hobson (48), who conducted a detailed study on the relationship between tomato softening and cellulase, used acetone precipitates of tomato extracts and a reductometric rather than a viscometric assay. In agreement with other workers, he found high cellulase in young tomatoes. The activity decreased steadily, as fruit size increased, until the mature green stage and then began to increase. Cellulase continued to increase up to the red stage, but not into the overripe stage. In contrast to other workers, Hobson found higher cellulase in the firmer of two varieties, and firmness was not correlated with cellulase in fruit in the sub-genera *Eriopersicon* and *Eulycopersicon*. He concluded that the contribution of cellulase to softening is of minor importance in tomatoes.

An explanation for the difference in opinion concerning the role of cellulase in tomato softening may lie in the recently revealed complexity of tomato cellulase. Pharr and Dickinson (49) identified two enzymes in locular contents of ripening tomatoes. One enzyme reduced the viscosity of CMC solutions and generated reducing groups, but it did not attack insoluble cellulose. The second enzyme hydrolyzed cellobiose rapidly, and the rate of cleavage decreased as the chain length of the substrate increased. Sobotka and Stelzig (50) provided evidence for four celluloytic enzymes in tomatoes. Two of the enzymes were endocellulases that degraded both CMC and insoluble cellulose. The endocellulases differed in pH optimum, degradation of short chain cellodextrins, and the size of product released from insoluble cellulose. One of the other enzymes was characterized as a nonspecific β-glucosidase. It hydrolyzed cellobiose rapidly; the rate of cleavage decreased as the chain length increased. The fourth enzyme also hydrolyzed cellobiose, but the rate of reaction decreased very slowly with increasing chain length. This enzyme produced glucose as the main product, and therefore can be classified as an exocellulase.

Cellulase, as measured viscometrically with CMC, has also been studied in relation to the softening of dates and peaches. Activity was absent in dates at the green stage but began to develop between the green and early red stages (51). Activity increased sharply at the early red stage and reached a maximum at the late red stage. The development of cellulase closely paralleled the loss of firmness, suggesting a role for this enzyme in date softening. Similarly, cellulase was not detected in green peaches, but activity developed and increased during ripening (52). The increase in cellulase was greatest before peach firmness decreased significantly, suggesting that cellulase may not only be involved in peach softening, but also in initiation of the process.

2. Pectic Enzymes. The pectic substances have received more attention in relation to fruit softening than any other cell wall component. The reasons for the concentration on pectin are its predominance in the middle lamella, the relatively high level of pectin in most fruits, and above all, the appearance of water-soluble pectin that accompanies the softening of many fruits. Pectin solubilization occurs in apples (40), pears (10), peaches (53, 54, 55), tomatoes (56), and other fruits. The soluble pectin content of apples increased more than 3-fold during a change in firmness of 4.8 to 3.4 kg (40). The pectic changes in ripening pears are quite similar to those in apples. The proportion of soluble pectin is very low in unripe pears, but increases markedly with ripening (10). Some earlier workers (56) found a direct relation between ripening and soluble pectin in pears and proposed the use of soluble pectin as an index of maturity. There is controversy over whether pectin is solubilized in tomatoes, however. Woodmansee et al. (57) were not able to confirm earlier claims of pectin solubilization in tomatoes.

Changes in pectin solubility are most pronounced in freestone peaches. Postlmayr et al. (54) found that 5% of the dry matter of unripe Fay Elberta peaches was pectin, a fourth of which was water-soluble. On ripening, the pectin content decreased to 4.1%, but the proportion of water-soluble pectin increased to 71% of the total. In contrast to the changes in the freestone variety, ripening of Halford clingstone peaches was accompanied by a decrease in total pectin from 4.4 to 3.8%, but the soluble pectin content remained relatively unchanged. Shewfelt (55) studied three pectin fractions (water-soluble, Versene-soluble, and Versene-insoluble) in five freestone and one clingstone varieties. In the clingstone peaches, the proportion of the three pectin fractions remained essentially constant. But in freestone peaches, the water-soluble pectin increased rapidly at the expense of the other two fractions. Only 20% of the pectin in unripe peaches was water-soluble, whereas the level increased to 80% in ripe peaches. Shewfelt et al. (58) subsequently isolated and characterized the three pectin fractions. They demonstrated that peach ripening is accompanied not only by pectin solubilization but also by reduction in molecular weights, as determined viscometrically. In yet another study on peaches, Pressey et al. (59) found that the total pectin content of Elberta and Red Haven peaches did not change significantly during fruit softening while the soluble pectin increased sharply. They analyzed the soluble pectin by gel filtration and found that the molecular weights of the pectin decreased progressively during ripening.

The complexity of pectin seems to offer a number of possibilities for enzyme action leading to solubilization. At one extreme, release of pectin molecules may represent cleavage of its linkages to other cell wall components. Cleavage could involve terminal residues of the galacturonan chains or hydrolysis of the polymers attached to the chains. Evidence is accumulating to show that this may be the case in apples because the pectin does not appear to be degraded during solubilization. On the other hand, pectin may be solubilized by hydrolysis of long, insoluble molecules. In this case, hydrolysis could occur in the galacturonan chains or at the linkages with rhamnose. Two enzymes that could be involved in the mechanism are pectinesterase and polygalacturonase.

a. Pectinesterase. Pectinesterase catalyzes the hydrolysis of methyl esters in pectin. It is highly specific for the methyl ester and the ester as it occurs in pectin, and does not hydrolyze other esters and the methyl ester in short chain galacturonans. The enzyme is activated by divalent cations or monovalent cations at high concentrations. The optimum pH range is 5 to 8 and is affected by cations. The action of pectinesterase on pectin produces blocks of free carboxyl groups (60), indicating that deesterification occurs in a linear manner. Solms and Deuel (61) suggested that ester hydrolysis occurs only adjacent to free carboxyl groups. In detailed studies on the mode of action of tomato pectinesterase, Lee and Macmillan (62) demonstrated that it acts at both the reducing ends and interior loci on highly esterified pectin chains.

Pectinesterase occurs commonly in various parts of higher plants, including the fruit. Kertesz (63) found that tomatoes are a particularly rich source of the enzyme. The activity is high in green tomatoes and, according to Kertesz, increases about 4-fold during ripening. This suggests that softening may be related to high pectinesterase activity. However, Hamson (64) reported that pectinesterase is higher in firm than in soft varieties. In a study on six breeding lines and one variety, Hall and Dennison (65) did not find a significant correlation between firmness and pectinesterase. Furthermore, Hobson (66) observed only a 40% increase in pectinesterase during ripening of tomatoes. In contrast to the findings of Hamson, the firmer of the two varieties contained lower pectinesterase.

Part of the confusion about pectinesterase in tomatoes may be due to the complexity of the enzyme. The activity can be resolved into a number of components (67). The number of pectinesterases and their relative levels vary with both tomato variety and stage of ripeness. For example, green Marion tomatoes contained two pectinesterases. On ripening, one enzyme decreased while the other nearly tripled, and a third form of pectinesterase appeared. One of the enzymes appears to be the major component in most varieties. The same three enzymes were present in Homestead tomatoes in the same proportions. The dwarf variety Pixie contained two of the above enzymes and a different enzyme. The four tomato pectinesterases differed in molecular weight, stability to heat, pH optimum, and activation by cations.

An even greater heterogeneity of tomato pectinesterase has been demonstrated by thin-layer isoelectric focusing (68). At least eight enzymes were separated, although one component predominated in all samples. The pectinesterases appear to have similar molecular weights, and to differ mainly with respect to charge properties.

Pectinesterase has been detected in many other fruits, although the activities are usually much lower than in tomatoes. There have been conflicting reports on the presence of the enzyme in apples. Apparently it is present in low amounts and does not change during apple ripening (69). Nagel and Patterson (70) reported that pectinesterase in pears decreases during maturation. The level of pectinesterase in peaches varies with the variety, and the changes in activity show no distinct trend with advancing ripeness (55).

The degree of esterification of pectin in avocados gradually decreases during ripening (71). Pectinesterase is high in young avocados (72). A pronounced decrease in activity was observed during the intensive growth stage, and still further reduction during fruit maturation. Similarly, pectinesterase decreased sharply in stored avocados through the softening period. The more mature the fruit at harvest, the more pronounced was the decrease in pectinesterase during storage (72).

Hultin and Levine (73) separated the pectinesterase activity in bananas into three fractions by differential extraction. The enzymes differed in pH optimum, thermal stability, and sensitivity to SDS, as well as in solubility. In contrast to the decrease in pectinesterase in avocados, the activity in bananas increased about 10-fold during ripening. One of the pectinesterases increased continually during ripening, whereas the other two enzymes were maximal after about 2 days of ripening. The greatest increase in all three enzymes occurred as the bananas changed from green to yellow.

Thus, the total pectinesterase activity increases in bananas and decreases in avocados during ripening, but in most fruits the enzyme is present at the immature stage and appears to remain relatively constant during ripening. A role for this enzyme in fruit softening, therefore, is not apparent. Multiple forms of pectinesterase are present in bananas and tomatoes, and a specific form of enzyme might be involved in softening.

Also, the direct role of pectinesterase in pectin solubilization is not obvious because a decrease in esterification alone would not be expected to increase solubility. To the contrary, an increase in free carboxyl groups would lead to greater interaction with Ca^{2+} and to decreased solubility. But pectinesterase action must precede degradation of pectin by polygalacturonase, and in this way pectinesterase could exert regulation on the process of fruit softening. Judging from the lack of a consistent pattern of fluctuation of the enzyme, regulation of pectinesterase probably involves a mechanism other than changes in enzyme concentration. The fact that pectin is never completely deesterified in tomatoes, for example, even though pectinesterase is always present, suggests that other regulatory mechanisms are active.

b. Polygalacturonase. Pectolytic enzymes are classified on the basis of mode of action as either hydrolases or lyases. The latter degrade galacturonans by a transelimination mechanism in a random or terminal manner with preference for low or high ester substrates. The hydrolases also can function in a random or uniform manner, but all known enzymes in this group act only on deesterified substrates and therefore are called polygalacturonases. They are the only pectolytic enzymes known to occur in fruit tissues. Polygalacturonase was first detected in ripe tomatoes which remain the richest plant source. The list of fruits containing the enzyme is now fairly long, and the following discussion will deal with each commodity separately.

i. Tomatoes. In contrast to pectinesterase and cellulase, polygalacturonase is not present in green, immature tomatoes (63, 66). It appears in fruit near the onset of ripening and increases as the fruit softens. It is this apparent coincidence of polygalacturonase with softening that has suggested a key role for this enzyme in the process. Numerous studies have been directed at establishing such a relationship.

One approach is to study the enzyme level in relation to tomato firmness which differs considerably with a variety.

Tomato lines with high pigment characteristics tend to be firmer than other lines (74). Foda (75) found that high pigment lines contained more protopectin and less polygalacturonase than a commerical variety. Sobotka and Watada (76) also observed that two lines containing the high pigment characteristic had considerably lower polygalacturonase than lines without the characteristic. The activity was not only lower, but it increased at a slower rate than in other varieties. Hobson (77) reported that ripe fruit of the variety Potentate were both firmer and lower in polygalacturonase than those of the variety Immuna. A further study by Hobson (78) revealed a highly significant positive correlation between polygalacturonase and the compression of 6 different varieties at the commerical picking stage.

A fruit ripening inhibitor (rin) mutant of tomato can be produced by backcrossing (79). Tomatoes containing the recessive gene soften much more slowly than normal varieties. Hobson (80) observed that Potentate tomatoes backcrossed in this manner were associated with markedly reduced solubilization of pectin and considerably lower levels of polygalacturonase. This was confirmed by Buescher and Tigchelaar (81) who found that backcrossed Rutgers tomatoes remained firm and did not develop polygalacturonase. The physiological disorder "blotchy" ripening has also been evaluated in terms of polygalacturonase activity. The red areas of blotchy fruit contained less than two-thirds the activity in evenly red fruit, and the activity was even lower in non-red areas (77).

Polygalacturonase activity is highest in the outer locule wall of the pericarp tissue followed by the inner locule walls and the placental tissue (77). The enzyme is not present in the locular contents. The activity first appears in the placenta which usually shows incipient yellowness. It then develops in both the inner and outer locule walls as the change in color spreads to the pericarp. Hobson suggested that the parallel appearance of polygalacturonase and coloration is further evidence for a relationship between the enzyme and tomato ripening. McColloch et al. (82) also noted that deep red color in tomatoes was associated with high polygalacturonase.

Tomato polygalacturonase has been partially purified and characterized (83, 84). It cleaves pectate randomly first to oligogalacturonates and ultimately to galacturonic acid, but the rate of hydrolysis decreases rapidly with decreasing chain length. If the initial rate of pectate hydrolysis is 100, the rates of hydrolysis of tetra-, tri-, and di-galacturonate are approximatley 7, 1.6, and 1, respectively (84). Some of the properties of the activity in crude extracts suggest that the activity is produced by more than one polygalacturonase. McColloch and Kertesz (85) observed that most of the activity was destroyed by heating at relatively low temperatures, but part of it survived heating to 90°. Patel and Phaff (84) examined the heat-stable component and concluded that it was similar to the heat-unstable enzyme. But they observed double pH optima for the enzyme acting on tri-and tetra-galacturonates and attributed this to the presence of two enzymes. The observation that pectate inhibits the hydrolysis of small substrates at low pH, but only to the extent of 70%, is further evidence for a complex system of polygalacturonase in tomatoes (86).

McColloch (87) separated two polygalacturonases from tomatoes by moving boundary electrophoresis. One of the peaks was more active than the other, but the enzymes were not characterized. Pressey and Avants (88) separated the activity in tomato extracts into two peaks (PG I and PG II) by chromatography on DEAE-Sephadex A-50. The enzymes differed in molecular weight and stability to temperature and pH, although the more stable component did not survive heating to 90°. The pH optima for both enzymes were near 4.5. The activities of both enzymes shifted to the acid side with decreasing substrate size and increasing NaCl concentration, but PG I was less dependent on these factors. At enzyme levels producing equal numbers of reducing groups, PG II was much more effective than PG I in reducing the viscosity of pectate. However, the effect of PG I on the viscosity of pectate is too rapid for a mechanism of end group cleavage.

ii. Peaches. The changes in flesh firmness and pectin solubility are particularly pronounced in ripening freestone peaches. Shewfelt et al. (58) demonstrated that pectin is not only solubilized, but that the molecular weights of the soluble fraction gradually decrease, suggesting that peach ripening is accompanied by pectin degradation. Pressey et al. (59) confirmed that reduction in pectin molecular weights occurs during both tree and postharvest ripening. Although earlier attempts to detect polygalacturonase in peaches were unsuccessful, the presence of this enzyme was revealed by incubation of water-washed residues of peach tissue with polygalacturonate (59). Polygalacturonase activity was absent during the fruit enlargement stage; it appeared as the peaches began to ripen and then increased sharply. The appearance of the enzyme paralleled the formation of soluble pectin. Both processes were preceded by some fruit softening, but they coincided with the most rapid loss of firmness.

Pressey and Avants (89) subsequently solubilized and characterized the peach polygalacturonase. The procedure involved extraction of water-washed residues of ripe freestone peaches with 0.2 M NaCl followed by ultrafiltration of the macromolecules. Attempts to purify the activity by chromatography on Sephadex G-100 revealed two enzymes (PG I and PG II). The enzymes differed in a number of ways, including pH optimum, cation activation, and effect of substrate size. PG I functioned optimally at pH 5.5 and required Ca^{2+} for activity (89). It cleaved digalacturonate very slowly, but the rate of cleavage increased with substrate chain length to a maximum for substrates with a degree of polymerization of about 20. This enzyme exhibited the highest affinity for the largest substrates (90). PG II was most reactive with large substrates; its pH optimum was near 4.5 for the largest substrate (pectate), but it shifted to the acid side as the substrate size decreased (89).

The two peach polygalacturonases differed markedly in mode of action. PG II was a typical endopolygalacturonase; it rapidly reduced the viscosity of pectate relative to a slow release of reducing groups. This enzyme was very effective in solubilizing pectin from washed peach cell walls. In contrast, PG I had a very small effect on the viscosity of pectate. The product of its reaction was identified as galacturonic acid. The evidence presented showed that cleavage occurs at the nonreducing ends of the substrate chains, and that all chains are progressively shortened. The polygalacturonase in peaches therefore, consists of endo- and exo-cleaving enzymes. This was the first report of an exopolygalacturonase in fruit tissue, but we now know that this enzyme is widespread in higher plants.

iii. Dates. Hasegawa et al. (91) found a trace of polygalacturonase, as measured by the release of reducing groups from pectate, in green dates. The activity remained low during ripening of dates until the late red stage at which time the activity rose sharply to a maximum at the 50% soft stage. The greatest part of the increase in polygalacturonase, therefore, occurred immediately before date softening. The development of activity also followed softening within single dates, which commences at the apical end and progresses toward the stem end. They concluded from the close relation between polygalacturonase and softening that this enzyme may be involved in controlling the texture of dates. Properties of the date enzyme have not been determined.

iv. Avocado. McCready and McComb (92) first reported that polygalacturonase is absent in green avocados, but considerable activity develops during ripening. Reymond and Phaff (93) found that the enzyme first appears at the blossom end and spreads toward the stem end. In a more detailed study, Zauberman and Schiffmann-Nadel (72) did not detect polygalacturonase in fruit at various stages of development immediately after harvest. In harvested fruit, the enzyme developed earlier in more mature fruits, although the maximum levels of enzymes appeared to be independent of fruit maturity.

Avocado polygalacturonase exhibits optimal activity at pH 5.5 and in the presence of 0.14 M Na^+ (93). Phosphate and Calgon were inhibitory, but Ca^{2+} at 0.1 M was without effect. The enzyme appears to be an endopolygalacturonase; it hydrolyzes pectate to intermediate oligogalacturonates which are then slowly hydrolyzed to galacturonic acid (93, 94). Reymond and

Phaff (93) concluded that avocados contain an inhibitor of polygalacturonase because recovery of the enzyme during purification often exceeded 100%. Furthermore, the addition of extracts of immature avocados to purified polygalacturonase resulted in 48% inhibition, confirming the existence of an inhibitor. Precipitability of the inhibitor by 50% ammonium sulfate suggests that it is a protein.

v. Pears. McCready and McComb (92) presented evidence showing that ripening of pears is accompanied by degradation of pectic substances. During the ripening of pears, the solubility of pectin increased and precipitability with alcohol decreased. They did not detect polygalacturonase in unripe pears, but they found considerable activity in ripe pears. Pressey and Avants (95) have demonstrated that the activity consists of endo- and exo-polygalacturonases, similar to the system of polygalacturonases found in peaches. The exopolygalacturonase was optimally active at pH 5.5 and was activated by Ca^{2+} and Sr^{2+}. It was most reactive with polygalacturonate possessing a chain length of about 12 units. The endopolygalacturonase had a pH optimum at 4.5; it was most reactive with a substrate possessing a chain length of about 80.

vi. Cucumbers. Bell (96) detected pectolytic activity in cucumbers by using pectin at pH 4 as the substrate in the viscometric assay. The choice of pectin over pectate for the substrate may have been the reason for the low level of activity found. When Pressey and Avants (97) reexamined the polygalacturonase in cucumbers, they found that the activity consists solely of an exosplitting enzyme. This enzyme is similar to peach exopolygalacturonase in pH optimum (5.5), molecular weight (59,000), and mode of action. In contrast to the exopolygalacturonases, the cucumber enzyme was readily extracted by water. It was activated by Ca^{2+}, but not by Sr^{2+}. Substrates with chain lengths from about 6 to 12 were cleaved most rapidly.

vii. Cranberries. Patterson et al. (98) found polygalacturonase in cranberries that were breaking down in storage and softened by bruising, but not in sound berries. It is possible that the activity detected was of microbial origin. However, Arakji and Yang (99) isolated and characterized a polygalacturonase from McFarlin cranberries. They found that phenol binding agents increased the yields of activity. The enzyme was endo-cleaving with a pH optimum near 5.

c. Other Enzymes. Enzymatic cleavage of any linkage involved in maintaining the structure and rigidity of the middle lamella and cell wall could affect fruit firmness. Even the phenomenon of pectin solubilization may be due to enzymatic cleavage of linkages between pectin and other cell wall components rather than to degradation of the pectin chain. This may be the case in apples. Attempts to find polygalacturonase in ripe apples have been unsuccessful (100). Furthermore, soluble pectin is formed during apple ripening, but the pectin chains are not shortened (100).

The softening of apples is accompanied by not only pectin solubilization but also by the loss of galactose residues from the cell wall (101). A small decrease in arabinose also occurs, but the xylan, cellulose, and noncellulosic glucans are not hydrolyzed. The loss of galactose is substantial, decreasing to about a third of the initial level, and precedes the increase in soluble pectin and the decrease in apple firmness. This has prompted searches for galactan-degrading enzymes in apples. Bartley (102) has found that both soluble and cell wall preparations from apple cortex hydrolyze β-galactan and p-nitrophenyl derivatives of α- and β-galactoside. The β-galactosidase activity was present at a high level in firm apples and increased less than 2-fold during ripening. Bartley concluded that β-galactosidase may control pectin solubilization, but the fact that the loss of galactose precedes the release of pectin led him to suggest that the hydrolysis of other bonds may be involved (102).

Wallner and Walker (103) measured the enzymatic activity against a number of glycosides and polysaccharides to determine whether the cleavage of certain linkages varies with the stage of ripeness of tomatoes. Extracts of tomatoes hydrolyzed glucomannan, galactomannan, laminarin, arabinogalactan, xylan, and p-nitrophenyl derivatives of α-D and β-D-galactoside, α-D-glucoside, α-D-and β-D-xyloside, as well as cellulose and polygalacturonate. The major enzymes, in terms of reducing group formation, were laminarinase, polygalacturonase, and β-galactosidase. With the exception of glucomannanase, xylanase, and polygalacturonase, the enzymes were present at all stages of ripeness. During ripening, the β-galactosidase increased 4-fold in the placenta, but less in other parts of the fruit. The other enzymes did not change significantly. However, on the basis of susceptibility of cell walls to degradation by the mixture of tomato enzymes, Wallner and Walker concluded that changes in the cell wall occur before the appearance of polygalacturonase. An enzyme other than polygalacturonase may therefore be involved, but has not been identified.

Conclusions

Fruit softening probably is due to breakdown of the middle lamellae and cell walls. This desintegration may involve movement of calcium in the cell walls (26) or alteration of hydrogen bonding between cellulose and xyloglucan (38) in response to changes in pH. But evidence of covalent bond cleavage in cell wall components points to roles for several of the hydrolytic enzymes found in fruit tissues. There has been a tendency to emphasize the pectic enzymes because of the apparent relationship between pectin solubilization and fruit softening. The importance of polygalacturonase is supported by its widespread occurrence in fruits and its appearance during ripening. The discovery of exopolygalacturonase in fruits suggests that cleavage of terminal linkages in cell wall macromolecules may be involved in softening. But the process is complex and undoubtedly requires numerous enzymes, some of which remain to be identified.

Literature Cited

1. Nitsch, J. P. Ann. Rev. Plant Physiol. (1953), 4, 199.
2. Esau, K. "Plant Anatomy", Wiley and Sons, New York. 1965.
3. Mohr, W. P., and Stain, M. Can. J. Plant Sci. (1969), 49, 549.
4. Reeve, R. M. Amer. Jour. Bot. (1959), 46, 241.
5. Addoms, R. M., Nightingale, G. T., and Blake, M. A. New Jersey Agric. Exp. Sta. Bull. 507, (1930).
6. Smock, R. W., and Neubert, A. M. "Apple and Apple Products". Interscience Publishers, New York. 1950.
7. Nelmes, B. J., and Preston, R. D. J. Exp. Bot. (1968), 19, 496.
8. Porter, K. R., and Machado, R. D. J. Biophys. Biochem. Cytol. (1959), 7, 167.
9. Sifton, H. B. Bot. Rev. (1957), 23, 303.
10. Jermyn, M. A., and Isherwood, F. A. Biochem. J. (1956), 64, 123.
11. Williams, K. T., and Bevenue, A. J. Agric. Food Chem. (1954), 2, 472.
12. Knee, M. Phytochemistry (1973), 13, 2207.
13. Coggins, C. W., Knapp, J. C. F., and Richer, A. L. Date Growers Report. (1968), 45, 3.
14. Dennis, D. T., and Preston, R. D. Nature (1961), 191, 667.
15. Ranby, B. G. In "Encyclopedia of Plant Physiology" (W. Ruhland, ed.) 6, p. 268, Springer, Berlin (1958).
16. Frey-Wyssling, A. "Die pflanzliche Zellwand". Springer, Berlin (1959).
17. Bishop, C. T. Can. J. Chem. (1955), 33, 1521.
18. Sen-Gupta, U. K., and Rao, C. V. N. Bull. Chem. Soc. Japan (1963), 36, 1683.
19. Aspinall, G. O., and Canas-Rodrigruez, A. J. Chem. Soc. (1958), 4020.
20. Aspinall, G. O., and Fanshawe, R. S. J. Chem. Soc. (1961), 4215.
21. Barrett, A., and Northcote, D. Biochem. J. (1965), 94, 617.
22. Aspinall, G. O., Craig, J. W. T., and Whyte, J. Z. Carbohydrate Res. (1968), 7, 442.

23. Talmadge, K. W., Keegstra, K., Bauer, W. D., and Albersheim, P. Plant Physiol. (1973), 51, 158.
24. Rees, D. A., and Wight, N. J. Biochem. J. (1969), 115, 431.
25. Albersheim, P. In "Plant Biochemistry" (J. Bonner and J. E. Varner, eds.) Academic Press, New York, 1965, p. 313.
26. Doesburg, J. J. J. Sci. Food Agric. (1957), 8, 206.
27. Hirst, E. L., and Jones, J. K. N. J. Chem. Soc. (1939), 454.
28. McCready, R. M., and Gee, M. J. Agric. Food Chem. (1960), 8, 510.
29. Hirst, E. L. J. Chem. Soc. (1942), 70.
30. Chanda, S. K., Hirst, E. L., and Perceval, E. G. V. J. Chem. Soc. (1951), 1240.
31. Das, A., and Rao, C. V. N. Aust. J. Chem. (1965), 18, 845.
32. Dekazos, E. D. J. Food Sci. (1972) 37, 562.
33. Bauer, W. D., Talmadge, K. W., Keegstra, K., and Albersheim, P. Plant Physiol. (1973), 51, 174.
34. Lamport, D. T. A., and Northcote, D. H. Nature (1960), 188, 665.
35. Lamport, D. T. A., Katona, L., and Roerig, S. Biochem. J. (1973), 173, 125.
36. Lamport, D. T. A. Nature (1967), 216, 1322.
37. Knee, M. Phytochemistry (1975), 14, 2181.
38. Keegstra, K., Talmadge, K. W., Bauer, W. D., and Albersheim, P. Plant Physiol. (1973), 51, 188.
39. Kertesz, Z. I., Eucare, M., and Fox, G. Food Res. (1959), 24, 14.
40. Bartley, I. M. Phytochemistry (1976), 15 625.
41. Nightingale, G. T., Addoms, R. M., and Blake, M. A. New Jersey Agr. Expt. Sta. Bull. 494 (1930).
42. Sterling, C. J. Food Sci. (1961), 26, 95.
43. Tracy, M. V. Biochem. J. (1960), 47, 431.
44. Dickinson, D. B., and McCollum, D. P. Nature (1964), 203, 525.
45. Hall, C. B. Nature (1963), 200, 1010.
46. Hall, C. B. Bot. Gaz. (1964), 125, 156.
47. Sobotka, F. E., and Watada, A. E. J. Amer. Soc. Hort. Sci. (1971), 96, 705.
48. Hobson, G. E. J. Food Sci. (1968), 33, 588.
49. Pharr, D. M. and Dickinson, D. B. Plant Physiol. (1973), 51, 577.
50. Sobotka, F. E., and Stelzig, D. A. Plant Physiol. (1974), 53, 759.
51. Hasegawa, S., and Smolensky, D. C. J. Food Sci. (1971), 36, 966.
52. Hinton, D. M. and Pressey, R. J. Food Sci. (1974), 39, 783.

53. Appleman, C. O., and Conrad, C. M. Maryland Univ. Expt. Sta. Bull. 283 (1926).
54. Postlmayr, H. L., Luh, B. S., and Leonard, S. J. Food Technol. (1956), 10, 618.
55. Shewfelt, A. L. J. Food Sci. (1965), 30, 573.
56. Kertesz, Z. I. "The Pectic Substances". Interscience, New York, 1951.
57. Woodmansee, C. W., McClendon, J. H., and Somers, G. F. Food Res. (1959), 24, 503.
58. Shewfelt, A. L., Paynter, V. A., and Jen, J. J. J. Food Sci. (1971), 36, 573.
59. Pressey, R., and Avants, J. K. J. Food Sci. (1971), 36, 1070.
60. Deuel, H., and Stutz, E. Advan. Enzymol. (1958), 20, 341.
61. Solms, J., and Deuel, H. Helv. Chim. Acata (1955), 38, 321.
62. Lee, M. and Macmillan, J. D. Biochemistry (1970), 9, 1930.
63. Kertesz, Z. I. Food Res. (1938), 3, 481.
64. Hamson, A. R. Food Res. (1952), 17, 370.
65. Hall, C. B., and Dennison, R. A. Proc. Amer. Soc. Hort. Sci. (1960), 75, 629.
66. Hobson, G. E. Biochem. J. (1963), 86, 358.
67. Pressey, R. and Avants, J. K. Phytochemistry (1972), 11, 3139.
68. Delincee, H. Phytochemistry (1976), 15, 903.
69. Pollard, A., and Kieser, M. E. J. Sci. Food Agric. (1951), 2, 30.
70. Nagel, C. W., and Patterson, J. E. J. Food Sci. (1967), 32, 294.
71. Dolendo, A. L., Luh, B. S., and Pratt, H. K. J. Food Sci. (1966), 31, 332.
72. Zauberman, G., and Schiffmann-Nadel, M. Plant Physiol. (1972), 49, 864.
73. Hultin, H. O., and Levine, A. S. J. Food Sci. (1965), 30, 917.
74. Thompson, A. E. Science (1955), 121, 986.
75. Foda, Y. H. Ph.D. Dissertation, University of Illinois (1957).
76. Sobotka, F. E., and Watada, A. E. Proc. W. Va. Acad. Sci. (1970), 141.
77. Hobson, G. E. Biochem. J. (1964), 92, 324.
78. Hobson, G. E. J. Hort. Sci. (1965), 40, 66.
79. Rich, C. M., and Butler, L. Advan. Genet. (1956), 8, 267.
80. Hobson, G. E. Phytochemistry (1967), 6, 1337.
81. Buescher, R. W., and Tigchelaar, E. C. Hortscience (1975), 10, 624.
82. McColloch, R. J., Keller, G. J., and Beavens, E. A. Food Tech. (1952), 6, 197.

83. Patel, D. S., and Phaff, H. J. Food Res. (1960), 25 37.
84. Patel, D. S., and Phaff, H. J. Food Res. (1960), 25, 47.
85. McColloch, R. J., and Kertesz, Z. I. Arch. Biochem. (1948), 17, 197.
86. Pressey, R., and Avants, J. K. J. Food Sci. (1971), 36, 486.
87. McColloch, R. J. Ph.D. thesis, Kansas State College, (1948).
88. Pressey, R., and Avants, J. K. Biochem. Biophys. Acta. (1973), 309, 363.
89. Pressey, R., and Avants, J. K. Plant Physiol. (1973), 52, 252.
90. Pressey, R., and Avants, J. K. Phytochemistry (1975), 14, 857.
91. Hasegawa, S., Maier, V. P., Kaszycki, H. P., and Crawford, J. K. J. Food Sci. (1969), 34, 527.
92. McCready, R. M. and McComb, E. A. Food Res. (1954), 19, 530.
93. Reymond, D., and Phaff, H. J. J. Food Sci. (1965), 30, 266.
94. McCready, R. M., McComb, E. A., and Jansen, E. F. Food Res. (1955), 20, 186.
95. Pressey, R., and Avants, J. K. Phytochemistry (1976), 15, 1349.
96. Bell, T. A. Bot. Gaz. (1951), 113, 216.
97. Pressey, R. and Avants, J. K. J. Food Sci. (1975), 40, 937.
98. Patterson, M. E., Doughty, C. C., Graham, S. O., and Allan, G. Proc. Amer. Soc. Hort. Sci. (1967), 90, 498.
99. Arakji, O. A., and Yang, H. Y. J. Food Sci. (1969), 34 340.
100. Doesburg, J. J. I.B.V.T. Commun. No. 25 (1965).
101. Knee, M. Phytochemistry (1973), 12, 1549.
102. Bartley, I. M. Phytochemistry (1974), 13, 407.
103. Wallner, S. J. and Walker, J. E. Plant Physiol. (1975), 55, 94.

11

Uses of Endogenous Enzymes in Fruit and Vegetable Processing

ROBERT C. WILEY

Department of Horticulture, Food Science Program, University of Maryland, College Park, Md. 20742

The quality of fresh and processed fruits and vegetables may be improved by utilizing naturally occurring enzymes which can alter appearance, flavor, nutrition, texture and functional properties. Research with endogenous enzymes as well as immobilized enzymes is finding renewed emphasis as consumers become more concerned about food additives.

Management of endogenous enzyme systems to develop quality attributes of desirable characteristics is an extremely complex operation. Generally, it is difficult to modify a single enzyme system at a predetermined site and on a specific substrate without disturbing or activating other natural enzyme systems. This secondary activation may have an undesirable effect on the food product in question. Also Schwimmer (1) has indicated that some cell decomposition or disruption is continually taking place during maturation, ripening, handling, pre-processing and processing. These occurrances break natural compartmentalization and increase the difficulty of proper enzyme utilization.

To a great extent, enzyme utilization rather than end-point control requires precise knowledge of enzyme systems, their substrates, cofactors, and inhibitors. These biocatalysts can probably best be managed in the pre-processing sector of the food chain. For example, Wiley and Winn (2) found polyphenol oxidase activity and subsequent browning in green beans and grapes could be controlled with field treatment by using a modified mechanical harvester. The products were treated at the moment of harvest with SO_2 gas in the presence of CO_2 and N_2 to reduce oxygen levels.

Because of the obvious complexities and tremendous number of plant products, the endogenous enzymes selected for study must be ranked for importance according to dominance of the plant species, importance of plant structure, i.e. type of plant organ, and possible economic impact of the enzyme utilization on resultant improvement of quality and functional properties. There seems to be little question that emphasis on enzyme management in fruits and vegetables has centered on texture control, although some researchers have studied enzyme utilization in relation to appearance factors such as color-hue, value and chroma, gloss and defects, flavor enhancement and nutritional quality.

This paper emphasizes the influence of endogenous enzymes on textural factors and will briefly review natural enzyme systems used to modify appearance, flavor and nutritional quality.

APPEARANCE, FLAVOR AND NUTRITIONAL QUALITY

The enzymes in fruits and vegetables which affect their appearance are generally associated with negative and/or undesirable reactions such as browning, loss of chlorophyl, loss of intensity of the anthocyanin pigments and the like. Most enzymic browning reactions in plant products are related to the polyphenol oxidases. Polyphenol oxidases which catalyze a number of reactions involving phenolic compounds have engendered a great amount of research interest (3). Other enzymes such as peroxidases and catalases also contribute to the undersirable brown color reactions although their presence has traditionally been used as an indicator of inadequacy of blanch in frozen vegetables. Dehydrated prunes, raisins, dates and related fruits are about the only plant products which are considered beneficiaries of the browning reactions. This browning requires little or no regulation and occurs generally during normal processing operations.

Chlorophyllase which is specific for catalyzing the hydrolysis of the phytyl ester linkage of chlorophyll is involved in green hue shifts in many green vegetables and chlorophyll-continuing plant material. Braverman (4) indicates information is unclear concerning the use of endogenous chlorophyllase to preserve the green color in this type of tissue. Perhaps the best perspective relating to green color and chlorophyllase is the work of Van Buren et al. (5) who suggested that the conversion of chlorophyll to pheophytin in blanched green beans at 60-80°C was primarily due to pH levels in the pods which were shifted from 6.5 to 5.5 by endogenous pectinesterase (PE) activity.

Anthocyanase enters plant products through bruises and insect damage and is thought to be mainly from fungal origins. The use of endogenous anthocyanase activity has been limited although Van Buren et al. (6) have suggested that an anthocyanin degrading enzyme is present in bruised sour cherries. Natural color can be reduced in berry juices and wines by deliberate or inadvertent addition of anthocyanase.

Color and gloss have been improved in reconstituted dehydrated sweet potato flakes, according to Hoover (7), by activating the natural amylolytic enzyme systems (mainly β amylase). The pureed tissue was held at 70-85°C for 2-60 minutes to get this desirable conversion.

In the area of flavor, it is clear that few of the taste sensations such as sweet, sour, bitter, salty, or even pain, heat or cold are developed by utilization of endogenous enzyme systems. This probably relates to the fact that most taste ingredients can be added economically to food products and consequently there is no practical reason to activate specific enzyme systems.

Although a great amount of research has been conducted on odor development by the use of enzymes, recently attention has been on the Cruciferae family and the Allium genus. Odors appear to be a normal physiological process in plant materials with development during the entire life of the plant through pre-precursors, precursors and their odor components. A large percentage of these odors are developed by both chemical and/or enzymes means from amino acids, isoprene groups, acetates, and break-down products of many more complex moities. Konigsbacher (8) has visualized that odor formation results from a series of chemical compounds in a chain of reactions which probably begin with water, sunlight, and carbon dioxide. It is obvious that each step is chemical and enzymatic, with each enzyme system probably requiring special substrates and cofactors to make the reaction move forward.

The problem in triggering an aroma reaction during a processing operation is the possible disappearance of the aroma before it reaches the consumer. The most satisfactory way to protect aroma factors in plant tissues for later activation is to preserve pre-precursors and precursors, maintain compartmentalization, and reduce as much as possible heat shock to the tissues which may destroy cell integrity or utilize precursor substances. Some ideas along this line were suggested by Puangnak, Wiley, and Kahlil (9) studying heat activated dimethyl sulfide (DMS) in sweet corn. They found this compound evolves mainly from the pericarp complex. Microwave blanching may protect this desirable corn-like aroma in cooked frozen samples by reducing heat treatment to flavor precursors in the pericarp area.

In the nutritional quality area, it appears that enzymes have been infrequently used in fruit and vegetable tissue to make food products more nutritive. The pectinases, hemicellulases, and cellulases all have the potential to break down polymers into more soluble and more digestible food substances, but these techniques have not been used extensively to provide more nutritious food. Amylases have been used, however, to increase soluble solids in sweet potatoes for better functional properties rather than improved nutrition (7).

Ascorbic acid oxidase, enzymes involved in carotene production, and the action of lipoxgenase for destruction of lipids, are of concern in the handling and processing of fruit and vegetable products, but probably cannot be involved in worthwhile utilization programs. Lipase, which converts lipids to fatty acids and glycerol, and polyphenol oxidases, which can destroy vitamins, must be carefully controlled in all fruit and vegetable products.

TEXTURAL FACTORS

The textural factors emphasized in this section will relate mostly to softness and firmness of fresh apple fruits and processed apple slices. The studies of cell-wall polysaccahrides, pectic enzymes, and calcium firming conducted by several colleagues in this department will be presented. There will be no attempt to include other textural and tactile sensations such as viscosity, consistency, chewiness, stringiness, brittleness, cohesiveness, oilness, wateriness, etc.

Polysaccahrides Involved in Softening

Natural apple softening during the post-harvest period has almost universally been considered a breakdown of pectins, which by strict definition are galacturonans, built up of 1-4 linked D-galacturonic acid residues. Other investigators have felt that non-galacturonide material is attached to the polyuronide chain and is also involved in fruit softening. Speiser et al (10) were among the first to indicate that ballast materials, or hemicelluloses, are probably attached to the polyuronide chain by ester linkages. Later, Tavakoli and Wiley (11), in considering the structure of apple cell walls, found pectic substances comprising units of D-galacturonic acid, hemicelluloses comprising units of D-xylose, L-arabinose, D-galactose, D-glucose, and cellulose comprising residual material were 30%, 40% and 30% respectively on a dry weight basis. These cell wall components, except for cellulose, were evaluated by gas liquid chromtography (GLC) as trimethyl silyl (TMS) derivatives of monomers formed after fungal pectinase and hemicellulase hydrolysis of ethanol precipitates. These precipitates were prepared from serial extractions of sugar-free alcohol insoluble solids (AIS). Results showed the hexosans and the pentosans in the hemicelluloses consisted of 60% glucose, 27% galactose 6.5% xylose and 6% arabinose moieties in about a 10:5:1:1 ratio. This GLC technique did not detect rhamnose which has been found in the cell wall of apples, but did show substantial amounts of the various glucose moieties, even in the more mature and starch-free fruits.

Tavakoli and Wiley (11) using apples of a wide range of maturity and ripeness found, as shown in Table 1, through regression analysis of objective firmness measurements and cell wall components, that galactans, glucans, and water insoluble pectinic acids were responsible for approximately 80% of the variation in firmness changes of the three major Appalachian apple varieties used in processing - Golden Delicious, Stayman, and York Imperial. These workers also reported, that according to their results the hemicelluloses and the pectic substances which occur with cellulose in apple parenchyma tissue in a very close knit cellular matrix, should be regarded as the major chemical constituents which gradually decompose as fruit undergoes ripening. The

Table 1. Multiple regression and correlation analysis of firmness (lbs. force) values with the cell wall constituents (per cent fresh weight basis) of Golden Delicious, Stayman, and York Imperial, n = 36.

Variable	Single correlation coefficients	Partial correlation coefficients	Cumulative regressions	
			R^2	R
Galactose	0.867	0.294	0.741	0.867
Glucose	0.836	0.298	0.772	0.879
Water insoluble extract-galacturonic acid	0.857	0.221	0.788	0.888
Residue	−0.080	−0.175	0.797	0.893
Arabinose	0.524	0.113	0.802	0.896
Xylose	0.622	0.049	0.803	0.896
Water soluble extract-galacturonic acid	−0.164	−0.047	0.804	0.896
Total galacturonic acid	0.767			

Proceedings of the American Society for Horticulture Science

Table 2. Effect of variety on firmness and on galactose, water insoluble extract galacturonic acid, and the residue content of apple cell wall materials[a] (per cent fresh weight basis).

Variety	Means			
	Shear-press (lb. force)	Water insoluble extract galacturonic acid (Pectinic acids)	Galactose (Galactan)	Residue (Cellulose)
Golden Delicious	321	0.30	0.19	0.50
Stayman	322	0.30	0.18	0.58
York Imperial	480	0.46	0.26	0.68
L.S.D. at 1% level	25	0.08	0.05	0.08

[a]Combined data for maturation and storage treatments.

Proceedings of the American Society for Horticulture Science

portion of apple fruit firmness which diminishes during storage life should be referred to as "dynamic firmness" since it is greatly affected by changes in the intermicellar polysaccharides. The starch content of fruit in a very immature stage and the osmotic properties of the cells may also produce some effects on firmness. Cellulose, however, was not influenced by maturities and storages. Therefore, cellulose should be regarded as a static constituent of the apple fruit cell wall which provides the cells with relatively fixed or base-line levels of strength. This strength may be referred to as "static firmness" which prevails throughout the storage life of apples.

The data in Table 2 by Tavakoli and Wiley (11), shows that York Imperial, a relatively firm variety, contained a significantly greater amount of water-insoluble pectinic acids, galactans, and cellulose than either Golden Delicious or Stayman. The levels of these substances should be used to characterize the differences in firmness among these varieties.

It appears that a number of polysaccharides are involved in apple fruit softening. Rombouts (12) has recently reviewed pectic enzyme research and presented several theories concerning the linkages of the various oligisaccharides to the pectic substances. Further, Keegstra et al (13) have suggested that the various macromolecular components of sycamore cell walls are covalently cross-linked. It is possible that the same linkages occur in apple parenchma tissue.

Enzymes Systems Related to Softening

The enzymes systems involved in apple softening and overall textural changes are still under question. There have been few reports of either endogenous polygalacturonases (PG) or pectin lyases(PL) in disease free apple parenchma tissue. The presumed absence of these important enzymes in this tissue places in further jeopardy the theory that apple fruit softening is due to galacturonan 1-4 linkage breaks and/or translimination in polygalacturonic acids. Mannheim and Siv (14) have also reported that PG was absent in oranges.

PE, which removes methyl esters from methylated pectic substances, is the remaining endogenous pectic enzyme that is thought to be involved in physiological softening changes of apple fruit. Rombouts (12) pointed out, Schultz et al (15) were the first to suggest that the mode of action of PE resembles that of a zipper. Zipper action may account for the major softening in apple fruit up to and including the first 60-90 days of cold storage, by separating the pectin and hemicellulose complexes from the cellulose matrix, or by causing internal separation of the "dynamic" constituents such as the galacturonans, pentosans, and hexosans.

Some cold storage work (1°C) conducted by Lee (16) on Golden Delicious and York Imperial varieties showed PE activity increased in storage up to 60 days and then leveled off at a high plateau. This plateau may have been due to accumulation of products or limitations in substrate level, which can have an inhibiting effect on enzyme activity. A negative correlation coefficient of -0.85 (n=40) was shown between PE activity in acetone powder from raw slices and shear press readings on raw slices. Results indicated that as PE activity increases in raw fruit, presumably in a Ca^{2+} ion deficit, the fruit became softer. He also showed that PE activity was approximately 2 times higher in the soft variety Golden Delicious as compared with the firm variety York Imperial.

In considering the problems associated with processing apple slices, it is clear that a substantial amount of PE activity has taken place in the apple cell wall during maturation and ripening of the fruit and the pH intregity of the cell wall has been maintained by the slow demethylation of the pectic substances. Lee (16) suggested that PE activity is very low in apple tissue which normally has a pH of about 3.4. He predicted PE specific activity at this pH could be 5-10 μg CH_3OH per mg protein in a 30 min period at 35°C. Cold storage further reduces this PE activity and might account for the very slow softening changes which take place in controlled atmosphere (CA) and refrigerated storage. Apple fruit received by processors then would be quite variable in terms of PE activity which had taken place during maturation and ripening and subsequent storage.

Calcium Firming Plant Tissue

Locenti and Kertesz (17) were among the first to indicate treatment of fruit tissue with Ca^{2+} salts would firm tissues and that the tissue firming compound was calcium pectate. Archer (18) and Collins and Wiley (19) reported Ca^{2+} salts were not used to any great extent to firm apple slices because of problems in getting a uniform firming response. They found that calcium lactate was a good firming agent and gave a more desirable flavor response than either calcium chloride or calcium gluconate

Van Buren et al (20), working with calcium firming of green beans, found PE activity resulted in firmness of tissue and a decreased solubility of the pectic substances, particularly in the presence of Ca^{2+} salts. They suggested use of the endogenous PE system, under proper processing conditions, to allow the enzyme to act on the pectins.

Penetration of Ca^{2+} salts into heat processed apple tissues using $^{45}CaCl_2$ as shown in Figure 1 was studied by Collins and Wiley (21). The dark areas in the autoradiograms of the apple pieces show $^{45}CaCl_2$ distribution. Section A shows only very slight amount of Ca^{2+} penetration occurs if the fruit is submerged 15 min in a $^{45}CaCl_2$ solution at 20°C. Section B presents submerged fruit which was then canned and heat processed for 8 min at 100°C. These samples do not show complete penetration of ^{45}Ca to the center of the piece. Section C shows a more complete penetration of the piece, when previously submerged slices were subjected to 2 min vacuum treatment for deaeration, steam blanched for 30 sec. at 10 psi, canned and then processed for 8 min at 100°C.

Figure 2 shows that canning and then processing of apple slices in boiling water for 8 min using $^{45}CaCl_2$ as a packing medium, did not result in full penetration of $^{45}CaCl_2$ to the middle of the pieces. These slices received the usual vacuum and steam treatments prior to heat processing

Results indicate Ca^{2+} ions are fairly effective in firming apple fruit if handled properly, however, very mature fruit and certain soft varieties, such as overmature Golden Delicious, Red Delicious, and Romes continue to present texture control problems. Van Buren (22), working with green beans, also found blanching treatments in the presence of Ca^{2+} did not always effectively firm the pods.

The continuing problems of firming plant tissues during processing has stimulated the idea that endogenous PE could be used to activate sites for Ca - bonding. However, considerable research information is required so that the PE can be measured easily under simulated processing conditions. These conditions should include wide ranges in temperature, pH, soluble solids and the like.

PE Measurement and Partial Characterization

Lee and Wiley (23) developed a dialysis procedure to characterize apple PE activity. The dialysis and GLC procedure to measure methanol produced by PE demethylation involved using a unitized dialysis cell under agitated conditions. Each dialyzer comprised 2 polymethyl-methacrylate half cells separated by a Visking membrane with 4.8 mμ pore diameters. Each half cell has a circular cavity (4.0 cm diameter, volume 7 ml) and a filling hole.

One side of the compartment, designated the front was charged with 5 ml of the reaction mixture. The charge contained 0.5% of apple acetone powder and 0.15M NaCl in 1% phosphate buffered pectin solutions adjusted to various pH levels. The other side of the compartment, designated the back, was charged with an equal volume of distilled water. The filling holes were sealed with tape and the cells were shaken in a temperature controlled oven. These phosphate buffers ranging from pH 4.5-9.0 allowed satisfactory measurement of PE activity over this pH range. The methanol produced was taken as PE activity and the dialysis procedure allowed simultaneous production and separation of methanol from the pectin substrate into the distilled water.

After dialysis and depending on the methanol concentration in the pectin free, distilled water side of the dialysis cell, a 5 μℓ sample of the methanol and distilled water mixture was injected directly into the gas chromatograph.

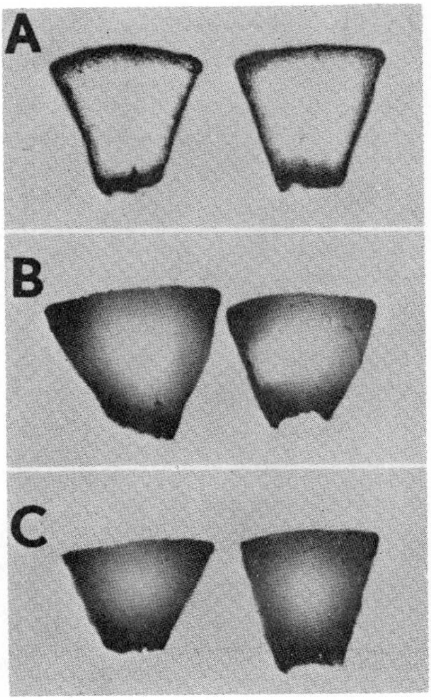

Journal of Food Science

Figure 1. Autoradiograms of transverse sections of mature Stayman apple slices submerged in a solution of $Ca^{45}Cl_2$ for 15 min at 68°F. A, *submerged only;* B, *submerged and processed; and* C, *submerged, subjected to vacuum and steam, and processed.*

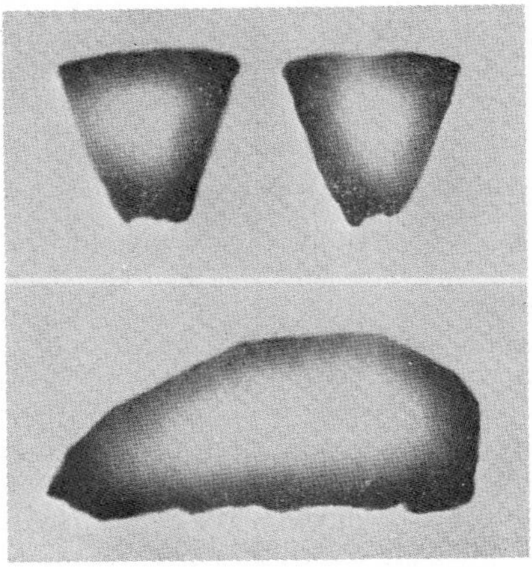

Journal of Food Science

Figure 2. Autoradiograms of longitudinal and transverse sections of mature Stayman apple slices subjected to vacuum and steam. Slices were processed in cans containing $Ca^{45}Cl_2$ solution.

The concentration equilibrium for the closed system dialyser with compartments charged with equal volumes of solution were as follows:

Initial concentration of methanol,
in front: $C_f = C_i$;
C_i is designated as concentration of methanol
as a result of PE activity.
At equilibrium concentration, if all methanol
is diffusible, $C_f = C_b = 0.5\ C_i$.
R equals the ratio of $C_b/0.5_i$. At equilibrium, R = 1
if all methanol is diffusible; if some of the methanol is
not diffusible, R will be less than 1.

Table 3 shows a substrate of 1% pectin and reaction times of about 3-4 hrs (at 35°C) gave satisfactory results using this method

Apple PE exhibits the characteristic bell-shaped curve showing optimum activity between pH 6.5 and 7.5. Figure 3 shows the PE curves for Golden Delicious apple fruit of 4 different maturities. The GO code indicates Golden Delicious apple samples were harvested and tested immediately without further storage. GOA, GOB, GOC, and GOD depict maturity levels. The GOA fruit was harvested Sept. 11, the GOD fruit Oct. 3 indicating a spread of 23 days between fruit of the first and last harvest. It seems clear that PE activity is significantly higher in the more mature fruit or fruit from later harvests.

Lee and Wiley (23) found the optimal reaction temperature for apple PE is 55°C and the activation energy (E) calculated by the graphic procedure is 5800 calories. Activation energy should be regarded as the amount of energy needed to place the substrate molecules in a reactive state. If E is large, the rate of reaction will increase rapidly for a fixed increase in temperature. The substrate molecules must first absorb that amount of E. They then can react and be converted to products. This means that PE activity can be expected to be 2.5 times higher at 25°C as compared with 0°C. This compares favorably to data from storage research where firmness of apples held at 0°C for 2 months was about twice as firm as those held at 20°C for the same period.

Apple PE shown in Table 4 is completely inactivated at 80°C for 10 min or 90°C for 5 min.

According to published figures, the temperature of inactivation of plant PE shows a wide range. Heating for 5 min at 70°C gave almost complete inactivation of purified PE, whereas a similar preparation from orange was inactivated by 30 min at 56°C. Other less purified extracts have been found more resistant to heat; for example, McColloch and Kertesz (24) found 50% activity left in tomato extracts after one hour at 70°C, Joslyn and Sedky (25) found a marked difference in the thermal stability of PE in citrus juices according to species and variety.

Table 3. Diffusibility of methanol[a] through a cellulose membrane[b] under agitated conditions.

Concentration of methanol	R values[c]			
	Reaction times (hours)			
	1	2	3	4
100 ppm	0.65	0.95	0.96	0.95
200 ppm	0.66	0.98	0.95	0.99
300 ppm	0.54	0.96	0.95	0.98
400 ppm	0.38	1.00	0.99	1.02

[a]Methanol was mixed in 1% pectin solution at 35°C.
[b]A Visking membrane obtained from Arthur H. Thomas Co., Phila., Pa., 4465-A2 dialyzer tubing.
[c]R value represents the ratio of methanol concentration between front and back sides of membrane in a dialyzer.

Journal of the American Society for Horticulture Science

Table 4. Temperatures for apple pectinesterase[a] inactivation.

Inactivation temperatures (°C)	Relative velocity (μg CH_3OH in 5 ml dialysate)	Inhibition (%)
	Control	
ambient	655	0
	5 min. heating	
100	0	100
90	0	100
80	50	92
70	80	88
60	325	50
	10 min. heating	
100	0	100
90	0	100
80	0	100
70	40	94
60	195	70

[a]Dialysis under agitation for 4 hr at 35°C in a pH 6.5 phosphate buffer.

Journal of the American Society for Horticulture Science

Table 5. Effects of buffer treatments on Ca in alcohol insoluble solids (AIS) of canned Stayman apple slices.[1]

Treatments: Buffer (1.0M) lactic acid—sodium lactate plus 1% $CaCl_2$	Tissue pH	Ca μg/g/AIS[2]
Control	3.48	3.70 a
Unbuffered plus Ca	3.27	12.79 b
pH 3.0	3.20	16.29 c
pH 3.5	3.57	17.51 d
pH 4.0	3.94	18.49 ef
pH 4.5	4.27	18.95 f
pH 5.0	4.52	17.94 de
pH 5.5	4.88	32.59 g

1 All fruit stored (34°F) 50 days.
2 Means (average 10 observations) not followed by the same letter are significantly different at the 5% level.

Food Technology

According to Lee and Wiley (23) PE can also be inhibited by sucrose and the percent of apple PE inhibition progressively increased in solutions containing from 5 to 15% sucrose. Approximately 5, 25, and 40% PE inhibition were found in the presence of 5, 10, and 15% sucrose concentrations, respectively. These results are similar to those of Chang et al (26) who reported that a 13% sucrose concentration inhibited papya PE. The actual mechanism by which sucrose inhibits PE activity is now known. However, an environment of lower water activity due to sucrose may be the most important factor involved, since PE requires water for its reaction. Sugar solids, added to fruit, should be carefully managed if it becomes desirable to activate PE during pre-processing operations.

Endogenous PE Utilization

Once a method is developed to determine enzymatic activity under simulated pre-processing and processing conditions, the utilization of that endogenous enzyme system may commence. Wiley and Lee (27) have pointed out that texture modification of apple slices during processing is necessary to accomodate customer's specifications. Soft texture is desirable in no-cook pies, whereas some products such as apple-turnovers and the like may need firmer fruit to withstand baking. As indicated earlier, there is great variation in firmness of raw apple products and even though heat processing tends to make for more textural uniformity, the Ca-firming reaction has been difficult to control because of unknown or undesirable handling practices used on fruit prior to processing. Modifying apple texture by activating the PE enzyme system during the relatively short pre-processing periods of slicing, trimming, vacuumizing, steaming and/or blanching and during the early heating phases of thermal processing appears possible.

Wiley and Lee (27) have investigated PE activation and have vacuumized apple slices in a wide range of buffer solutions prepared from 0.1 M citric acid-disodium phosphate mixtures containing 0.5% calcium chloride. This procedure was based on earlier research of Collins (28) who showed an improved up-take of calcium in apple AIS as the pH of the tissue increased from about 3.5 to 5.0 (Table 5). These results were probably due to increased PE activity at the higher pH levels.

Figure 4 shows the relationship between PE activity in the acetone powder of raw Golden Delicious tissue and the firmness of a similar sample of Golden Delicious slices which were vacuumized over a wide pH range in the presence of 0.5% $CaCl_2$. The slices were then blanched in steam at 10 psi for 2 minutes, cooled and measured in the Kramer shear press. It appears that maximum firmness in the Ca^{2+} treated blanched slices, held in a concentration of 0.5% $CaCl_2$ over the wide pH range, occurred at

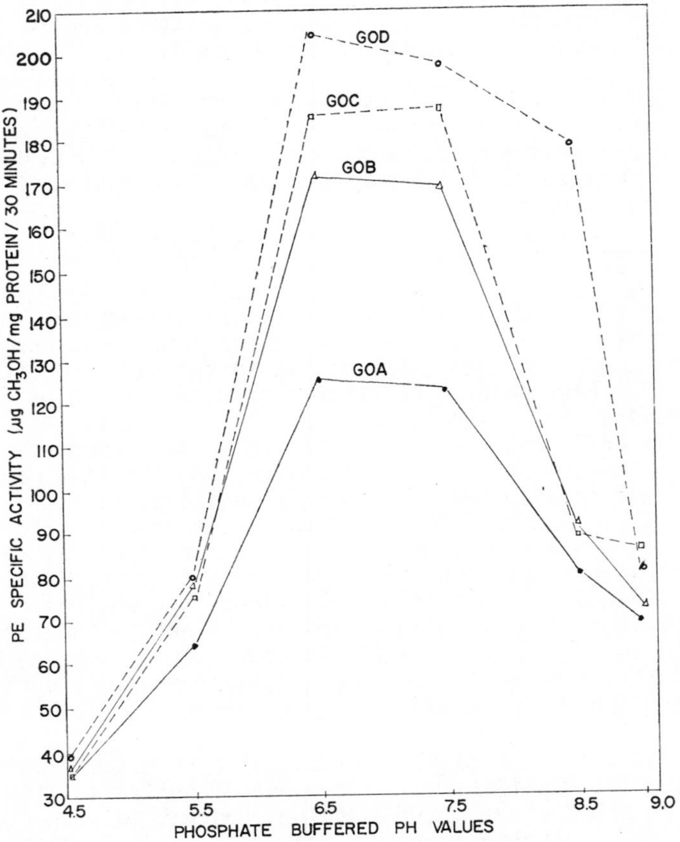

Journal of the American Society for Horticultural Science

Figure 3. Effects of pH on PE specific activity

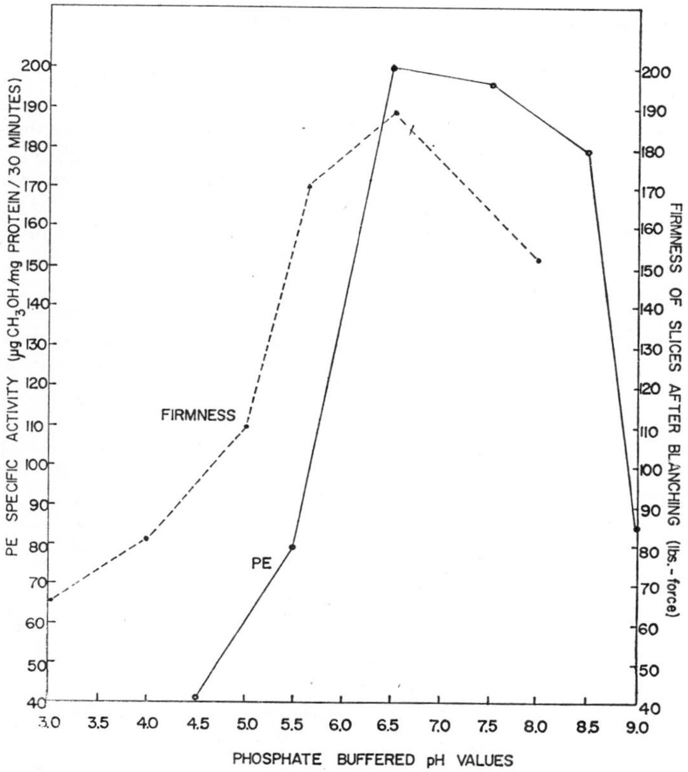

Food Technology

Figure 4. Effects of pH on PE activity and firmness of calcium-treated blanched Golden Delicious slices

approximately the same pH level which is optimum for pectin-esterase activity.

Apple slices can also be treated with $Ca(OH)_2$ to firm the tissue and modify pH. Golden Delicious slices impregnated with a 1% $Ca(OH)_2$ solution gave a firmness value of 158 lb/force. This was much firmer than control which was 78 lb/force. The samples firmed with the $Ca(OH)_2$ treatment were then pH reversed with a 1% citric acid solution to about 3.95. This reduction in pH did not significantly reduce the firmness of the tissue which had been previously induced at the higher pH level in the presence of $Ca(OH)_2$. The $Ca(OH)_2$ treated slices showed a dull brownish coloration and had to be acidified to return the color of slices to normal. Case hardening was noticed in the 1% $Ca(OH)_2$ treated slices and concentrations below this level should be used in commercial operations.

Possible Processing Applications

Apple processors should consider utilizing endogenous PE systems to standardize the textural quality of their products. Techniques for enzyme management have already been approved in principle by the Food and Drug Administration. Recent research by Wagner et al (29) and resultant regulations (30) for tomato puree and paste allow acidification by HCl and neutralization with NaOH prior to straining. This technique, to improve the consistency of tomato products, controls and utilizes characteristics of endogenous PE and PG enzyme systems.

Apple slice manufacture presents a somewhat more complex set of operations. The fruit may be utilized in a thermally processed system or as a frozen item requiring different enzyme inhibiting operations than for canned slices. For example, many apples for freezing are sulfited without the use of high blanching temperatures. Most generally problems occur with slightly soft fruit and/or fruit almost too soft and mushy to be utilized. Utilization of endogenous PE is primarily for soft fruit. Although it is recognized that apple slices become soft and mushy during thawing after freezing, the following examples deal with canned slices.

A general schematic flow sheet used by apple slice processors is as follows: classify fruit according to firmness and variety, size grade, automatically peel, core, slice, trim, dry or wet flume, vacuum blanch, convey, fill and heat process.

The PE utilization procedure should be used after the fluming step and would incorporate a variable speed steam or hot water treatment and a 2nd vacuumizing unit with suitable controls to moniter temperature, time of hold, pH, soluble solids, and the like. For the various catagories of soft fruit the variable speed treatment chamber would be used to activate PE at about 55°C. The 1st vacuum unit would be used to impregnate slices with Ca^{2+} salts and/or Ca^{2+} bases. The 2nd vacuum unit would only be used where pH revision was necessary to return firmed slices to their normal pH, color and flavor.

CONCLUSIONS

It is evident that only a few endogenous enzyme systems have been used to alter quality characteristics and/or functional properties of processed fruits and vegetables. The most important studies can be summarized as follows:

(1) Heat activated β amylase is used to improve functional properties of sweet potato flakes.

(2) Heat and pH activated PE can be used to provide sites for calcium firming of thermally processed apple slices.

Activation of PE during maturation and ripening considered in the light of the following obervations:

(a) Apple softening during maturation and ripening appears to be due to the cumulative changes of several macromolecular components. They are galactans, glucans and insoluble pectinic acids. Therefore linkages between these components must be considered in any Ca^{2+} related-PE utilization firming operation.

(b) Cellulose tends to lend base-line or background strength to mature or ripe apple tissue and is not an important factor during enzyme utilization operations.

(c) PE activity shows a highly significant negative correlation (-0.85) with firmness changes in apple fruit and thus appears to be a factor in apple tissue structural integrity.

(d) A dialysis - GLC procedure to measure methanol produced by PE activity provides an analytical tool to provide information for the management of the PE system during pre-processing and processing procedures.

(e) PG is presumed to be absent in disease free apple parenchma tissue thus may be disregarded during PE activation operations. PG, if present, would be activated by Ca^{2+} ions and decrease effective firming.

Further research is needed to develop and utilize endogenous enzyme systems in fruit and vegetable products.

Literature Cited

1. Schwimmer, J. Food Sci. (1972) 37: 530.
2. Wiley, R.C. and Winn, P.N., N.Y. State Proc. Veg. Conf. (1970)
3. Whitaker, J.R. "Principles of Enzymology for the Food Science" 636P Marcel Dekker Inc. N.Y. 1972.
4. Braverman,J.B.S."Introduction ot the Biochemistry of Foods, 336p Elsevier Pub. Co. Amsterdam, 1963.
5. Van Buren, J.P., Moyer, J.C. and Robinson W.B. Food Technol (1964) 18: 1204.
6. Van Buren, J.P., Scheiner, D.M. and Wagenknecht, A.C. Nature (1960) 185: 165.
7. Hoover, M.W. Food Technol. (1967) 21: 322.
8. Kongisbacher, K.S. "Enzymes Use and Control in Foods" Institute of Food Technologist Short Course pG-1 Chicago, 1976.
9. Puangnak, W., Wiley, R.C. and Kahlil, T. 1976 Private Communication 12p.
10. Speiser, R., Eddy, C.R. and Hills, C.H. J.Phys, Chem. (1945) 49: 563.
11. Tavakoli, M. and Wiley, R.C. Proc. Amer. Soc. Hort. Sci. 1968) 92: 780.

12. Rombouts, F.M. "Occurrence and Properties of Bacterial Pectate Lyases" Ph.D. Dissertation Wageningen Netherlands p. 132 1972.
13. Keegstra, K., Talmadge, K.W., Bauer, W.D. and Albersheim, P. Plant Physiol (1973) 51: 188.
14. Mannheim, C.H. and Siv, S. Fruchtsaft Industrie (1969) 14: 98.
15. Schultz, T.H., Lotzkar, H., Owens, H.S. and Maclay, W.D. J. Phys. Chem. (1954) 49: 554.
16. Lee, Y.S. "Measurement, Characterization and Evaluation of Pectinesterase in Apple Fruits" Ph.D. Dissertation University of Maryland, College Park 105p. 1969.
17. Loconti, J.D. and Kertesz, Z.I. Food Res (1941) 6: 449.
18. Archer, R.W. Canner and Packer (1962) 131: 28.
19. Collins, J.S. and Wiley, R.C. Maryland Agric. Exp. State. Bul. (1963) A-130 Influence of added calcium salts on texture of thermal-processed apple slices 62p.
20. Van Buren, J.P., Moyer, J.C. and Robinson, W.B. J. Food Sci. (1962) 27: 291.
21. Collins, J.L. and Wiley, R.C. (1967) J. Food Sci. 32: 185.
22. Van Buren, J.P. Food Technol. (1968) 22: 132.
23. Lee, Y.S. and Wiley, R.C. J. Amer. Soc. Hort. Sci. (1970) 95: 465.
24. McColloch, R.J. and Kertesz, Z.I. Food Technol (1949) 3: 94.
25. Joslyn, M.A. and Sedky, A. Food Res (1940) 5: 223.
26. Chang, L.W.S., Morita, L.L. and Yamamoto, H.Y. J. Food Sci. (1965) 30: 218.
27. Wiley, R.C. and Lee, Y.S. Food Technol (1970) 24: 126.
28. Collins, J.L. Effect of Calcium Ion-Structural Polysaccahride Interaction on Texture of Processed Apple Slices. Ph.D. Dissertation, University of Maryland, College Park 99p 1965.
29. Wagner, J.R. and Miers, J.C. Food Technol (1967) 21: 920.
30. Anon "Code of Federal Regulations" Food and Drugs Parts 10-129 (1974) Part 53:20 General Services Admin., Washington, D.C.

12

Oilseed Enzymes as Biological Indicators for Food Uses and Applications

JOHN P. CHERRY

Southern Regional Research Center, Agricultural Research Service, U.S. Department of Agriculture, New Orleans, La. 70179

Research utilizing enzymes as indicators of physiological and biochemical changes in living organisms is not new (25,38). Enzyme synthesis is genetically controlled, however, their specific task(s) and feedback control mechanisms are also influenced by other constituents in the cellular environment; i.e., cofactors, substrates and products associated with metabolic processes. Scientists have developed the technology to partially imitate in vivo conditions in order to study the active enzymes extracted from their natural environment. Nevertheless, enzymes do falter when abused, becoming excellent indicators of several types of cellular change.

Gel electrophoresis is a method developed to qualitatively study enzymes. The background and theory of electrophoretic techniques were discussed by Ornstein (46), and applications of these procedures to analyze and compare proteins and enzymes have been presented by a number of investigators (6,8,24,29,48,55,57). Basically, the method applies an electric charge to separate aqueous extracts of proteins in a gel matrix, such as polyacrylamide or starch. Protein mobility depends upon a combination of factors, including net charge, molecular size, and conformation. Histochemical staining procedures have been developed to detect the location of enzymes in gels based on their catalytic activity (39).

Electrophoretically detected enzyme changes have been used to characterize Aspergillus-peanut interrelationships; in the fungus, development and differentiation occurs while the peanuts undergo senescence (14,18). Electrophoretic zymograms of enzymes extracted from various plant and animal tissues were used in chemotaxonomic studies of genetic speciation and genera taxonomy (21,38). These studies showed that a great amount of genetically controlled molecular diversity, or enzyme multiplicity, exists in nature as isozymes; i.e., genetically related enzymes with similar substrate specificities but different electrophoretic mobilities. Zymograms can also vary because of tissue ontogeny. These reports suggest that, in all types of electrophoretic studies, care should be taken to insure that standard known zymograms are developed for the materials to be examined prior to any evaluation of treatment effects.

Complexity of Identifying Isozymes as Multiple Enzyme Forms

Hunter and Markert (32) revealed a vast multiplicity of electrophoretically distinct bands with enzyme activity. The term "isozyme" was proposed by Markert and Moller (42) to refer to enzymatically active proteins that are separated electrophoretically and that catalyze the same biochemical reaction. Markert (39) later proposed modifying the word isozyme with terms, such as allelic, nonallelic, homopolymeric, conformational, hybrid, and conjugated (words which deal with genetic and chemical terminology). The principal types of molecular multiplicity that generate isozyme patterns were reviewed by Markert and Whitt (40). It is theorized that enzyme multiplicity is created at the molecular level by various combinations of the following: (a) different polypeptides coded by allelic and nonallelic genes; (b) polymers of various sizes; (c) homo- and heteropolymers; (d) polypeptides secondarily modified in various ways; and (e) different conformations produced by permutations of polymer subunits, or alternate tertiary and quaternary configurations of proteins. Organisms may utilize both genetic and nongenetic mechanisms to form properties of enzymes necessary to fit special metabolic requirements.

Shaw (54) classified isozymes into primary types, which include distinct molecular entities produced from different genetic sites, and secondary types, which result from a significant alteration in the structure of a single polypeptide. Scandalios (52) has employed the term isozyme to define the heterozygous state of genetically variant enzymes. He showed a multiplicity of alleles (genes) carrying information for a specific type of enzyme activity. Crosses between several lines of maize with different catalase phenotypes revealed six alleles, whereas leucine aminopeptidase isozymes were controlled by two separate genetic loci. It was further noted that different isozymes were present in different tissues and in extracts of tissues at different stages of development.

In 1971, the IUPAC-IUB Commission on Biochemical Nomenclature (34) recommended that multiple forms of an enzyme possessing the same activity, separated and distinguished by suitable methods (preferably by electrophoretic techniques but also by chromatography, kinetic criteria, chemical structure, etc.) and occurring naturally in a single species, should be generally defined as "multiple forms of the enzyme...". Multiple forms of enzymes that have been specifically characterized by genetic differences in primary structure should be defined as isoenzyme or isozyme. These include genetically independent proteins, heteropolymers (hybrids) of two or more noncovalently bound polypeptide chains, and genetic variants (allelic). Enzymes not falling into this latter category include proteins conjugated with other groups, proteins derived from one polypeptide chain, polymers of a single subunit, and conformationally different forms.

Detecting Multiple Forms of Enzymes

Esterases Investigations were completed to standardize techniques (protein and enzyme extraction, purification, and gel electrophoresis) used to examine enzymes (2,3,30,31,41). This provided reproducible qualitative data for comparative purposes; i.e., uniformity in intensity of gel staining and repeatable spatial arrangement of enzyme bands between experiments. This also established electrophoretic patterns which serve as standards to which enzymes from damaged, processed, or mold-infected substances might be compared.

A typical example of on-gel electrophoretic characterization of enzyme activity was shown in rabbit erythrocytes (Fig. 1; 13), using polyacrylamide gel electrophoresis. This example revealed a complex array of nonspecific esterases in samples of hypotonic extracts and fractions. Although the rabbits were considered to be isogenic, some genetic variability in esterase bands was noted (cf., region 1.5-2.5 cm of gels A and B in both membrane and cytoplasmic extracts). Nonspecific esterases were detected by incubating gels in a mixture of α- and β-naphthyl acetate. By using on-gel techniques, the esterases were characterized on the basis of their substrate specificities and susceptibilities to inhibitors. Group I esterases were inhibited by diisopropylphosphorofluoridate (DFP), phenylmethanesulfonyl fluoridate (PMSF) and eserine, but not by mercuric chloride or p-chloromercuribenzenesulfonic acid (CMBSA) (Fig. 1). These esterases showed specificity toward acetylthiocholine iodide and thiophenyl acetate and thus appeared to be composed of acetylcholinesterases. Group II (acetylesterases) consisted of enzymes that hydrolyzed esters of acetic acid and β-naphthyl laurate and thiophenyl acetate, and was not inhibited markedly by any of the inhibitors tested. The enzymes in Groups III a,b, and c, IV and VI, consisting of slow, intermediate, and rapid migrating carboxylesterases, respectively, were highly active toward α-naphthyl butyrate, and were inhibited partially or completely by DFP and PMSF but not by mercurials or tri-o-tolyl phosphate. Only Groups I and VI were inhibited by eserine. Groups V and VII showed specificity typical of arylesterases, being partially inhibited by DFP, PMSF, and the mercurial reagents.

These data revealed that the nonspecific esterases in hypotonic cytoplasmic or membrane fractions from erythrocytes which could be detected using nonspecific substrates were really heteromorphic (specific esterases). Some of the definitions for describing the term isozyme may be applied to these multiple forms of esterases detected in erythrocytes. However, until each of the electrophoretic bands of esterase activity is related to a specific gene(s) by breeding studies, a true definition of isozyme cannot be used to describe them.

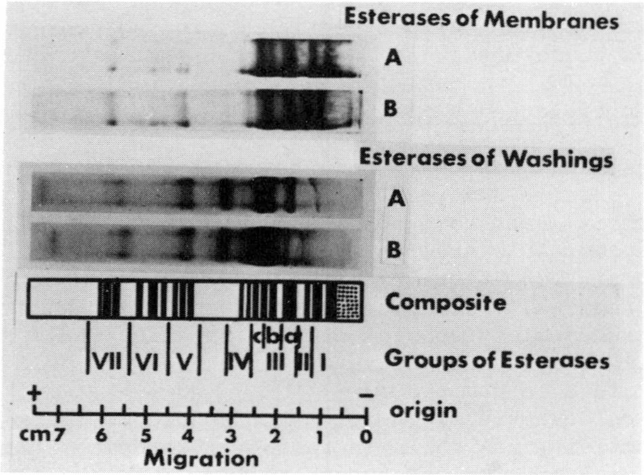

*Figure 1. Polyacrylamide gel electrophoretic patterns of nonspecific esterase activity in membrane and cytoplasmic fractions of rabbit erythrocytes. Methods for preparation of erythrocyte fractions are described by Cherry and Prescott (12). A composite shows the classes of esterase activity (*I: *acetylcholinesterases;* II: *acetylesterases;* III a, b, *and* c, IV *and* VI: *carboxylesterases and;* V *and* VII: *arylesterases. The bands in region 3.0–3.5 cm of gels of the cytoplasmic fraction are hemoglobin.*

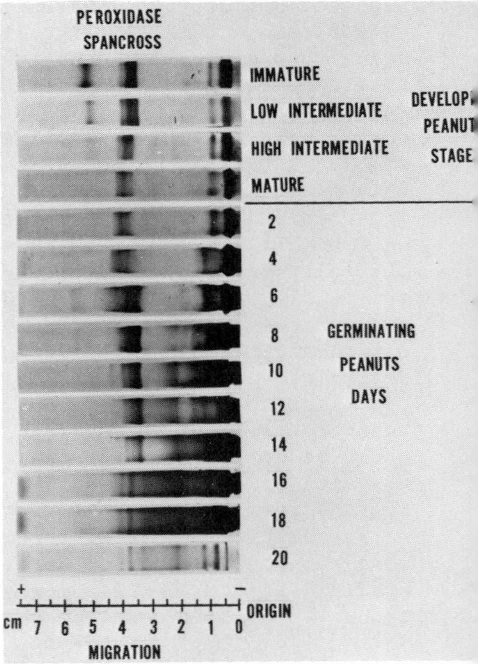

Figure 2. Polyacrylamide gel electrophoretic patterns of peroxidases in peanut cotyledons at various developing and germinating stages. Gel patterns of the cultivar Spancross are presented as an example of typical standard gels.

Application of Enzymes as Biological Indicators

Peroxidases Peroxidase is probably the most widely used enzyme as a biochemical indicator of disease, cellular injury, trauma damage, infection, etc., in plants (4,7,16,23,36,37,47, 51,53). In peanuts, most peroxidase activity has been located in the albumin fraction of the proteins (17). Peroxidase has been employed with peanuts as an index of blanching temperatures in food quality control and, with certain other enzymes, has been implicated in the development of off-flavors during storage (1,28,59). Since most peroxidases seem to be localized in a specific storage protein fraction of peanuts, analysis of this fraction may be a useful tool in quality control during preparation of protein isolates or concentrates. Alternatively, the removal of the albumin fraction during isolate preparation may prevent or retard the production of off-flavors caused by peroxidase activity.

Examination of many developing Spancross peanuts from individual plants grown in Georgia showed a decrease in peroxidase activity as seeds matured (Fig. 2). Zymograms containing one band at the origin, one in region 0.7 cm and three in region 3.0-4.0 cm, were typical of most patterns of mature seeds. A peroxidase band in region 5.0 cm distinguished immature and low intermediate seeds from high intermediate and mature peanuts. The presence of one major band in region 3.0-4.0 cm, the absence of isozymes at the origin and/or in region 0.7 cm, and the presence (or absence) of activity in regions 1.0-2.7 and 4.0-5.0 cm distinguished zymograms of mature seeds.

Germinating peanuts showed increased peroxidase activity in electrophoretic gels after 2 to 4 days (Fig. 2) that became especially prevalent in region 0-3.0 cm between 8 and 18 days germination. An increase in peroxidase activity was noted in region 3.0-4.0 cm at days 6 and 8, decreasing slowly thereafter to 20 days germination.

Esterases Esterases are popular in electrophoretic studies of isoenzymes in higher plants (8,10,15,19-21,23,43,44,52). Esterases from peanuts at immature and low intermediate stages of development showed much banding variation that differed for Spancross and Florunner cultivars (Fig. 3). After the low intermediate stage of peanut development, esterase activity on electrophoretic gels increased in regions 3.0-4.0 and 5.0-7.5 cm. The two cultivars could be distinguished by esterase bands in region 4.0-5.0 cm. Extracts from overmature seeds of Florunner showed decreased esterase activity on electrophoretic gels compared to those of mature seeds; no changes were observed in overmature seeds of Spancross. Intravarietal analysis of these peanut cultivars yielded both qualitative and quantitative esterase variations that were limited primarily to regions 4.0-5.0 cm of the zymograms. Variations within and between various other

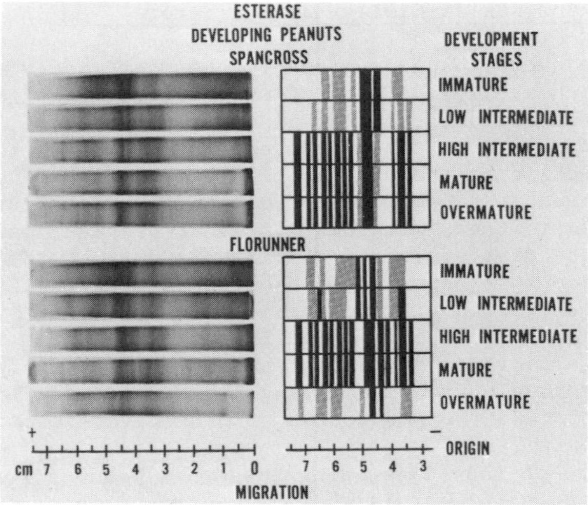

Figure 3. Typical polyacrylamide gel patterns of esterases in peanut cotyledons of Spancross and Florunner cultivars at various developing stages

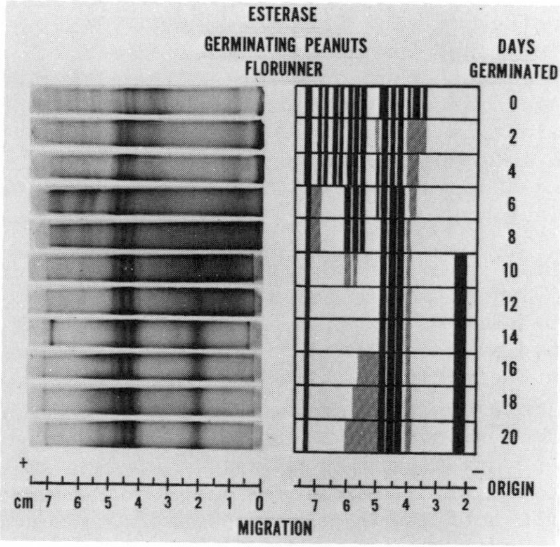

Figure 4. Polyacrylamide gels of esterase activity in Florunner peanut cotyledons germinating for various times from 0–20 days

peanut cultivars were limited to quantitative differences in region 4.4 cm. These studies suggest that esterase activity in peanuts may be useful in identifying particular stages of peanut development to maturity, as well as for distinguishing Florunner and Spancross cultivars.

Both qualitative and quantitative band variations appeared in region 2.0-7.0 cm shortly after Florunner seeds were germinated (Fig. 4). These changes increased markedly after 4 days, clearly distinguishing early germinated cotyledons from those developing between 6 and 20 days. Intensity of esterase activity in regions 2.0-2.5 and 4.5-5.0 cm of seeds germinated beyond 6 to 8 days increased quantitatively. Esterase bands of mature seeds present in other regions of the gel, decreased between 2 and 8 days germination.

Leucine Aminopeptidases Nearly all electrophoretic detection of variant peptidases has involved the use of synthetic substrates such as aminoacylnaphthylamides, especially leucyl-β-naphthylamide. As with the reaction media used for esterases, this synthetic substrate is not specific only for certain peptidases. Despite the vagueness concerning substrate specificity, the leucine aminopeptidases are very useful enzymes in electrophoretic studies of protein polymorphism, plant hybridization, and polypeptide hydrolysis (19,21,22,38,58).

Zymograms of leucine aminopeptidase activity distinguished high intermediate, mature, and overmature peanuts of Spancross and Florunner cultivars (Fig. 5). Cultivars and stages of seed development were similarly distinguished by leucine aminopeptidases, as was shown with esterases.

Changes in leucine aminopeptidase activity in Florunner cotyledonary extracts were examined on gels during extended seed germination up to 20 days (Fig. 6). The peptidase activity in these preparations remained consistently high during early stages of germination to day 6, after which it decreased. By day 12 most of it could not be detected. This observation suggested that peptidase activity may be partially responsible for the reserve protein breakdown during early stages of germination. Interestingly, the fast-moving bands (region 5.0-6.0 cm) showed very little activity up to 8 days of germination. However, after this time, most of the remaining activity was associated with these bands, suggesting that they may have been activated or synthesized de novo during this interval.

The breakdown of cotyledonary proteins was investigated qualitatively by gel electrophoresis (Fig. 7). Protein extracts from ungerminated seeds (0 day) consisted mainly of the monomeric (1.5 cm) and dimeric (1.0 cm) components of the major storage globulin of peanuts, arachin. Except for intensification of two bands in region 0-0.5 cm, no major changes were observed during the first 2 days of germination. Major changes in protein composition became evident at and after day 4. The arachin components and the bands in region 2.0-3.5 cm decreased quantitatively

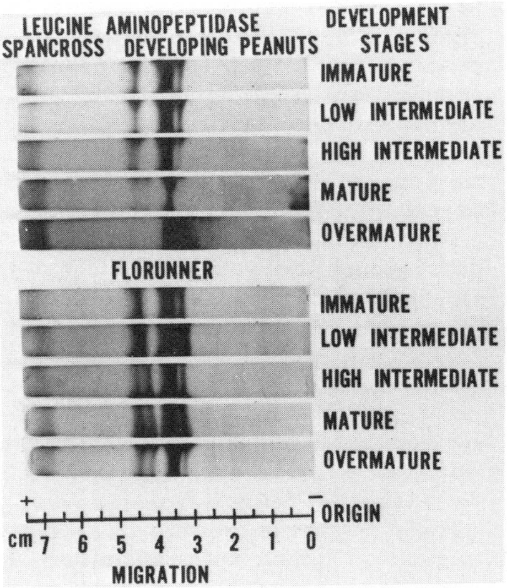

Figure 5. Polyacrylamide gel patterns of leucine aminopeptidase activity in developing peanut cotyledons of Spancross and Florunner cultivars

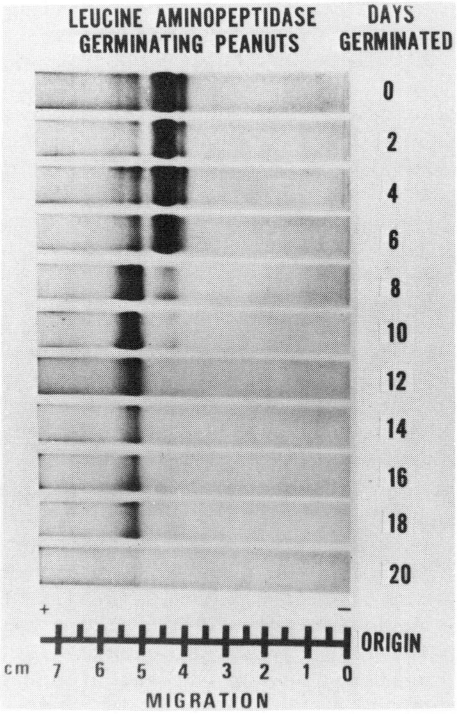

Figure 6. Polyacrylamide gels of leucine aminopeptidases in germinating Florunner peanut cotyledons

in the gel patterns, and simultaneously there appeared numerous proteins in region 2.5-7.0 cm. The appearance of small protein components in the gel patterns indicates hydrolysis of the major storage proteins of peanuts by proteases and peptidases to polypeptides of various sizes and to free amino acids, prior to their transport to the embryo. After day 14, no further major changes were noted in the few remaining proteins detectable by gel electrophoresis. This coincided with the observation that a consistently low quantity of protein and peptidase activity remained in the cotyledons between days 12 and 20 after germination began.

Fungal-Caused Seed Deterioration

Studies have shown that standard enzyme electrophoretic patterns of peanuts are distinctly modified by infection with various *Aspergillus* species (14,18). Biochemical transformations include deletion of some enzymes, intensification of others, and/or production of new bands. These changes in gel patterns of peanut enzymes indicate that the biochemical mechanisms operative in the saphrophyte-seed interrelationship function very efficiently and systematically for growth of the fungus at the expense of the seed. This is based on the observation that in addition to enzyme changes in the interrelationship, decomposition of storage globulins of large molecular weight to small protein components and free amino acids occurs in infected peanuts (9,11,14). The type of enzyme systems included in studies suggested a degradation of storage protein peptides by leucine aminopeptidases, hydrolysis of many ester linkages by esterases, hormonal interaction, and/or oxidation of organic substances with peroxides by peroxidases and oxidases, and decomposition of potentially toxic hydrogen peroxide by catalases. Many of the peanut enzymes remained active during *Aspergillus* invasion. In a number of cases, intensification of activities of some enzymes and the production of new bands were correlated with enzyme bands in mycelial tissue of the fungus collected from the seed surface. Quantitative and qualitative variations in gel patterns of certain enzymes also distinguished between fungal tissue from the exterior surface of peanuts and that grown on a synthetic medium, Czapek's solution (14,18). These variations in banding patterns of tissues collected from two separate sources may possibly be related to differences in nutritional needs of the fungus grown under the two conditions.

Phosphogluconate Dehydrogenase, Alcohol Dehydrogenase and Alkaline Phosphatase Examples of enzyme changes in peanuts infected for 4 days with various strains of *Aspergillus flavus* capable of producing no, intermediate, and high levels of aflatoxin are shown in Figure 8. In general, phosphogluconate and alcohol dehydrogenase banding activities detected on starch electrophoretic gels become difficult to discern in extracts of

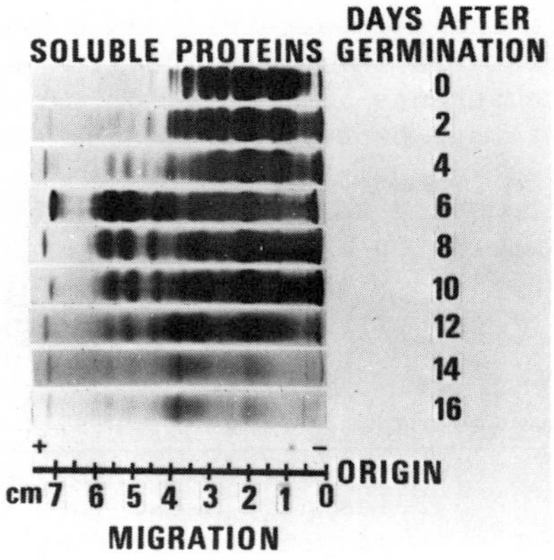

Figure 7. Polyacrylamide gel electrophoretic patterns of proteins in extracts from germinated Florunner peanut cotyledons

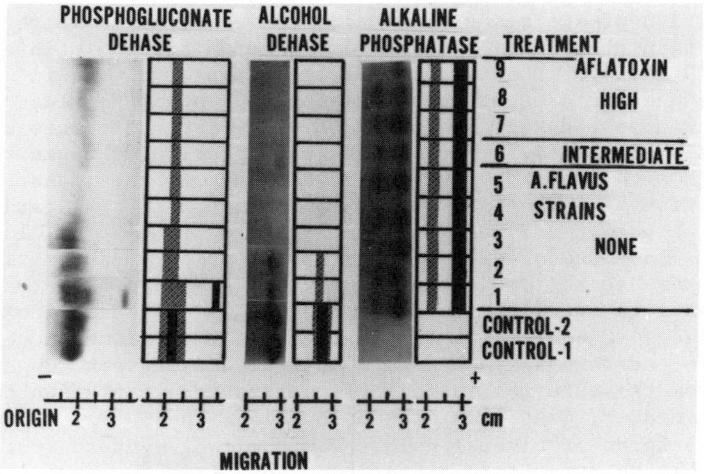

Figure 8. Starch gel electrophoretic patterns of phosphogluconate dehydrogenase, alcohol dehydrogenase, and alkaline phosphatase activity of noninfected and A. flavus-*infected peanuts after four days. Controls* 1 *and* 2 *represent peanuts at day zero and after incubation for four days at* 29°C *in a high humidity chamber, respectively.* A. flavus *strains from the Northern Regional Research Center, USDA-ARS, Peoria, Ill., were labeled as follows:* 1. *NRRL 5520,* 2. *NRRL 5518,* 3. *A-14152,* 4. *NRRL 3517,* 5. *A-62462,* 6. *NRRL 3353,* 7. *A-4018b,* 8. *NRRL 3251, and* 9. *A-13838.*

peanuts infected with *A. flavus*. Small amounts of activity were observed only in certain nonaflatoxin producing strains of *A. flavus*. A phosphogluconate dehydrogenase band in region 3.5 cm specifically distinguished peanuts infected with nonaflatoxin producing *A. flavus* strain 1 from all other contaminants.

In contrast, alkaline phosphatase was detected only in *A. flavus*-infected peanuts. Two bands of similar activity were noted in all contaminated peanuts, regardless of aflatoxin levels. Since alkaline phosphatase is not normally present in peanuts, this enzyme could be used as an indicator of *A. flavus* contamination in peanut protein preparations. This would be especially suggestive of compositional changes in proteins or other storage constituents and possible aflatoxin contamination of peanut products that could affect their functional and nutritional properties and their use in food products. Preliminary tests for alkaline phosphatase activity in peanut products prior to their utilization as food ingredients could be included as part of quality control measures normally used by food processors.

Esterases Region 3.0-7.0 cm contained detectable changes in esterase patterns after *A. flavus* infection of peanuts, compared with uninoculated or control seeds (Fig. 9). Four days after infection, esterase activity increased both quantitatively and qualitatively in regions 4.0-5.0 and 6.5-7.5 cm. Enzyme bands of most infected seeds in regions 3.0-4.0 and 5.0-6.0 cm decreased or were not readily detected in the gel patterns. Earlier studies with *A. parasiticus* showed similar intensification of esterase activity (14,18). These changes with *A. flavus* were similar to those of fungal tissue collected from the exterior surface of peanuts in the earlier work. Some of the esterase patterns distinguished *A. flavus* strains producing no, intermediate, or high levels of aflatoxins.

Leucine aminopeptidase Analysis for leucine aminopeptidase activity in polyacrylamide gels containing extracts from uninoculated peanut tissues yielded five bands in region 4.0-7.5 cm (Fig. 10). Four days after inoculation, each of the 9 *A. flavus* strains produced a zymogram that differed from that of the controls and was distinct from all other fungi. The number of bands in gels of infected peanuts ranged between 6 and 8, compared to 4 and 5 bands in control gels. In addition, band activity was more intense in many gels of infected seeds than in those of the controls. As suggested earlier in studies with developing and germinating peanuts (Figs. 5,6,7), changes in peptidase activity may be indicative of protein hydrolysis.

Gel electrophoretic patterns of proteins in soluble fractions showed that each *A. flavus* strain caused changes in these components during the infection period that were different from those of the control (Fig. 11); no protein changes were noted in gel patterns of noninfected seeds. In general, protein patterns of seeds infected by the various fungi showed new protein components

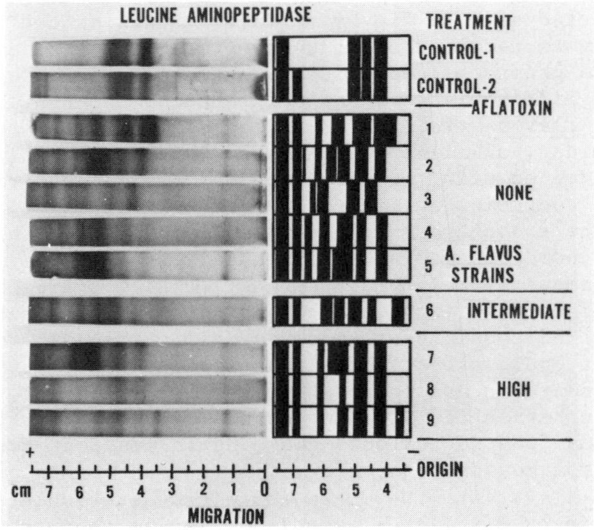

Figure 9. Polyacrylamide gel electrophoretic patterns of esterase activity in noninfected and A. flavus*-infected peanuts. Gel descriptions are given in Figure 8.*

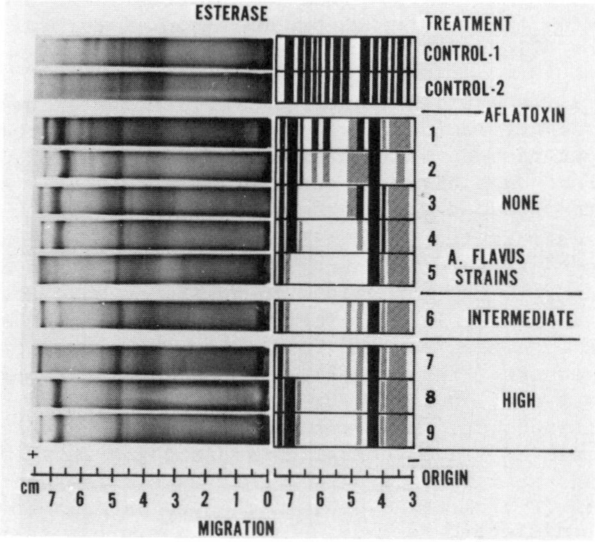

Figure 10. Polyacrylamide gel electrophoretic patterns of leucine aminopeptidase activity in noninfected and A. flavus*-infected peanuts. Gel descriptions are given in Figure 8.*

in region 0-1.0 cm plus increased mobility and poor resolution of the bands in region 1.0-2.5 cm as the infection progressed to day 4. At the same time, bands normally located in region 2.0-3.5 cm disappeared, and a new group of polypeptides appeared in region 5.5-7.0 cm.

Speciation and Genetic Crossability

Substantial evidence has been accumulated suggesting that the genetic base, or gene pool, of cultivated plants does not have the reserve germ plasm needed to resist many of the new problems brought on by pollution, dwindling water supplies, insects, and plant pathogens. Fortunately, there are wild species of most cultivated plants containing untapped genetic resources for future breeding programs. Wild species contain old sources of germ plasm which can be used to decrease "genetic erosion" of commercial crops by broadening the genetic base in cultivated varieties (45). Once the crossability of species within a genus is understood, studies can begin to determine "bridge species" for the introduction of new genetic information into cultivated varieties (56).

Gel electrophoresis of enzymes offers a biochemical approach to the evolutionary aspects of plant speciation. Amino acid changes within a protein caused by genetic mutational changes can result in altered migration rates when the proteins are compared in a matrix system of polyacrylamide or starch gel. Since numerous electrophoretic analyses have shown that species differ from one another in band frequencies, the individuality of each plant species can usually be expressed according to its enzyme banding patterns (9,10,19-22,52).

The genus Gossypium is comprised of about 30 diploid and three natural allotetraploid species (5,27,33,50). The diploid (2n) plants were divided into six groups or genomes labeled A to F; tetraploid (4n) species were described 2(AD)n since they were hypothesized to have originated from a cross between plants from the A and D genomes. Analysis of species relationships within Gossypium by polyacrylamide gel electrophoresis of proteins and enzymes in extracts of dormant seeds was reported by Cherry et al. (19-22). The chemotaxonomic comparisons of protein and enzyme patterns for species within and between genomes supported the present classification of Gossypium and the origin of the natural allotetraploids. In addition, the esterase studies showed that much more variation in banding patterns existed within the wild species than in new cultivars.

Figure 12 presents an example of species within the D genome which have esterase banding patterns that are more similar to each other than to members of other genome groups (A-C, E-F). Endrizzi (26) showed cytologically that G. lobatum and G. aridum were closely related. Gossypium harknessii was shown to cross freely with G. armourianum(35). The highest chiasma frequency

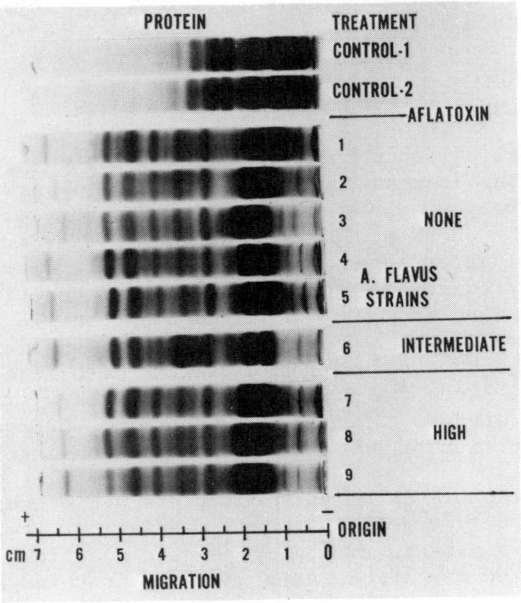

Figure 11. Polyacrylamide gel electrophoretic patterns of proteins in noninfected and A. flavus-*infected peanuts. Gel descriptions are given in Figure 8.*

among the D genome species was produced in crosses between G. klotzschianum and G. davidsonii (49). These comparisons were further supported by the isozyme patterns of the esterases; i.e., the more closely related the species were genetically, the more similar were their banding patterns. On the other hand, G. raimondii appeared to produce esterase patterns uniquely different from other species of the D genome.

Much esterase variability is observed within and between species of the D genome. For example, six different zymograms (A-F), were observed for the G. thurberi seeds (268 as compared to 24 for all other species). The presence of six esterase gel patterns indicates that notable genetic variability exists within species at the molecular level, and intraspecific variability increases when many seeds are examined.

Of major interest in this study with G. thurberi are the esterases located in a relatively narrow region of the gels, 6.5-9.2 cm. Within this region, distinct variations (both qualitative and quantitative) occur in the bands. A comparison of the relative frequencies of seed expressing a specific pattern, from each of the four populations analyzed, is shown in Table I. In general, zymograms A-D occur much more frequently in natural populations than do zymograms E and F. These data suggest that banding variation may be expressed both within and between populations.

The origin of this banding variation noted for G. thurberi (and other species; Fig. 12) may be due to one or more genetic possibilities in the plants: (a) isozymic variants of a specific esterase gene in homozygous or heterozygous pairing with its paired gene; (b) a number of different and similar polypeptide subunits in various combinations derived from closely related nonallelic genes; (c) enzymes completely unrelated and not from a common source; and (d) a selective combination involving all of these possibilities. As to which of these alternatives is most likely, it can be noted from the data for G. thurberi that the band in region 8.0 cm is the only one present in all of the zymograms and it is also present in greater amounts than any of the other esterases of this species. Thus, all of the latter bands may have arisen from this main esterase through mutation within a single gene (intragenic variation) or through duplication of a single gene that then underwent mutation (intergenic variation). Hybridization studies of pure plants that produce each of these esterase bands, however, would be needed to test this hypothesis.

Much variability of the esterases can also be noted between species of the D genomes. In a number of cases, specific banding patterns occurred with a low frequency in one species and a high frequency in another, whereas the reverse situation could be obtained with other gel patterns. For example, in G. harknessii, a zymogram with three esterase bands occurs with a high frequency of 67%. Within this species, a zymogram with four esterase bands

TABLE I. Frequency of occurrence of nonspecific esterase patterns.*

Location (Tucson, Arizona area)	Total Seeds Analyzed	Zymograms					
		A	B	C	D	E	F
Santa Catalina Mountains	78	0.72	0.00	0.18	0.09	0.03	0.00
Rincon Mountains	72	0.13	0.26	0.08	0.49	0.04	0.00
Santa Rita Mountains	72	0.81	0.08	0.00	0.11	0.00	0.00
Baboquivari Mountains	45	0.18	0.07	0.36	0.22	0.11	0.07
TOTALS	268	0.49	0.10	0.13	0.22	0.04	0.01

*Zymograms A–F of the natural populations of G. thurberi (Fig. 12).

occurs in low frequency of 21%. In G. armourianum, the frequencies of these two gel patterns are reversed (25% versus 50%). Similar comparisons can be made among gel patterns of G. aridum, G. lobatum, G. harknessii and G. gossypioides. This indicates that different combinations of genes coding for esterases are possibly present within the species of this group. The varying zymograms may be a reflection of natural selection pressures operating to produce the most stable genetic composition for each species. These isozyme patterns may also be due to differential genetic control of seed esterase production and composition in the plant caused by environmental pressures during a particular growing season.

Allopolyploids as Bridge Species

Examination of enzyme patterns of seeds from allotetraploids formed by genetic breeding studies showed that these gels compared closely to the additive zymograms of the parent species when the latter were combined in a synthetic mixture (19,21). This technique can be used to determine parental species of naturally occurring hybrids. Where wild hybrid species cross with cultivated varieties and form viable seed, the parentals or bridge species could be determined for introduction of valuable genetic information into new cultivars.

An example of such a study is shown in Figure 13. Esterases from seeds of a man-made viable allotetraploid $2(A_2D_1)_4$ species (AZ 239) were examined by gel electrophoresis. This allotetra-

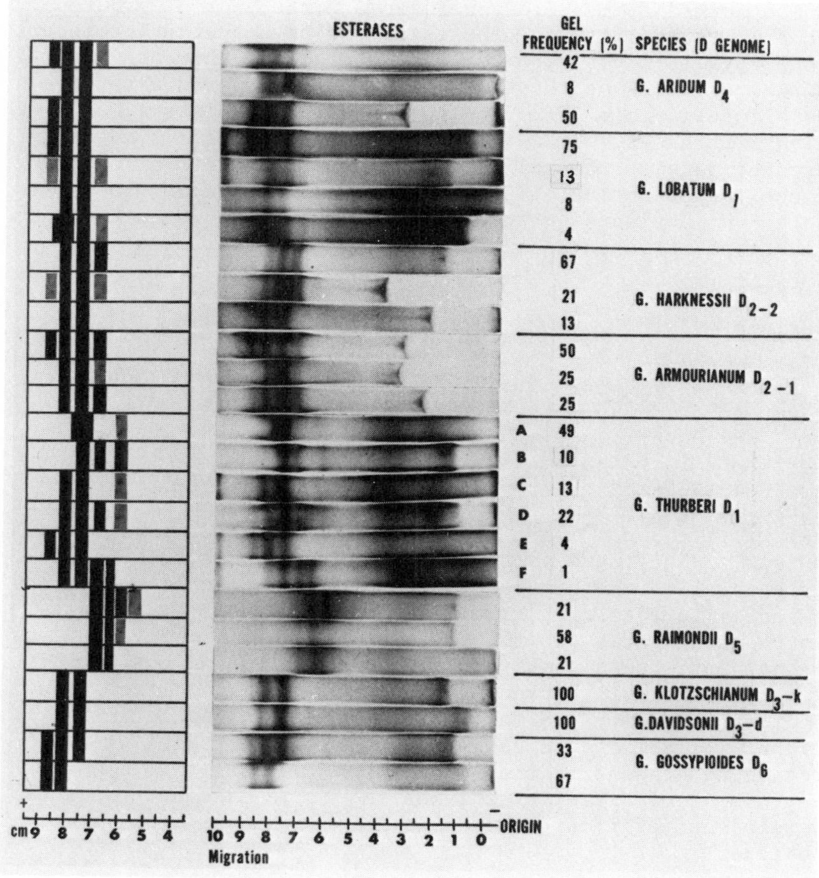

Figure 12. Polyacrylamide gel electrophoretic patterns of seed esterases from nine species wtihin the D *genome of the genus* Gossypium. *Included are the frequencies of zymograms formed from seeds of each species.*

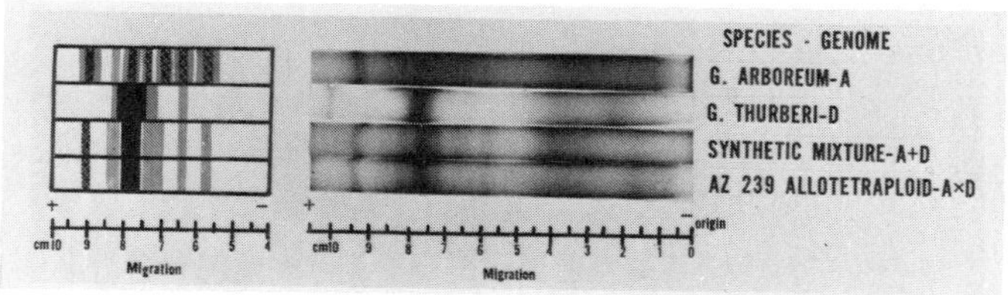

Figure 13. Polyacrylamide gel electrophoretic patterns of seed esterases in a comparison between parentals, their synthetic mixture, and the allotetraploid. Comparisons include G. arboreum *var G24* (A_2 *genome*), G. thurberi (D_1 *genome*), *their synthetic mixture, and their synthetic allotetraploid:* $2(A_2D_1)_4$, AZ 239.

ploid was formed from a cross between a variety of *G. arboreum* (A_2) and *G. thurberi* (D_1). The esterases of a synthetic mixture of seed extracts formed from *G. arboreum* and *G. thurberi* and the individual extracts of these parentals were compared with that of the allotetraploid AZ 239. In the synthetic mixture, an additive esterase zymogram was formed. The additive zymogram compared closely to the gel pattern of AZ 239. This allotetraploid has the potential of serving as a bridge species to cross with other wild plants shown by classical genetic techniques and gel electrophoresis to be closely related to *G. arboreum* and *G. thurberi*.

Conclusions

Technology has enabled scientists to take enzymes out of their *in vivo* environments to study them in test tubes. Enzyme activity during isolation, characterization, and identification procedures is a biological indicator of these catalytic proteins. Nevertheless, enzymes do change in activity and/or are altered structurally when abused. Because of this they can be excellent indicators of changes related to food processing or physiological mechanisms. Specific task(s) performed by enzymes are genetically controlled and can be used to evaluate genetic speciation and taxonomy to determine bridge species in breeding studies for introducing genes of wild species into cultivated varieties. Fungi-host interrelationships can be determined and evaluated, using enzymes to follow changes in development and differentiation of the former and a rapid form of senescence in the latter. Functional molecules such as enzymes should be studied with an understanding of genetic and ontogenetic variability. A sufficient number of different enzymes should be examined to insure the development of proper standards for reaching appropriate conclusions.

LITERATURE CITED

1. Acker, L., and H. O. Beutler, Fette Seifen Anstrichm., 67:430 (1965).
2. Augustinsson, K. B., B. Axenfors, I. Andersson, and H. Ericksson, Biochim. Biophys. Acta 293:424 (1973).
3. Augustinsson, K. B., Ann. N.Y. Acad. Sci. 94:844 (1961).
4. Baptist, J. N., C. R. Shaw, and M. Mandel, J. Bacteriol. 99:180 (1969).
5. Beasley, J. O., Genet. 27:25 (1942).
6. Berry, J. A., and R. G. Franke, Amer. J. Bot. 60:976 (1973).
7. Bhatia, C. R., and J. P. Nilson, Biochem. Genet. 3:207 (1969).
8. Brewbaker, J. L., M. D. Upadhya, Y. Makinen and T. MacDonald. Physiol. Plant. 21:930 (1968).

9. Cherry, J. P., and L. R. Beuchat, Cereal Chem. 53:750 (1976).
10. Cherry, J. P., Peanut Sci. 2:57 (1975).
11. Cherry, J. P., C. T. Young, and L. R. Beuchat, Can. J. Bot. 53:2639 (1975).
12. Cherry, J. P., and J. M. Prescott, Proc. Exp. Biol. & Med. 147:418 (1974).
13. Cherry, J. P., and J. M. Prescott, (Unpublished data, 1974).
14. Cherry, J. P., R. Y. Mayne, and R. L. Ory, Physiol. Pl. Path. 4:425 (1974).
15. Cherry, J. P., and R. L. Ory, Phytochem. 12:283 (1973).
16. Cherry, J. P., and R. L. Ory, Phytochem. 12:1581 (1973).
17. Cherry, J. P., J. M. Dechary, and R. L. Ory, J. Agric. Food Chem. 21:652 (1973).
18. Cherry, J. P., R. L. Ory, and R. Y. Mayne, J. Amer. Peanut Res. Ed. Assoc. 4:32 (1972).
19. Cherry, J. P., F.R.H. Katterman, and J. E. Endrizzi, Theoret. & Appl. Genet. 42:218 (1972).
20. Cherry, J. P., and F. R. H. Katterman, Phytochem. 10:141 (1971).
21. Cherry, J. P., "Comparative Studies of Seed Enzymes of Species of Gossypium by Polyacrylamide and Starch Gel Electrophoresis," Ph.D. Dissertation, University of Arizona, Tucson (1971).
22. Cherry, J. P., F. R. H. Katterman, and J. E. Endrizzi, Evolution 24:431 (1970).
23. Cubadda, R., A. Bozzini and E. Quattrucci, Theoret. & Appl. Genet. 45:290 (1975).
24. Davis, B. J., Ann. N.Y. Acad. Sci. 121:404 (1964).
25. Dixon, M., and E. C. Webb, "Enzymes," Academic Press Inc., New York (1964).
26. Endrizzi, J. E., J. Hered. 48:221 (1957).
27. Fryxell, P. A., Adv. Front. Plant Sci. 10:31 (1965).
28. Gardner, H. W., G. E. Inglett, and R. A. Anderson, Cereal Chem. 46:626 (1969).
29. Gottlieb, L. D., BioSci. 21:939 (1971).
30. Holmes, R. S., and C. J. Masters, Biochim. Biophys. Acta 151:147 (1968).
31. Holmes, R. S., and C. J. Masters, Biochim. Biophys. Acta 132:379 (1967).
32. Hunter, R. L., and C. L. Markert, Science 125:1294 (1957).
33. Hutchinson, J. B., R. A. Silow, and S. G. Stephens, "The Evolution of Gossypium and the Differentiation of the Cultivated Cottons," Oxford University Press, London (1947).
34. IUPAC-IUB Commission on Biochemical Nomenclature, J. Biol. Chem. 246:6127 (1971).
35. Kearny, T. H., Leaflets Western Bot. 8:103 (1957).
36. Kruger, J. E., and D. E. LaBerge, Cereal Chem. 51:578 (1974).

37. LaBerge, D. E., J. E. Kruger, and W.O.S. Meredith, Can. J. Plant Sci. 53:705 (1973).
38. Manwell, C., and C. M. A. Baker, "Molecular Biology and the Origin of Species: Heterosis, Protein Polymorphism and Animal Breeding, University of Washington Press, Seattle, Washington (1970).
39. Markert, C. L., Ann. N.Y. Acad. Sci. 151:14 (1968).
40. Markert, C. L., and G. S. Whitt, Experientia 24:977 (1968).
41. Markert, C. L., and R. L. Hunter, Histochem. Cytochem. 7:42 (1959).
42. Markert, C. L., and F. Moller, Proc. Natl. Acad. Sci. 45:753 (1959).
43. Marshall, D. R., and R. W. Allard, J. Hered. 60:17 (1969).
44. Menke, J. F., R. S. Singh, C. O. Qualset, and S. K. Jain, Calif. Agric. 27:3 (1973).
45. Miller, J., Science 182:1231 (1973).
46. Ornstein, L., Ann. N. Y. Acad. Sci. 121:321 (1964).
47. Pawar, V. S., and V. K. Gupta, Ann. Bot. 39:777 (1975).
48. Payne, W. J., J. W. Fitzgerald, and K. S. Dodgson, Appl. Microbiol. 27:154 (1974).
49. Phillips, L. L., Amer. J. Bot. 53:328 (1966).
50. Phillips, L. L., Evolution 17:460 (1963).
51. Reddy, M. M., and S. F. H. Threlkeld, Can. J. Genet. Cytol. 13:298 (1971).
52. Scandalios, J. G., Biochem. Genet. 3:37 (1969).
53. Schipper, A. L., Phytopath. 65:440 (1975).
54. Shaw, C. R., Intern. Rev. Cytol. 25:297 (1969).
55. Shaw, C. R., Science 149:936 (1965).
56. Simpson, C. E., and R. L. Haney, Texas Agric. Prog. 19:10 (1973).
57. Smithies, O., Adv. Prot. Chem. 14:65 (1959).
58. Sopanen, T., and J. Mikola, Plant Physiol. 55:809 (1975).
59. Wagenknecht, A. C., Food Res. 24:539 (1959).

13

Enzymes and Oxidative Stability of Peanut Products

ALLEN J. ST. ANGELO, JAMES C. KUCK, and ROBERT L. ORY

Southern Regional Research Center, Agricultural Research Service, U.S. Department of Agriculture, New Orleans, La. 70179

Investigations of the causes of oxidative deterioration of food lipids are being conducted in all types of foods -- marine, dairy, meat, and vegetable products. This research is a continuous effort to improve nutritional value of foods and to increase shelf-life stability. Also, because of medical implications of peroxidized lipids in food products, one can readily understand the reason for the great interest in this area of research.

Peanuts (groundnuts) are grown primarily in India, China, Africa, South America, and the United States. In most of these countries, peanuts are extracted and refined, the oil is processed for food uses, and the meal is sold for animal food. The United States is the only country that consumes the whole nut directly in food products. Here, peanuts are used in many different food products -- peanut butter, candy, confections, bakery products, salted nuts -- and as a cooking oil. In newer food applications, peanuts are being used as defatted peanut flour and as full-fat flour (1,2). However, defatted meals still contain significant amounts of residual oil that can give rise to lipid peroxides.

Peanuts contain approximately 50% oil and are a good source of unsaturated fatty acids (80%), of which 20%-37% is polyunsaturated. The unsaturated fatty acids can be oxidized, affecting both the quality and flavor of peanut products. Peroxidation of fatty acids has been attributed to a number of causes: light, air, heat, trace metals, enzymes, microorganisms, and even the presence of free fatty acids that can act as free radical catalysts. The quality of raw peanuts during storage and processing, therefore, can be seriously affected by activation of endogenous enzymes that oxidize fatty acids or induce other undesirable reactions.

Peanuts are also one of the most handled crops; they require digging during harvesting, then combining, curing, shelling, drying, blanching, and packaging for storage. In each of these steps, they are susceptible to damage that can activate enzymes and ultimately affect the quality of the final product.

The three causes of peroxidation that are of interest to us are autoxidation in air and catalysis by metals and by enzymes. Autoxidation of unsaturated fatty acids in raw peanuts is a slow process, sometimes taking weeks or months. In this reaction, oxygen is added across a double bond presumably forming a cyclic peroxide that cleaves to a fatty acid hydroperoxide. This hydroperoxide further decomposes into acids, aldehydes, ketones, and other substances that form during storage or processing. These byproducts are a major cause of undesirable odors and flavors. They can also damage and lower the nutritive value of the final food product by complexing with amino acids and proteins.

Assay Methods for Peroxidation of Unsaturated Fatty Acids

Several methods are used to follow the progress of rancidity, or to determine shelf-life stability. These methods include peroxide value (PV) determination, thiobarbituric acid (TBA) indication of malonaldehyde formation, measuring volatile carbonyl products, and oxygen consumption. In our studies, the conjugated diene hydroperoxide (CDHP) method described previously (3,4) is employed to study shelf-life stability of peanut products. This method is rapid, simple, requires only a drop of oil, and the results correlate well with those of the peroxide value method (5). Briefly, it involves extracting a weighed sample of either peanut butter or chopped peanut kernels with hexane, centrifuging, then measuring the absorption of the hexane supernatant at 234 nm. Figure 1 illustrates the direct relationship of CDHP

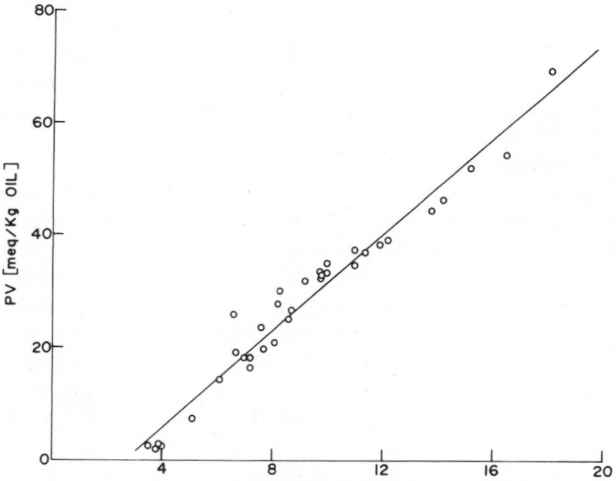

Figure 1. Comparison between peroxide value (PV) and conjugated diene hydroperoxide (CDHP) values in peanut butter stored 90 days at 25°C

to peroxide values in 34 peanut samples. CDHP values are calculated in μMoles/g sample, using 28,000 as the extinction coefficient instead of the 24,500 used in the original method (5). The equation for the conversion is:

$$PV = -11.44 + 4.302(CDHP);$$

the correlation coefficient for the two methods is 0.9812.

Peanut Butter

The most important use of peanuts in the United States is peanut butter. Millions of pounds are used in this product each year. Extending the shelf life of peanut butter is a major concern. When peanuts are homogenized for processing into peanut butter, they become susceptible to oxidation because of their high polyunsaturated fatty acid content.

The first step to eliminate peanut deterioration is to find its cause. Several minor constituents of peanut butter were investigated as possible catalysts of fatty acid oxidation (6). These included salt, enzymes, and metalloproteins that are present in peanuts. In each series of tests, the constituents were added to peanut butter and stored at room temperature (25°C) for 28 days. CDHP values were determined before, during, and after storage. The most significant increases in oxidation occurred in samples containing ferric or cupric ions, lipoxygenase, tyrosinase, peroxidase, or chlorophyll. The degree of oxidation depended upon their microenvironment, i.e., whether or not the material was added in an aqueous or nonaqueous solvent.

In a more complete investigation that included storage of the peanut butter samples for 3 months at room temperature, copper and iron salts and three indigenous metalloproteins, tyrosinase, peroxidase, and lipoxygenase, were found to be major catalysts of peroxide formation (Figure 2). Tyrosinase contains copper, peroxidase has iron, and lipoxygenase contains iron (7,8).

Since preliminary evidence suggested that chelating agents such as citric acid and EDTA could retard the catalytic oxidative effect of the metalloproteins (5,9), an additional study was conducted on the effects of these agents on fatty acid oxidation in peanut butter.

Peanut butter was prepared in the laboratory with and without citric acid, EDTA, or ascorbyl palmitate. After initial CDHP values were determined, the peanut butters were stored in sealed glass jars up to 180 days (5 months). Duplicate samples were periodically removed and analyzed for CDHP contents. As shown in A of Figure 3, peanut butter containing citric acid showed an increase of 5.7 μMoles of CDHP after 140 days storage, compared to 11.3 for the control. When EDTA was added to peanut butter (Figure 3B), the increase was only 5.9 μMoles CDHP, compared to 10.3 for the control, after storage for 156 days. The

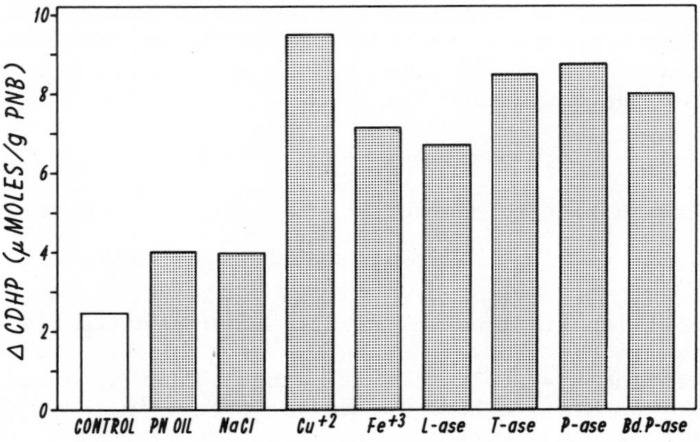

Journal, American Peanut Research and Education Association

Figure 2. Effects of various additives suspended in peanut oil on lipid oxidation during storage for three months at 25°C (6)

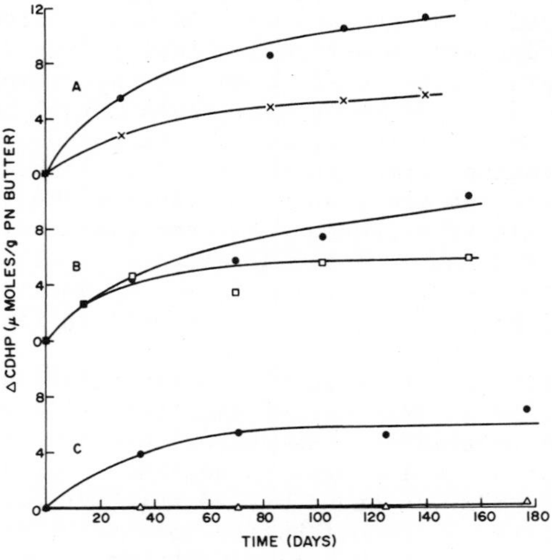

Figure 3. Effect of 0.01% citric acid (A), ethylenediaminetetraacetic acid, 0.01% (B), and ascorbyl palmitate, 0.1% (C) on oxidation of peanut butter during storage at 25°C

effect of ascorbyl palmitate (Figure 3C) was even more striking; it completely inhibited further peroxidation, whereas there was an increase of 7 μMoles CDHP for the untreated control. These data reflect a reduction of 50% peroxidation for the peanut butter containing citric acid, a 43% decrease for the sample containing EDTA, and a 100% reduction for the sample with added ascorbyl palmitate. Citric acid and EDTA are known chelating agents and would presumably bind those metals present in peanut butter.

It is generally accepted that linoleic acid is one of the primary substrates in fatty acid peroxidation. The linoleic/oleic acid ratio is frequently used as an index for predicting the susceptibility of peanut oil towards oxidation (10,11,12). However, Young and Holley (13) have reported a negative correlation between oil-keeping time and nitrogen content in peanuts. In other words, the higher the protein content, the lower the stability. While they suggested that the negative correlation was coincidental, our data indicate that in peanut butter, proteins (metalloproteins, in particular) appear to act as catalysts and thus become an intricate part of the oxidative mechanism.

Ascorbyl palmitate was shown by Cort (14) to be a better antioxidant than butylated hydroxyanisole (BHA) and butylated hydroxytoluene (BHT) for increasing the shelf life of vegetable oils and potato chips. As demonstrated in Figure 3C, ascorbyl palmitate significantly increased the shelf life of peanut butter, a high protein/oil-containing food.

To investigate the possible effects of different handling and processing steps, several different samples of freshly prepared commercial peanut butters were analyzed for initial CDHP contents. Results shown in Figure 4 indicated that no two samples had the same initial peroxide content. The CDHP contents of nine different samples of peanut butter varied from 3.2 to as high as 15.5 (6), suggesting that the quality of the peanuts before roasting and processing varied considerably, and that some were already in different stages of peroxidation. These findings stress the importance of quality checks for peroxide content of peanuts before processing.

Lipoxygenase in Raw Peanuts

The results in Figure 4 indicate that reactions taking place in the peanuts prior to processing may affect the quality of processed peanuts. The biological catalyst considered to be the principal cause of peroxidation of unsaturated fatty acids in many raw or unprocessed plant products is the enzyme lipoxygenase. The occurrence of this enzyme is rather widespread in legumes and cereal grains, but its existence in animal tissue is still questionable. The enzyme is highly specific for the peroxidation of unsaturated fatty acids that contain the *cis-cis*

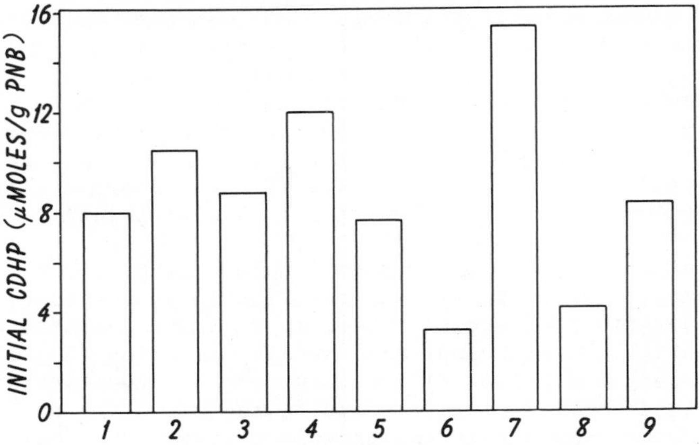

Figure 4. Initial CDHP content of nine different peanut butters

1,4-pentadiene systems; i.e., linoleic, linolenic, and arachidonic acids, esters, and triglycerides.

The presence of lipoxygenase in raw peanuts has been reported by several investigators over the past two decades (15-18). St. Angelo and Ory (17) partially purified the enzyme by ammonium sulfate fractionation and proceeded to separate it from a coexisting enzyme, presumably linoleic acid hydroperoxide isomerase. This latter enzyme, which was first identified by Zimmermann in flaxseed (19), was recently reported to be a lyase (20). Figure 5 represents an elution pattern on Sephadex G-100 that we employed to separate lipoxygenase, "hydroperoxide isomerase," and peroxidase.

The 40% ammonium sulfate fraction described previously (17) was chromatographed in phosphate buffer. Peak 1 contains the lipoxygenase and the "isomerase." Peak 2 contains an active peroxidase. No activity was found in Peak 3.

Aqueous lipoxygenase preparations are pH dependent and are not very stable, even when stored below freezing for several days. Approximately 30% of the activity is destroyed by incubating the enzyme at 36°C for 30 minutes. At temperatures above 40°, all lipoxygenase activity is lost (17). In Figure 6, activity of the enzyme in relation to pH is presented. The enzyme was extracted with phosphate buffers from pH 5.8 through 8.4; one preparation was extracted with deionized water. The enzyme preparations were allowed to stand at 4°C for 8 days. Over the 8-day period, activity was lost in each solution. However, these results indicated that the enzyme was the most stable for about 3 days when kept in phosphate buffer at pH 7.4.

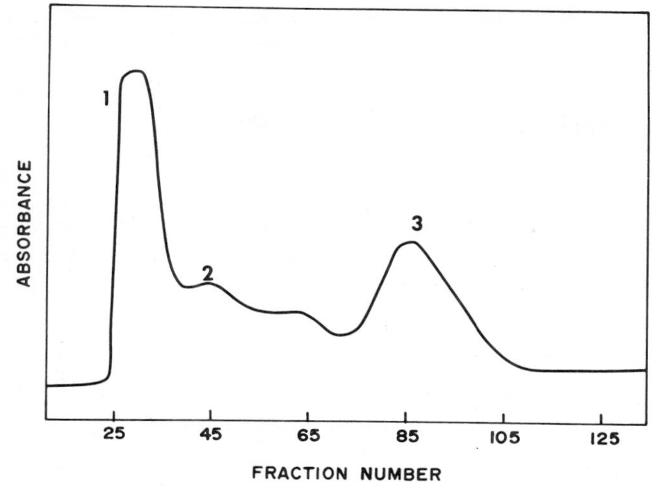

Figure 5. Separation of peanut lipoxygenase and peroxidase activities from peanut extracts on Sephadex G-100, 0.1M phosphate buffer pH 7.8. Absorbance measured at 280 nm.

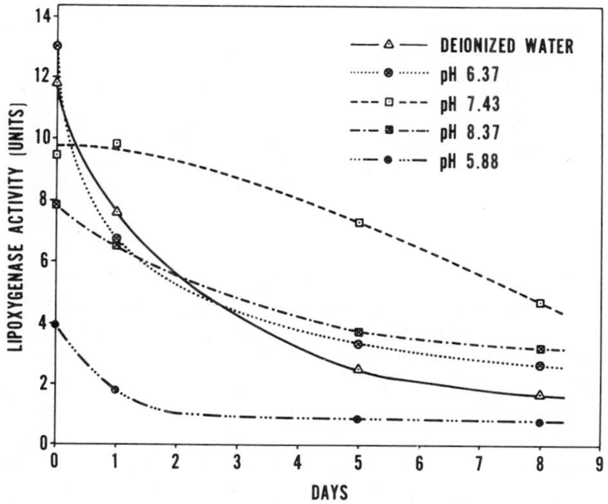

Figure 6. Relationship between pH and lipoxygenase stability

Lipoxygenase isolated from peanuts with 40% ammonium sulfate was active over a pH range from 4 through 7, with the optimum at 6.1. Cyanide added at the usual 1 mM inhibitor concentration did not inhibit the enzyme (17). Recently, Sanders et al. (18) reported separating three lipoxygenase isozymes in raw peanuts. Two had pH optima of 6.2 and the other had an optimum at 8.3. They also reported that higher concentrations of cyanide inhibited the acidic isozymes. Because the cyanide was added to the reaction mixture before addition of the enzyme, it was difficult to establish the actual reaction conditions and pH of the cyanide-substrate solution. Therefore, we reinvestigated the effect of cyanide on peanut lipoxygenase.

Since aqueous solutions of sodium and potassium cyanide are usually strongly alkaline, it was necessary to first investigate the effect of cyanide on the pH of assay solutions. The results (Table I) show that, as the CN concentration increases, even

Table I
pH of KCN Dissolved in Phosphate Buffer and Water

Phosphate Buffer (pH)	Phosphate Buffer (M)	KCN (mM)	Solution (pH)
6.1	0.01	30	9.5
6.1	0.10	30	6.8
6.1	0.10	300	9.8
8.3	0.01	30	10.8
8.3	0.10	30	9.7
5.7	0.20	300	8.9
4.7*	0.10	300	9.4
5.5, deionized water		30	10.5

*NaH_2PO_4

though added in acidic buffer, the pH of the final solutions becomes very alkaline. This suggested that the pH of the cyanide solutions could be critical for lipoxygenase studies, and lead to erroneous conclusions.

In our more recent peanut isozyme-cyanide experiments, we prepared the isozymes by extracting raw peanuts with 0.5 M NaCl, 0.05 M Tris buffer, pH 7.2, and centrifuging twice at 29,000 g for 10 minutes at 0°C. The supernatant was dialyzed overnight against deionized water at 4°C, then recentrifuged. A pH 6 isozyme was found in the resuspended precipitate, and the alkaline isozyme was in the clear supernatant fraction. In each assay, the reaction mixture contained 1 ml of the substrate solution, 1.95 ml of buffer, and 0.05 ml of the enzyme preparations. When cyanide was investigated, a microvolume (25-100 µl) of cyanide solution was added to buffer containing the enzyme, which was allowed to stand for 5 minutes. Then substrate was added and the mixture was immediately assayed for lipoxygenase activity as previously described (17). The pH of the reaction mixture was also determined immediately following the assay. When cyanide solutions were incubated with the acidic enzyme-buffer for 5 minutes prior to addition of substrate, activity still followed the pH curve, indicating that increasing amounts of cyanide buffered within the pH optimum range of the lipoxygenase do not inhibit the pH 6.2 isozyme. If not strongly buffered, cyanide can cause an increase in pH that will effect lipoxygenase activity. Cyanide had no effect on the alkaline isomerase.

Linoleic Acid Hydroperoxide Lyase

A sequential enzyme in the lipid oxidation chain in germinating watermelon seedlings was recently described by Vick and Zimmermann (20). This enzyme, a lyase that cleaves linoleic acid hydroperoxide, was originally thought to be an isomerase (19). The lyase also appears to be present in dormant peanuts. Direct GC analysis of a crude peanut lipoxygenase reaction mixture, as shown in Figure 7, demonstrated that methyl linoleate was cleaved by the peanut enzyme preparation (21). Cleavage occurs at the C-13 bond releasing two components; hexanal and at least one other component, presumably a methyl ester of a C-12 fatty acid, but not confirmed with a standard compound.

Effect of Lipid Peroxidation on Protein Solubility

As noted earlier, oxidative degradation of unsaturated fatty acids produces hydroperoxides and their secondary breakdown products, acids, alcohols, aldehydes, and ketones. These compounds are known to damage proteins, enzymes, and amino acids (22,23). Recently, St. Angelo and Ory (24,25) investigated the effects of lipid peroxides on raw and processed peanuts during storage. The oil and protein were separated from peanuts according to the diagram shown in Figure 8. CDHP contents were determined on hexane extracts of the oils. The deoiled meal, the salt-soluble proteins, and the salt-insoluble residues were analyzed for protein contents. The results of these studies (24) showed that hexane extracts from roasted whole peanuts, after 12 months storage at 30°C, contained an increased CDHP value of 32.4 μMoles/g.

The development of peroxides in several varieties of raw and roasted peanuts stored for 12 months at 4°C also was investigated, see Figure 9. Raw peanuts, plus corresponding roasted samples, were stored in sealed glass jars. Data obtained with twelve different samples fit into three general curves: (1) those that did not develop peroxides in the first 28 weeks and whose CDHP values remained at 3 μMoles/g of peanuts throughout the storage period; (2) those that were undergoing peroxidation, but whose initial CDHP values were also 3 μMoles/g; and (3) those that showed peroxidation taking place, with slightly higher CDHP values, initially 8.5 μMoles/g, and increasing to 13 μMoles/g after

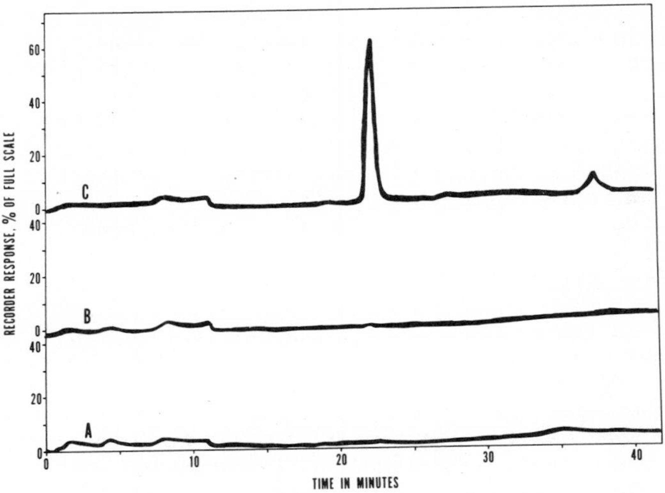

Figure 7. Direct GC analysis of reaction products from oxidation of methyl linoleate by peanut extract. Curve A *represents the profile obtained when the extract was deleted from the reaction mixture; Curve* B, *minus the substrate; Curve* C, *total reaction mixture: 0.1 mL substrate, 0.04 mL of Tween-20 emulsifier, 15 mL of 0.2*M.

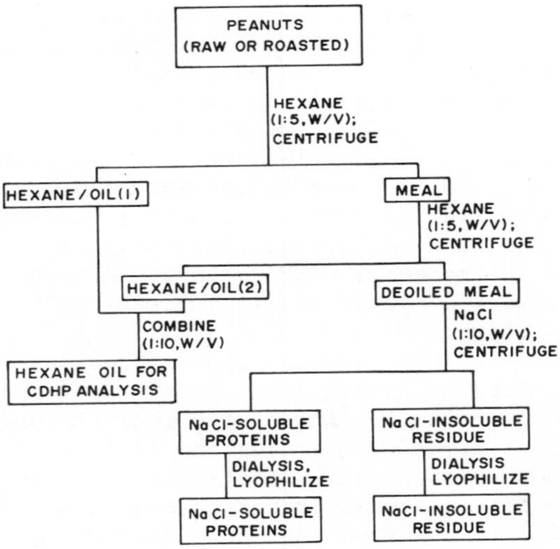

Figure 8. Scheme for fractionation of oil and protein from peanuts

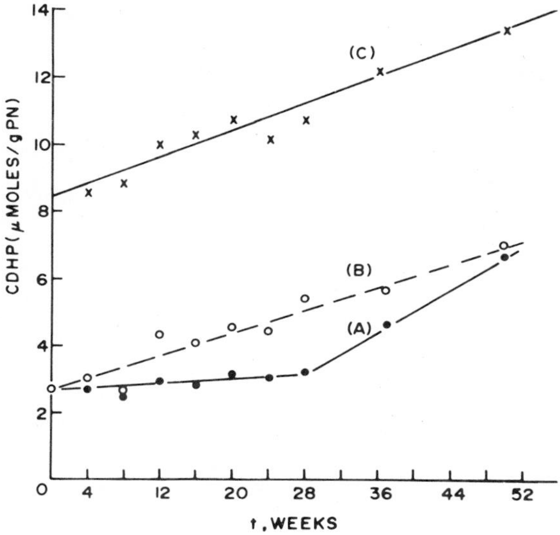

Figure 9. Effect of cold storage on peroxidation of fatty acids in raw peanuts. Peroxidation was measured as the increase in CDHP content for 12 different lots separated into three groups (25).

52 weeks. These data suggested that the peanuts yielding curves B and C were probably older than those depicted in curve A, and had already undergone some peroxidation prior to these storage experiments.

When the three peroxidation curves in Figure 9 were joined end-to-end, then the composite shown in Figure 10 was obtained. These data suggested that raw shelled peanuts could be safely stored at 4°C for at least 6 months before enzyme-catalyzed oxidation begins. Once this oxidation starts, storage at 4°C is unable to prevent it. Again, the most logical explanation for lipid oxidation in raw peanuts is lipoxygenase. Furthermore, Ericksson (26) has reported that pea lipoxygenase is active after storage for several months at -20°C.

When the corresponding roasted peanuts were assayed for CDHP content after storage for 1 year at 4°C, the rate of oxidation was 8 times faster than that in raw peanuts. CDHP units increased from an initial value of 8 μMoles/g to approximately 40 in 1 year and to 61 after 2 years storage. Since roasting destroys lipoxygenase, it is obvious that this lipid peroxidation is catalyzed by nonenzymic factors, most likely by metals or metalloproteins (e.g. inactivated lipoxygenase, peroxidase, or tyrosinase) present in the peanuts, as shown in the experiments with peanut butter (3,4).

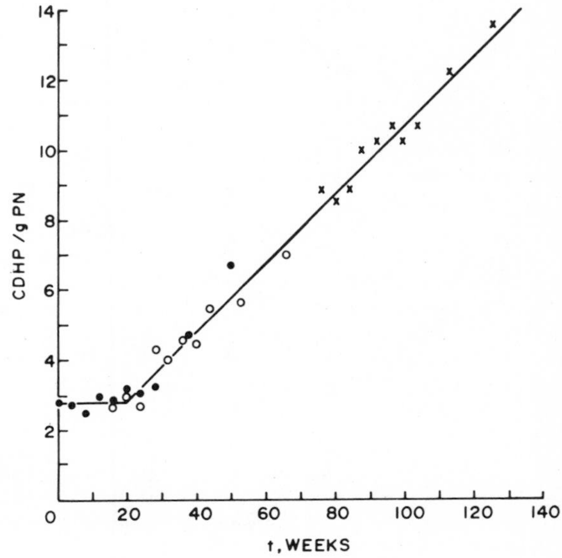

Figure 10. Composite of curves A, B, and C from Figure 9 placed end-to-end. Time is hypothetical age of stored seeds, as described in text.

St. Angelo and Ory (24) have shown that with increasing lipid peroxidation in raw peanuts, recovery of salt-soluble fractions increased, and salt-insoluble residues decreased correspondingly. This phenomenon was investigated further with intentionally damaged (by crushing) peanuts stored under 3 different conditions: sample 1 stored in sealed jars at 4°C; sample 2, in jars covered with cheesecloth (access to air) at 30°C; and sample 3, ground and blended with 10% rancid oil to promote lipid-protein interaction, then stored at 30°C while covered with cheesecloth. After 4 1/2 months, the samples were deoiled with hexane and the absorbance measured at 234 nm. Defatted meals were then extracted with sodium chloride, and the yields of soluble and insoluble fractions determined (24). The results showed that sample 1, stored at 4°C, had an absorbance of 0.55; 480 mg of salt-soluble material was recovered from 1.5 g. For samples 2 and 3, peroxidation and salt-soluble material increased; absorbance readings were 1.23 and 1.90 and weights of soluble material recovered were 550 and 588 mg, respectively. Conversely, as peroxidation increased, the salt-insoluble residues recovered from the three corresponding samples decreased (562, 537, and 515 mg, respectively). It appears, therefore, that enzyme-catalyzed lipid peroxidation in raw peanuts also promotes interactions between the peroxides and proteins in the seed. Subsequent changes in solubility of the proteins could possibly affect potential use of defatted peanut flour as protein supplements in protein-fortified food products.

Effects of Blanching on Quality

Another factor that can affect quality of peanut products is blanching, the process of removing the skins (testae) that contain tannins that may contribute to off-flavors. There are several methods for blanching peanuts, but we examined only two, water and spin blanching (27).

In spin blanching, the peanuts are first passed through a series of finely adjusted razor blades to slit the skins and then through a dryer at 71°C to lower the moisture content to 5%. The kernels are then passed through a series of rollers which remove the skins. The nuts are air separated and sorted.

In water blanching, the kernels are slit as in spin blanching, sprayed with 86°C water for 1 1/2 minutes, and the skins are then mechanically removed. Finally, the kernels are dried to 5% moisture.

The effects of water and spin blanching on oxidative stability of raw and roasted peanuts were compared, using unblanched controls, water blanched, and spin blanched peanuts that were divided into two portions each. One portion of the samples were stored in a sealed glass jar at 25°C. The remaining portions were roasted at 175°C for 20 minutes in a rotisserie oven, then stored in sealed jars, as were the unroasted samples. All samples were examined for developing CDHP contents at monthly intervals for 210 days. There were no significant differences in shelf life of unblanched and spin blanched raw peanuts, but water blanched raw peanuts had a shorter shelf life and developed lipid peroxides faster than the spin blanched and control peanuts.

In roasted peanuts, the results were different. The unblanched controls had the shortest shelf life, and the roasted, water blanched peanuts had the longest shelf life of all three.

Results of the roasting experiments suggest that elevated temperatures applied during blanching may inactivate some enzymes, but the exact systems involved are still not understood.

The foregoing results indicate, that selection of peanut storage conditions should be based upon their pretreatment and intended use to insure the maintenance of fresh quality in peanuts and in their roasted products. If peanuts are to be stored without vacuum pack conditions after roasting, they should be water blanched. On the other hand, if they are to be stored raw for long periods before roasting, they either should be unblanched or spin blanched.

Concluding Remarks

Oxidation of unsaturated fatty acids in peanut products form compounds that can adversely affect flavor and quality. Lipoxygenase has been suspected as the major cause of lipid oxidation, but roasting conditions in commercial processes should inactivate this enzyme. In roasted peanut products, metalloproteins are probably responsible for lipid oxidation. In raw peanuts stored for long periods before roasting, handling practices that can damage seeds could also activate enzymes to form peroxides, aldehydes, ketones, and alcohols, and shorten the shelf life of the final products. It is evident that improvements must be made in each of these handling and processing techniques if quality of the final product is to be improved.

Literature Cited

1. Harris, H., Davis, E. Y., Van de Mark, M. S., Rymal, K. S., Spadaro, J. J., Auburn Univ., Agr. Res. Exp. Sta. Bull. 431, Auburn, Ala., 1972.
2. Mitchell, J. H., U. S. Patent 3,689,287 (Sept. 5, 1972).
3. St. Angelo, A. J., Ory, R. L., Brown, L. E., J. Amer. Peanut Res. Ed. Assoc. (1972) 4, 186.
4. St. Angelo, A. J., Ory, R. L., Brown, L. E., Oleagineux (1973) 28, 351.
5. St. Angelo, A. J., Ory, R. L., Brown, L. E., J. Amer. Oil Chem. Soc. (1975) 52, 34.
6. St. Angelo, A. J., Ory, R. L., J. Amer. Peanut Res. Ed. Assoc. (1973) 5, 128.
7. Chan, H. W. S., Biochim. Biophys. Acta (1973) 327, 32.
8. Roza, M., Francke, A., Biochim. Biophys. Acta (1973) 327, 24.
9. St. Angelo, A. J., Ory, R. L., J. Amer. Oil Chem. Soc. (1975) 52, 38.
10. Holley, K. T., Hammons, R. O., Univ. of Georgia, Agr. Res. Sta. Bull. 32, Tifton, Ga., 1968.
11. Worthington, R. E., Hammons, R. O., Oleagineux (1971) 26, 695.
12. Young, C. T., Mason, M. E., Matlock, R. S., Waller, G. R., J. Amer. Oil Chem. Soc. (1972) 49, 314.
13. Young, C. T., Holley, K. T., Univ. of Georgia, Agr. Res. Sta. Bull. 41, Experiment, Ga., 1965.
14. Cort, W. M., J. Amer. Oil Chem. Soc. (1974) 51, 321.
15. Siddiqi, A. M., Tappel, A. L., J. Amer. Oil Chem. Soc. (1957) 34, 529.
16. Dillard, M. G., Henrick, A. S., Koch, R. B., Food Res. (1960) 15, 544.
17. St. Angelo, A. J., Ory, R. L., in "Symposium: Seed Proteins," Inglett, G. E., Ed., Avi Publishing Co. Inc., Westport, Conn., 1972, p 284.

18. Sanders, T. H., Pattee, H. E., Singleton, J. A., Lipids (1975) 10, 681.
19. Zimmermann, D. C., Biochem. Biophys. Res. Commun. (1966) 23, 398.
20. Vick, B. A., Zimmermann, D. C., Plant Physiol. (1976) 57, 780.
21. St. Angelo, A. J., Dupuy, H. P., Ory, R. L., Lipids (1972) 7, 793.
22. Karel, M., Schaich, K., Roy, R. B., J. Agr. Food Chem. (1975) 23, 159.
23. Matsushita, J., J. Agr. Food Chem. (1975) 23, 150.
24. St. Angelo, A. J., Ory, R. L., J. Agr. Food Chem. (1975) 23, 141.
25. St. Angelo, A. J., Ory, R. L., Peanut Sci. (1975) 2, 41.
26. Ericksson, C. E., J. Agr. Food Chem. (1975) 23, 126.
27. St. Angelo, A. J., Kuck, J. C., Amorim, H. L., Amorim, H. V., Ory, R. L., J. Amer. Peanut Res. Ed. Assoc. (1975) 7, 77.

14

Enzymes in Soybean Processing and Quality Control

JOSEPH J. RACKIS

Northern Regional Research Center, Agricultural Research Service, U.S. Department of Agriculture, Peoria, Ill. 61604

Urease, lipoxygenase, and peroxidase activity analyses are used to measure the degree of heat treatment which is a primary determinant of the functionality of soy proteins. Lipoxygenase, in raw soy flour, can replace chemicals to improve dough and crumb quality and bleach pigments during breadmaking. This is the only commercial application of the enzymes in soybeans. The functional importance of lipoxygenases are associated with the formation of lipohydroperoxides coupled with secondary oxidation of the hydroperoxides. Enzyme-active soy flours are used at the 0.7 to 1% level (wheat-flour basis). At higher levels, lipoxygenase activity is responsible for the formation of objectionable flavors in food systems.

Enzymes are assuming greater importance in modifying the functionality of soy proteins. Proteolytic modification shows promise in development of a new generation of modified soy proteins for food use. Pepsin-hydrolyzed protein isolates are used as whipping agents for egg white replacement. Another development that may have application in food processing is the "plastein reaction." Proteolytic enzymes are used to partially hydrolyze soy protein isolates, and then the hydrolyzate in the presence of supplemental amino acids is incubated with enzymes to reverse the process and reconstitute protein-like substances (plasteins) with different composition and functional properties. The use of alpha-galactosidases to eliminate soybean flatulence by hydrolysis of the oligosaccharides, raffinose and stachyose, has been described.

Soybean protein products are ingredients widely used in processed foods and compete with other protein sources. Economic factors and unique functional properties have helped to establish present markets for soybean proteins. Properties of the proteins have been altered by processing such as heat treatment, alcohol extraction, and texturization by extrusion. Enzyme treatment affords the possibility to develop a broader spectrum of functional properties.

A functional property of a food ingredient is one that imparts a desirable change or contributes some favorable quality to a fabricated food during processing or in the finished product. A functional ingredient should also inhibit factors that create undesirable changes with a subsequent loss in quality. Functionality includes: nutritional value, flavor and color characteristics, protein solubility, water-fat absorption, texture, improved baking performance, and regulation of enzyme activity.

The present paper will be concerned with the use of soybean enzymes in food processing and the application of other sources of enzymes to modify protein and carbohydrate constituents in soybeans. Measurement of enzyme activity as a quality control specification for the manufacture of edible-grade soy protein products will also be discussed.

Enzyme-Active Soy Flours

A full range of full-fat and defatted soy flours are used in processed foods (1). These products have widely different functional properties dependent upon the temperature, time, and moisture conditions used in their manufacture. Such moist heat treatment is referred to as toasting. Enzyme-active full-fat and defatted soy flours are produced by omitting the toasting operation. Raw soy flour is a good source of enzymes, including amylases, lipases, urease, and lipoxygenases. However, the lipoxygenase enzyme system is the only one of commercial importance to the food processing industry. Raw soy flour is one of the richest sources of lipoxygenase.

Lipoxygenase. The importance of enzyme-active soy flour is its use as a bleaching agent and dough improver for the production of breads, buns, and other fermented bakery goods. The lipoxygenases are the functional components of these flours.

Lipoxygenase (E.C. 1.13.1.1.3) catalyzes oxidation of unsaturated fatty acids in lipids containing *cis*,*cis*-1,4-pentadiene system by molecular oxygen to hydroperoxides. Linoleic and linolenic acids are the most common fatty acid moieties in unsaturated lipids that act as substrates. The name lipoxidase was first coined by Andre and Hou (2) for an enzyme in soybeans that oxidized fat, but now the preferred name is lipoxygenase. Soybeans contain multiple forms of lipoxygenase (isoenzymes) as well as a complex system of other enzymes that further decompose lipid hydroperoxides. The decomposition of the lipohydroperoxides can be classified into six characteristic pathways (3). Although Kies *et al.* (4) conclude that carotene oxidase and lipoxygenase are two distinct enzymes, it appears that one of the lipoxygenase isoenzymes possesses similar activity.

The importance of lipoxygenase is in coupled secondary oxidations which have a desirable effect in breadmaking. Hydroperoxides are also known to damage membranes, affect biological functions, oxidize protein sulfhydryl groups (5), and produce undesirable grassy/beany flavors (6). Therefore, the level and extent of lipoxygenase activity should be carefully regulated when used in a food system.

Lipoxygenase in Breadmaking. Cotton (7) has reported that about 71 million pounds of full-fat and defatted soy flour is used in baked goods in the United States. However, most of the flours are processed with moist heat which readily destroys enzymes and the primary use of these flours is for functional purposes, other than that associated with enzymatic activity. The amount of enzyme-active soy flour produced in the United States is unknown.

A series of patents by Haas and Bohn (8) and Haas (9) describe a process for the preparation of a stabilized agent by mixing raw defatted soy flour with 4 parts of corn flour. The addition of 0.75 to 2% of the corn-soy mixture (based on wheat flour weight) is sufficient for bleaching and dough improvement. Higher amounts can lead to flavor problems in bread. Detection levels are summarized in Table I. Flavor problems in bread occur when enzyme-active soy flour is incorporated at higher levels than that required to improve dough characteristics in order to increase the protein content for nutritional purposes (10).

Table I. Flavor Detection Level of Soy Flours in Bread

Workers	Detection level of soy, %
Haas (1934)	>0.4
Finney et al. (1950)	<4
Ofelt et al. (1952)	>5
Ehle and Jansen (1965)	>4
Cotton (1974)	2

Wolf (10)

In Europe, production of enzyme-active flour has become a specialized business (11). There, raw full-fat soy flour is used as a dough improver in the production of bread by two processes: (a) traditional fermented bread and (b) mechanically developed bread. Production figures are not available, but it is estimated that over 90% of the bread produced in the United Kingdom contains soy flour (12) and that it is rare to find a bread recipe in England which does not contain enzyme-active soy flour (11).

Lipoxygenase Function in Bread. In the traditionally fermented bread process, oxidative reactions of lipohydroperoxides generated by the soy lipoxygenase enzyme system are reported to bleach wheat pigments, condition and/or oxidize gluten for maximum softness of crumb, reduce the rate of staling, decrease binding of added shortening, and reduce production costs (12). Because of its high protein content, enzyme-active soy flour helps to retain water after baking, thereby keeping the loaf soft and fresh longer.

The recommended levels of addition of soy flour is from 0.7% i.e., 2 lb per sack (280 lb) of flour with 1-1/2 to 2 times its weight in extra water. There are also commercially available, composite bread improvers based on soy flour incorporating yeast foods and crumb softening agents.

Soy lipoxygenase activity also exerts a beneficial effect on the quality of bread made by the mechanically developed bread process. The bulk fermentation time is replaced by intense mechanical action, referred to as the Chorleywood bread process. A high level of oxidizing agent is required (75 ppm of ascorbic acid) and an addition of 0.7% triglyceride shortening. The optimal level of enzyme-active full-fat soy flour is 0.7% based on flour weight.

The beneficial effect of 0.7% enzyme-active soy flour added to a Standard Chorleywood recipe is illustrated in Figure 1. Loaves containing soy flour become softer even after 50-hr storage. Many other advantages are obtained with soy additives (12).

Soy lipoxygenase reduces the binding of added shortening with protein during dough mixing (13), increases gluten strength (14), and improves baking performance and keeping quality (12). These effects are also attributed to the coupled peroxidation of polyunsaturated lipids mediated through the formation of hydroperoxides (13).

In the presence of air, high-energy mixing leads to a release of bound lipid through a lipoxygenase-coupled oxidation of lipid binding sites of the proteins in the dough. Full-fat enzyme-active soy flour is a valuable source of enzymes required for this oxidation. The lipoxygenase present in wheat flour, in common with purified preparations of soybean lipoxygenase (3), is specific for free linoleic and linolenic fatty acids. However, soy flour contains several lipoxygenases which oxidize free and esterified fatty acids.

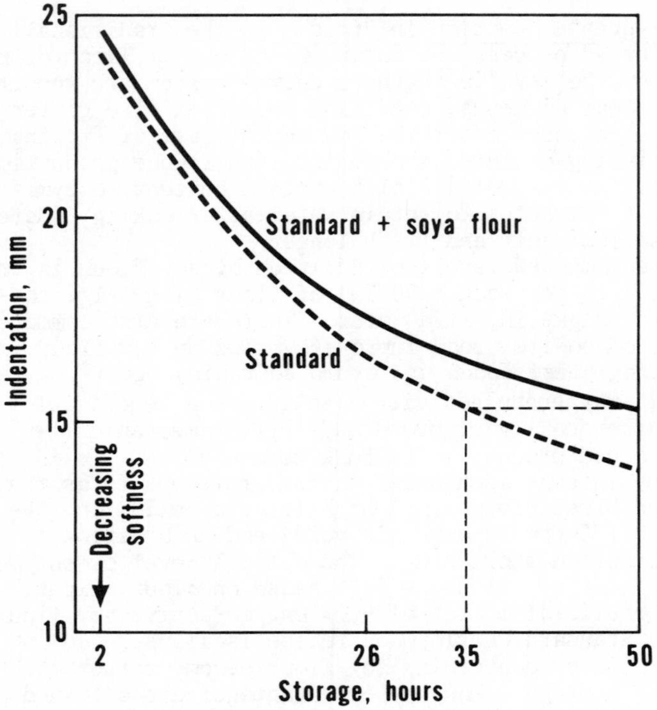

Figure 1. Effect of enzyme-active, full-fat soy flour on softness and storage stability of bread: Curve A, *standard Chorleywood process; Curve* B, *standard plus soy flour* (12)

Results of studies (13) which differentiate wheat and soy flour lipoxygenases in their effect on lipid binding are summarized in Table II. In the absence of free linoleic acid, neither the wheat flour lipoxygenase nor the purified soy lipoxygenase were able to induce the coupled oxidation required to release bound lipid in an air-mixed dough. About 80% of the total lipid remained bound. However, with the addition of a raw soy flour, capable of oxidizing free and esterified linoleate, the amount of lipid binding was greatly reduced. In the presence of added free linoleic acid, maximum release of bound lipid occurs with both the crude or purified soy lipoxygenase extracts. Wheat lipoxygenase was much less effective.

When soy flour is added to dough, there is maximum oxidation of linoleic and linolenic acids in all the lipids of wheat flour. In control doughs, wheat lipoxygenase oxidized only free fatty acids and monoglycerides (15, 16).

Table II. Effect of Soy Lipoxygenases on Lipid Binding[a]

	Bound lipid, % of total lipid		
Added enzyme	None	Crude soy lipoxygenase	Purified soy lipoxygenase
Shortening fat alone	81.0	58.5	80.0
Shortening plus 5% linoleic acid	66.5	39.0	36.2

[a]Fat-extracted flour supplemented with triglyceride shortening either alone or with 5% linoleic acid.

Daniels *et al.* (13).

Kleinschmidt *et al.* (17) have developed a process which uses lipoxygenase activity in soy flour to develop flavor in continually mixed bread processes. A dispersion of enzyme-active defatted soy flour and peroxidized cottonseed or soybean oil is added to the liquid ferment during baking. When used at the 0.5% level (based on flour), an oil with a peroxide value of 0.03 moles per kilogram of fat is required to produce an acceptable wheaty/nutty flavor. Depending upon conditions used in baking, additional flavors are generated by a Maillard-type reaction.

Enzymes in Soybean Processing

Precise control of heat treatment is critical in that the degree of enzyme inactivation, improvement in nutritive value, and other functional properties of soy protein products depend upon temperature, time, and moisture conditions (1). A number of official and unofficial methods have been used to evaluate the degree of heat treatment which is the primary determinant of functionality. Some of the measurements used for the control of processing are summarized in Table III.

Table III. Property Measurement Used to Control Processing of Edible Soy Flour

Control measurement	Property affected: Flavor	Property affected: Nutritive value
Lipoxygenase	+	-
Peroxidase	+	-
NSI[a]	-	+
PDI	-	+
Urease	-	+
TI	-	+
Available lysine		+

[a] NSI = nitrogen solubility index, PDI = protein dispersity index, and TI = trypsin inhibitor. Rackis *et al.* (18).

Protein Dispersibility Index (PDI) is the percent of total protein (N X 6.25) dispersed under control conditions of extraction and measures the effect of heat treatment on protein solubility (AOCS method BA10-65 (19). Nitrogen Solubility Index (NSI) is the percent of total nitrogen which is soluble in water under controlled conditions of extraction, AOCS method BA11-65 (19).

Live Steam Process. The PDI of raw unprocessed soybean meal is in the range of 90 to 95% and that of toasted products, 10 to 20%. A continuous spectrum of PDI values and of enzyme activity can be obtained through controlled heat processing (20). The relationship between changes in PDI and enzyme activity is shown in Table IV. Of the enzymes tested, lipoxygenase was the most sensitive and urease was the most resistant to heat processing. Urease has no effect on the functionality of edible-grade soy products. However, in feed manufacturing, measurement of urease activity is required to insure sufficient destruction of urease activity when soybean meal is used in animal rations containing urea.

Table IV. Enzyme Activity in Soy Flour Treated with Moist Heat

Enzyme	Negligible PSI[a] 80-100	Light cook PSI 50-80	Moderate cook PSI 20-50	Toasted PSI 0-20
Lipoxidase	+	-	-	-
Urease	+	+	+	-
Diastase (beta)	+	+	-	-
Lipase	+	+	-	-
Protease	+	+	-	-

[a] PSI = Protein Solubility Index. Hafner (20).

Lipoxygenase is an important factor in the generation of flavor compounds from the lipids when soybeans are processed under high-water activity such as in the preparation of soy beverages. To what extent lipoxygenase is responsible for the lipid oxidation at low moisture levels (8-14%) that occurs during processing of soybeans into defatted soy flour, concentrates, and isolates is still not known. A recent review discusses the relationship between lipoxygenase and flavor of soy protein products (10).

Several processes have been used to inactivate lipoxygenase in soybeans (Table V). The essential condition in all of these processes is heat which inactivates enzymes and insolubilizes most of the proteins and generates a cooked or toasted flavor.

Table V. Processes for Inactivating Lipoxygenase

Process
Grinding soybeans with hot water
Dry heating-extrusion cooking
Blanching
Grinding at low pH-cooking

Wolf (10)

Alcohol Processes. Extraction with aqueous alcohol removes many objectionable flavors and is used commercially to prepare soy protein concentrates. Hexane-alcohol azeotrope extraction of defatted soy flakes can also be used to prepare soy flours and concentrates with flavor scores approaching the blandness of wheat flour (21). A process patent issued recently incorporates both hexane-ethanol and aqueous alcohol extraction to prepare soy products with flavor qualities significantly higher than those prepared by present commercial practices (22).

Effect of azeotrope extraction on NSI, lipoxygenase, and peroxidase activity is given in Table VI (18). Over 99% of lipoxygenase activity was destroyed during azeotrope extraction at 56°C, whereas no destruction of peroxidase activity occurred. Since peroxidase is a more stable enzyme, measurement of its activity, rather than of lipoxygenase activity, may have greater significance in defining conditions for production of bland soy products. In addition, peroxidase is a very active enzyme capable of utilizing lipohydroperoxides which can be produced even in the absence of lipoxygenase activity (23). As shown in Table VII, peroxidase activity is much higher at pH 5.0 than 6.5.

Table VI. Enzyme Activity and Nitrogen Solubility of Processed Soy Flakes

Processing conditions	NSI	Lipoxygenase, µmoles O_2/min/g	Peroxidase[a]
Hexane-defatted, raw	92.9	249	1875
Azeotrope-extracted (3 hr)	88.8	1.8	1585
Azeotrope-extracted (6 hr)	83.5	0.4	2098
Hexane-defatted, toasted	41.3	0.5	244
Azeotrope-extracted (3 hr) (toasted)	41.0	---	---
Azeotrope-extracted (6 hr) (toasted)	31.2	0.3	70

[a] Absorption units/g, at pH 5.0. Rackis *et al.* (18).

Table VII. Peroxidase Activity in Maturing Soybeans

Days after flowering	Fresh weight, mg	Dry matter, %	Activity[a] pH 5.0	Activity[a] pH 6.5
35	283	25.8	---	14.2
40	376	31.3	46.2	16.4
42	452	35.3	40.5	13.5
44	470	36.8	42.1	18.9
48	516	37.4	45.2	17.4
51	577	39.6	48.5	15.6
55	500	44.6	48.5	16.1
68	300	74.5	44.0	12.1
72	235	89.0	19.1	8.0
Dehulled, defatted flakes			0.67	0.22

[a] Absorption units per g dry matter X 10^{-3}.
Rackis et al. (23).

The in situ inactivation of enzymes in soybeans that are wet-milled or steeped in various concentrations of aqueous alcohol is shown in Figure 2. Less than 1% of the original lipoxygenase activity remains after soaking soybeans in 30 to 80% aqueous ethanol at room temperature (24). Over 70% of the urease activity was destroyed by soaking the beans in 40 to 60% aqueous alcohol. Maximum flavor scores are obtained when soybeans are treated with 40 to 60% alcohol.

Lipoxygenase and peroxidase activities of soybeans as related to the beany and bitter flavor intensity values (FIV) of maturing soybeans are illustrated in Figure 3 (23). The beany intensity (Curve D) did not change much with increasing maturity as indicated by the small differences in FIV which varied from 2.0 to 2.7. A different pattern was observed for the bitter flavor (Curve C). The average FIV for bitter was 0.4 in the early stages of maturity. As the beans matured, the FIV for bitter increased to 1.6, an increase of about fourfold. There is a correlation ($r = 0.73$), significant at the 1% level, between lipoxygenase activity (Curve A) at pH 6.8 and the increase in bitter FIV (Curve C) in maturing soybeans. Lipoxygenase activity measured at pH 9.0 remained relatively constant and at a level much below that at pH 6.8 during maturation. Activity at pH 6.8 represents the isoenzyme, lipoxygenase-2, which oxidizes fatty acids in triglycerides as well as other lipids (3).

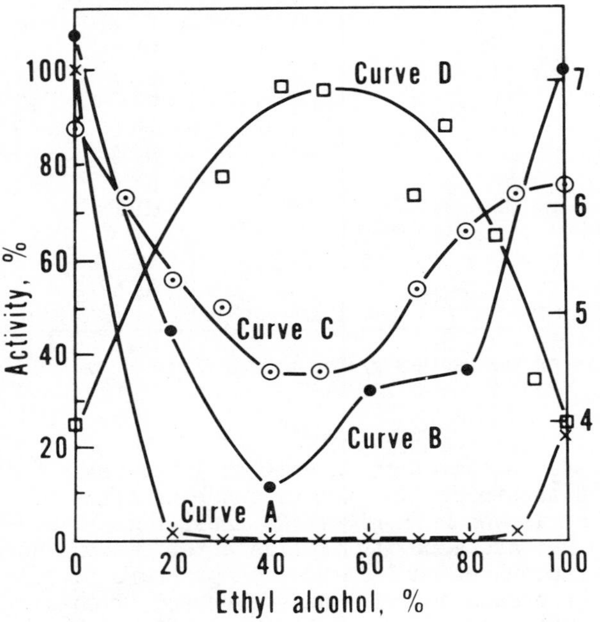

Cereal Chemistry

Figure 2. Effect of soaking soybeans for 24 hr in various aqueous ethanol solutions on urease, lipoxygenase, Nitrogen Solubility Index (NSI), and flavor score. Curve A, *lipoxygenase; Curve* B, *urease; Curve* C, *NSI; and Curve* D, *flavor score* (24).

The level of peroxidase activity (Curve B) in maturing soybeans is very high and remains at a relatively constant level throughout most of the maturation period. Based on our studies of maturing soybeans, we conclude that beany and bitter constituents preexist in immature soybeans. Whether the development of bitterness results from a coupled secondary oxidation reaction involving lipoxygenase and peroxidase remains to be determined.

Enzyme-Modified Soy Products

Proteolytic Modification. Although commercial soy protein products can fulfill a number of functional requirements in present markets for protein additives, enzymatic hydrolysis is also used to create functional properties for specific applications. Pepsin-digested soy isolates are used alone or in combination with egg albumin as aerating and whipping agents in confections, desserts, candies, and cakes, and as foaming ingredients in foam-mat drying of citrus powders (1). Pepsin-hydrolyzed soy isolates are approved as foaming agents in identity standards for soda water. Patents have been issued for enzyme-modified soy products as egg white replacers (25, 26). The use of enzyme treatment to increase solubility of heat denatured soy isolates (27) and to increase solubility in acid solutions have been described (28).

Modification of the functional properties of soy isolates with other proteinases has been investigated (29, 30). Enzymatic modification increased calcium tolerance, water absorption, emulsification, capacity, and foaming properties. However, emulsification and foaming stabilities were reduced. These functional properties were determined in model systems; usefulness of enzyme-modified soy isolates in actual food systems is yet to be evaluated.

Carbohydrate Modification. Raffinose, stachyose, and verbascose cause flatulence in man and animals (31). These oligosaccharides are related by having one or more α-D-galactopyranosyl groups, where the α-galactose units are bound to the glucose moiety of sucrose. α-D-Galactooligosaccharides escape digestion because there is no α-galactosidase activity (E.C. 3.2.1.22) in mammalian intestinal mucosa and because they are not absorbed into the blood. Consequently, bacteria in the lower intestinal tract metabolize them to form large amounts of carbon dioxide and hydrogen. Rate of gas production parallels the formation of monosaccharides resulting from enzymatic breakdown of the oligosaccharides and their intermediate products (Figure 4).

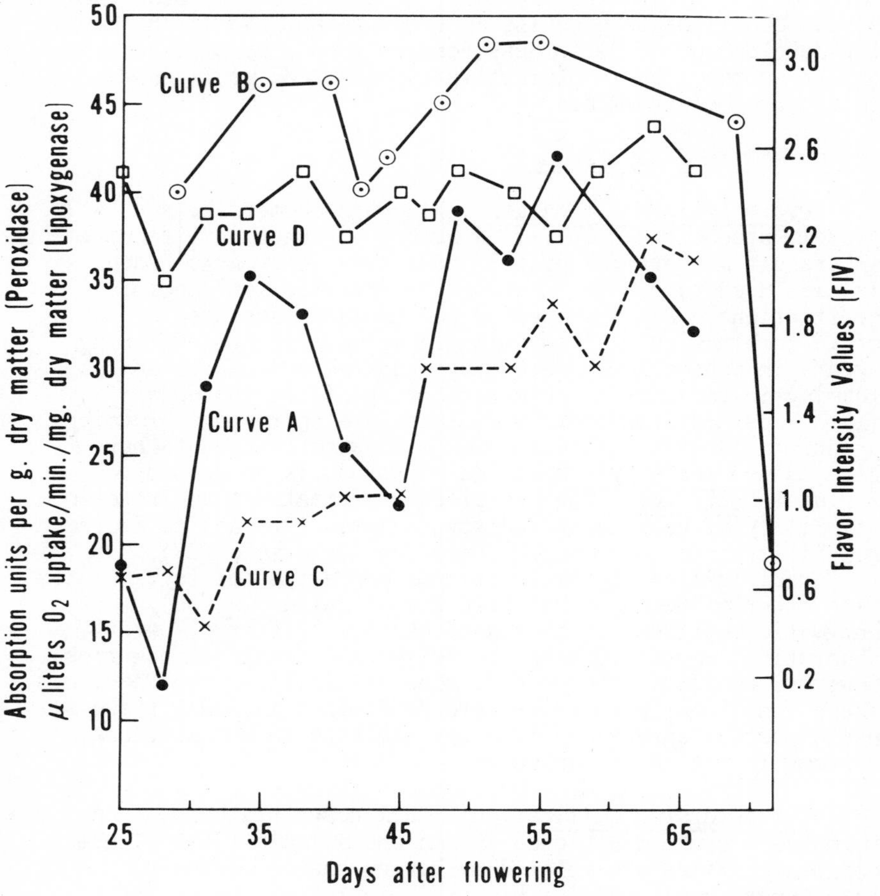

Cereal Chemistry

Figure 3. Relationship between lipoxygenase and peroxidase activities and Flavor Intensity Values (FIV) for bitterness and beaniness in maturing soybeans. Curve A, *lipoxygenase; Curve* B, *peroxidase; Curve* C, *FIV, bitterness; and Curve* D, *FIV, beaniness* (23).

Enzymatic processes that hydrolyze the oligosaccharides in full-fat and defatted soybean meal have been patented (32, 33). Use of immobilized enzyme systems to continuously convert raffinose and stachyose to monosaccharides has been described (34, 35, 36). As shown in Table VIII, about 4 hr was required to completely hydrolyze the oligosaccharides in soybean whey (36). Soybean oligosaccharides are also metabolized by lactic acid bacteria during the preparation of fermented soy milk (37).

Plastein Products. When soybean protein is partially hydrolyzed by proteinases (peptidyl-peptide hydrolases E.C. 3.4.4) and the hydrolyzate is then incubated with certain proteolytic enzymes under proper conditions, the hydrolysis is reversed and a "protein-like" substance is formed whose properties are different from that of the original protein. The new protein-like substance is called "plastein." The plastein reaction requires at least three specific conditions which are different from those for proteolysis (38). These conditions are summarized in Table IX. In addition, low molecular weight peptides are effective substrates for the plastein reaction. Of especial importance is the ratio of hydrophobic-hydrophilic groups of the peptide substrate (39).

Arai et al. (38) define plastein formation as a condensation-transesterification reaction giving rise to products of higher molecular weight. Other workers (40) have found that only slight changes in molecular weight distribution occur during plastein formation with enzymatic hydrolyzates of milk whey protein concentrates and fish protein concentrates. Apparently, transpeptidation and hydrolysis had a more important role than condensation reactions during plastein formation. Alternate theories for plastein formation have been reviewed (41).

Food Applications. Although the mechanism of the plastein reaction is still not well understood, plastein proteins have interesting properties which may be of value in food processing technology.

In a series of research reports, Fujimaki and coworkers conclude that the plastein reaction can be used to prepare bland protein-like substances free of a host of other impurities from crude soy protein isolates (38, 42) and single cell protein concentrates (43). A schematic diagram for preparing purified plastein from a crude protein material is illustrated in Fig. 5.

Table VIII. Glucose Production from Soybean Whey[a] by *A. awamori* α-galactosidase[b] Immobilized in an Amicon Hollow-Fiber Dialyzer[c]

Time, hr	Glucose, mg/ml
0.5	0.31
1.0	0.40
2.0	0.44
3.0	0.49
4.0	0.58
4.5	0.62
4.5 + soluble enzyme[d]	0.60

[a] Soybean whey (18 liters) was equivalent to 1% soybean slurry.

[b] Crude enzyme (900 ml) at 25 U/ml.

[c] At 45°C, recirculation rate was 185 ml/min.

[d] Consisted of 10 ml of the 4-5 hr sample plus 2.5 U of crude α-galactosidase. Incubated overnight at 25°C.

Smiley *et al.* (36).

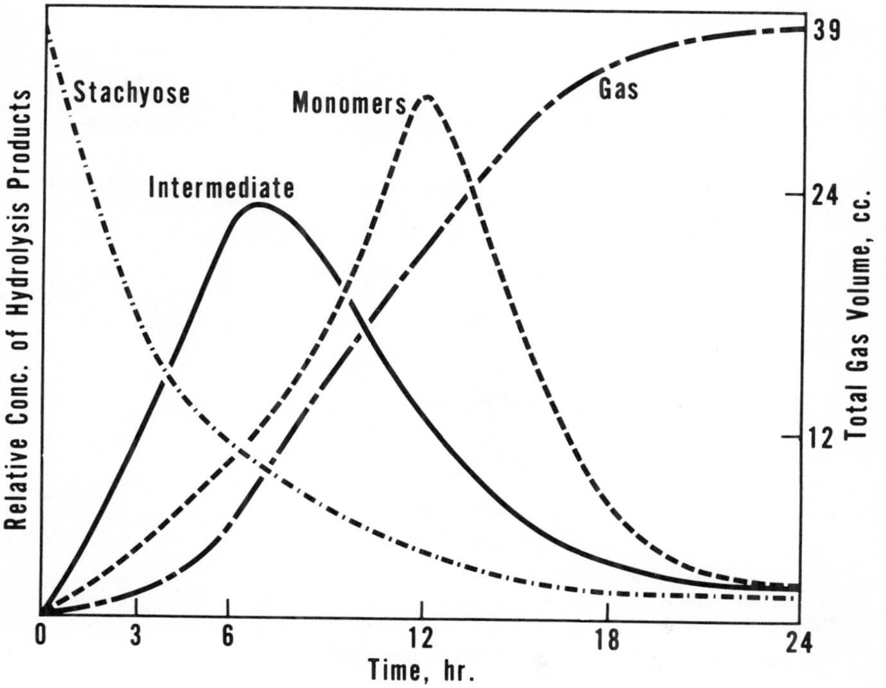

ACS Symposium Series

Figure 4. Relationship between the enzymatic hydrolysis of stachyose and gas production. Anaerobic culture isolated from the dog colon. Curve A, *stachyose; Curve* B, *disaccharides and trisaccharides; Curve* C, *monomer sugars; and Curve* D, *gas production* (31).

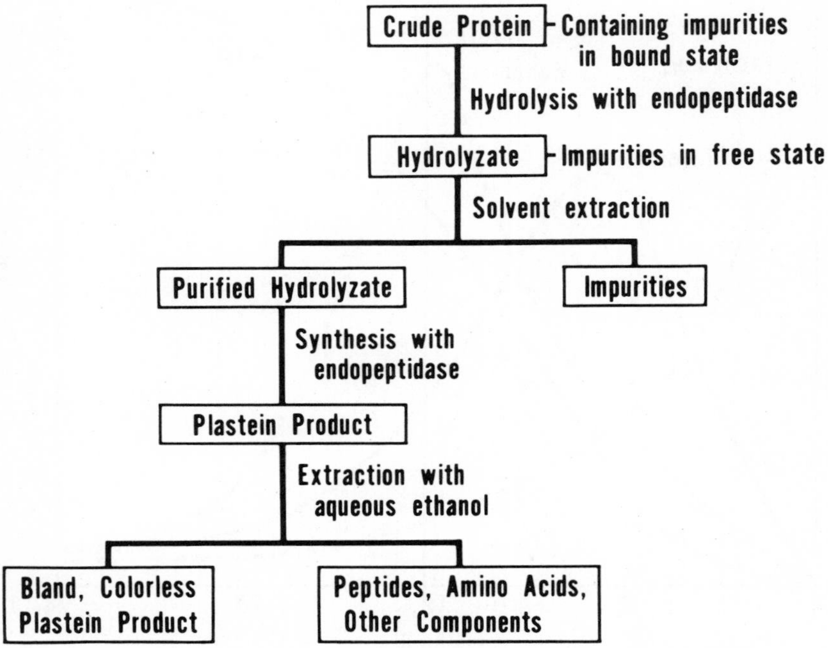

Cereal Foods World

Figure 5. Process for preparing purified, bland plastein products (38)

Table IX. Differences in Reaction Conditions Between Protein Hydrolysis and Plastein Synthesis

Condition	Hydrolysis	Synthesis
Substrate	Polypeptide	Oligopeptide
Concentration, %	<5	30-50
pH for acidic enzymes[a]	1-4	4-5
SH enzymes[b]	4-7	4-6
Alkaline enzymes[c]	7-10	5-6
pH range	Range 1-10	4-6
ES-ratio	1:100-3:100	1:100-3:100
Temperature, °C	37	37
Period, hr	24-48	24-96

[a] g pepsin: pH 1.6 in hydrolysis and pH 4.5 in synthesis.

[b] g papain: pH 5.5 both in hydrolysis and in synthesis.

[c] g α-chymotrypsin: pH 7.8 in hydrolysis and 5.3 in synthesis.

Arai *et al.* (38).

Table X. Amino Acid Enrichment Plasteins Prepared from These Pepsin-Hydrolyzed Substrates[a]

Amino acid	Soy protein isolate		Zein			
	Hydrolyzate	Met-enriched plastein	Hydrolyzate	Amino acid enriched plastein Try-	Thr-	Lys-
Methionine (Met)	1.18	7.98	---	---	---	---
Tryptophan (Try)	---	---	0.38	9.71	---	---
Threonine (Thr)	---	---	2.40	---	9.23	---
Lysine (Lys)	---	---	0.20	---	---	2.14

[a] Amino acid composition based on a weight percent. Consult source for changes in composition of its other amino acids for both hydrolyzates and plasteins.

Arai *et al.* (38).

In the hydrolysis phase, some of the low-molecular-weight flavor constituents and other impurities are removed by selective solvent extraction. Two methods are then used to remove the bitterness of the purified hydrolyzate: (a) further hydrolysis with an exopeptidase and (b) synthesis of plastein protein with an endopeptidase. Various enzymes have been used to debitter crude soy protein (42). The economic feasibility of such a process remains to be evaluated. In addition, no organoleptic evaluations have been made in comparing the acceptability of plastein soy protein and of protein isolates prepared from hexane:ethanol azeotrope-extracted soy flakes (21).

The use of the plastein reaction to improve the nutritional value of low-quality proteins by enrichment with essential amino acids has also been proposed (38). By incubating soy protein hydrolyzates with L-methionine ethyl ester, the methionine content of the soy plastein protein was increased nearly sevenfold (Table IX). Similarly, ethyl esters of L-tryptophan, L-threonine, and L-lysine were used to enrich the essential amino acid content of zein plastein protein.

Glutamic acid enrichment was used to form a soy plastein that exhibited high water solubility in the pH range of 1-13. Arai *et al.* (38) illustrates the differences in solubility of glutamic acid-enriched soy plastein and soy protein isolate. Such a product may find application in food processing where protein solubility in acid solutions is a requirement. Presently, only enzymatically hydrolyzed soy protein has such a solubility characteristic (28, 29).

Peptide gels with interesting rheological properties have been prepared (42). The plastein process may prove useful in preparing products having tailored compositional and functional properties, however, nutritional and consumer labeling problems are some of the many special considerations that will need to be evaluated (43).

Literature Cited

1. American Soybean Association. Proc. World Soy Protein Conf., Munich, Germany, November 11-14, 1973, *J. Am. Oil Chem. Soc.* (1974) 51, 49A.
2. Andre, E., Hou, K., *C. R. Acad. Sci.* (1932) 194, 645.
3. Gardner, H. W., *J. Agric. Food Chem.* (1975) 23, 129.
4. Kies, M. W., Haining, J. L., Pistorius, E., Schroeder, D. H., Axelrod, B., *Biochem. Biophys. Res. Commun.* (1969) 36, 312.
5. American Chemical Society, Symposium on Effects of Oxidized Lipids on Food Proteins and Flavor, *J. Agric. Food Chem.* (1975) 23, 125.

6. Kalbrener, J. E., Warner, K., Eldridge, A. C., Cereal Chem. (1974) 51, 406.
7. Cotton, R. H., J. Am. Oil Chem. Soc. (1974) 51, 116A.
8. Haas, L. W., Bohn, R. M. (to J. R. Short Milling Co.) U.S. Patents 1,957,333 and 1,957,336 (May 1, 1934).
9. Haas, L. W. (to J. R. Short Milling Co.) U.S. Patents 1,957,334 and 1,957,337 (May 1, 1934).
10. Wolf, W. J., J. Agric. Food Chem. (1975) 23, 136.
11. Pringle, W., J. Am. Oil Chem. Soc. (1974) 51, 74A.
12. Wood, J. C., Food Manuf. Ingredient Survey (January 1967) 11-15.
13. Daniels, N. W. R., Frazier, P. J., Wood, P. S., Baker's Dig. (1971) 45(4), 20.
14. Frazier, P. J., Leigh-Dugmore, F. A., Daniels, N. W. R., Russell-Eggitt, P. W., Coppock, J. B., J. Sci. Food Agric. (1973) 24, 421.
15. Mann, D. L., Morrison, W. R., Ibid. (1975) 26, 493.
16. Morrison, W. R., Panpaprai, R., Ibid. (1975) 26, 1225.
17. Kleinschmidt, A. W., Higashiuchi, K., Anderson, R., Ferrari, C. G., Baker's Dig. (1963) 37(5), 44.
18. Rackis, J. J., McGhee, J. E., Honig, D. H., J. Am. Oil Chem. Soc. (1975) 52, 249A.
19. American Oil Chemists' Society, "Official and Tentative Methods," 3rd Ed., The Society, Chicago (1973).
20. Hafner, F. H., Cereal Sci. Today (1964) 9, 164.
21. Honig, D. H., Warner, K. A., Rackis, J. J., J. Food Sci. (1976) 41, 642.
22. Hayes, L. P., Simms, R. P. (to A. E. Staley Mfg. Co.) U.S. Patent 3,734,901 (May 22, 1973).
23. Rackis, J. J., Honig, D. H., Sessa, D. J., Moser, H. A., Cereal Chem. (1972) 49, 586.
24. Eldridge, A. C., Warner, K., Wolf, W. J., Cereal Chem. (In press).
25. Turner, J. R., U.S. Patent 2,489,208 (1969).
26. Gunther, R. C., Canadian Patent 905,742 (1972).
27. Hoer, R. A., Frederiksen, C. W., Hawley, R. L., U.S. Patent 3,694,221 (1972).
28. Pour-El, A., Swenson, T. C., U.S. Patent 3,713,843 (1973).
29. Puski, G., Cereal Chem. (1975) 52, 655.
30. Pour-El A., Swenson, T. C., Cereal Chem. (1976) 53, 438.
31. Rackis, J. J., American Chemical Society Symposium Series, Number 15 (1975), Edited by A. Jeanes and J. Hodge, Chapter 13, 207.
32. Ciba-Giegy, A. G., French Patent 2,137,548 (1973).

33. Sherba, S. E., U.S. Patent 3,632,346 (1973).
34. Delente, J., Johnson, J. H., O'Connor, M. J., Weeks, L. E., Biotechnol. Bioeng. (1974) 16, 1227.
35. Reynolds, J. H., Ibid. (1974) 16, 135.
36. Smiley, K. L., Hensley, D. E., Gasdorf, H. J., Appl. Environ. Microbiol. (1976) 31, 615.
37. Mital, B. K., Steinkraus, K. H., J. Food Sci. (1975) 40, 114.
38. Arai, S., Yamashita, M., Fujimaki, M., Cereal Foods World (1975) 20(2), 107.
39. Arai, S., Yamashita, M., Aso, K., Fujimaki, M., J. Food Sci. (1975) 40, 342.
40. Hofsten, B., Lalasidis, G., J. Agric. Food Chem. (1976) 24, 460.
41. Eriksen, S., Fagerson, I. S., J. Food Sci. (1976) 41, 490.
42. Fujimaki, M., Yamashita, M., Arai, S., Kato, H., Agric. Biol. Chem. (Japan) (1970) 34, 483.
43. Fujimaki, M., Kato, H., Arai, S., Yamashita, M., J. Appl. Bacteriol. (1971) 34, 119.

15

Enzyme Methods to Assess Marine Food Quality

FREDERICK D. JAHNS and ARTHUR G. RAND, JR.

Department of Food Science and Technology, University of Rhode Island, Kingston, R.I. 02881

How fresh is a fresh fish? If we look at Figure 1, perhaps we can illustrate this point! In examining these fish, especially the one at the bottom, one might conclude "this fish is putrid." However, this evaluation would be incorrect. This fish may not be fresh, but it has been stored in ice at $0^{o}C$ for only a few days and certainly is not spoiled. The belly is almost completely digested in the lower fish but this fish is 6 days old postmortem. It is also possible to have fish only several hours postmortem in the same condition, because they weren't properly iced. While the tissue in the belly region on the lower fish was digested, a cooked fillet of this fish could be quite acceptable to many people. Note that the fish at the top of the picture was caught at the same time and was stored under the exact same conditions as the one at the bottom, yet somehow their appearance is different. It is this type of example that contributes to the elusiveness of evaluating the quality of seafood, particularly in trying to distinguish the fresh and edible stages. The inedible phase is usually quite clear to anyone.

There have been many organoleptic guidelines set for judgment of fish quality such as texture, appearance, color, eye transparency, and odor. In many cases this approach may give a general indication of quality, but they are very subjective (1) (2).

Realizing the drawbacks in using these criteria, and understanding the economic impact of properly assessing seafood quality, many investigators have set out to find a more reliable means. The most logical approach was to view the issue through chemical tests. In general most of the chemical tests appear to have variable success, with limitations being placed on them by innate biological differences from species to species. Some compounds such as trimethylamine, volatile nitrogen bases, ammonia, and volatile reducing substances appear mostly in the latter stages of refrigerated storage, which limits their value in that considerable period prior to this time, as indicated by

Dugal (3). There are a number of other techniques, such as an automated determination of volatile bases (4), the Torry Fish Freshness Meter and the Intelectron Fish Tester (5)(6)(7) which have attempted to simplify quality assessment and have met with some success. However, they are not without problems. For example, the Torry meter seems to be limited to lean fresh fish. Frozen or fatty fish produce somewhat variable results (8). The automated determination of volatile bases requires a laboratory situation.

The criteria outlined by Shewan & Jones (9) continues to be a sound basis for a marine food quality index. The method selected should detect a compound which:

1. Will be absent or present in constant amount in freshly caught food;
2. Should accumulate or disappear quickly at a steady rate during post-harvest aging and spoilage;
3. Has the capability of being simply and speedily estimated.

Enzyme methods which have the advantages of specificity and rapidity meet this basis quite well. The first enzyme methods to assess marine food quality were reported in 1964 (10)(11); this was an analysis for the compound hypoxanthine, a metabolic byproduct of ATP breakdown, employing an enzyme method developed almost two decades before by Kalckar (12). The initial work utilized ion exchange chromatography to separate the nucleotide bases in a fish tissue extract. This enzyme technique then employed xanthine oxidase (EC 1.2.3.2) to convert the hypoxanthine fraction to uric acid, followed by differential spectrophotometry in the UV for analysis. There has been ample demonstration of the usefulness of hypoxanthine as a measure of fish quality in several papers (3)(10)(13)(14)(15).

Burt *et al.* (16) proposed a method of automating the xanthine oxidase reaction for hypoxanthine analysis using the redox dye 2,6 dichlorophenolindophenol (DIP). This placed the method on a colorimetric basis, simplified the equipment requirements, and increased sample capacity. Figure 2 illustrates the interaction between hypoxanthine, DIP, and xanthine oxidase. The hypoxanthine concentration can be related to the extent of decolorization of the dye. The correlation between the automated colorimetric methods, measuring at 600 nm using DIP, and the manual enzyme method measuring uric acid appearance at 290 nm was excellent (16). Beuchat (15) utilized DIP in a manual enzyme assay to measure hypoxanthine directly in a deproteinated extract, and correlated this method with an organoleptic evaluation of catfish (*Ictalurus punctatus*). Jahns *et al.* (17) also used a slightly modified xanthine oxidase-DIP method to monitor hypoxanthine concentration in iced Winter Flounder (*Pseudopleuronectes americanus*). A steady increase in hypoxanthine concentration occurred with storage time until the values leveled off at about 5 μM/g after 10 days where organoleptic evaluation indicated the

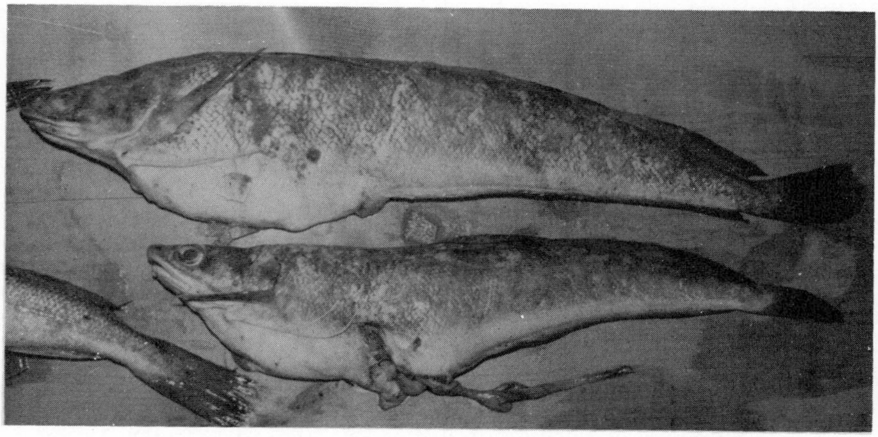

Figure 1. The appearance of red hake after six days of storage in ice (0°C)

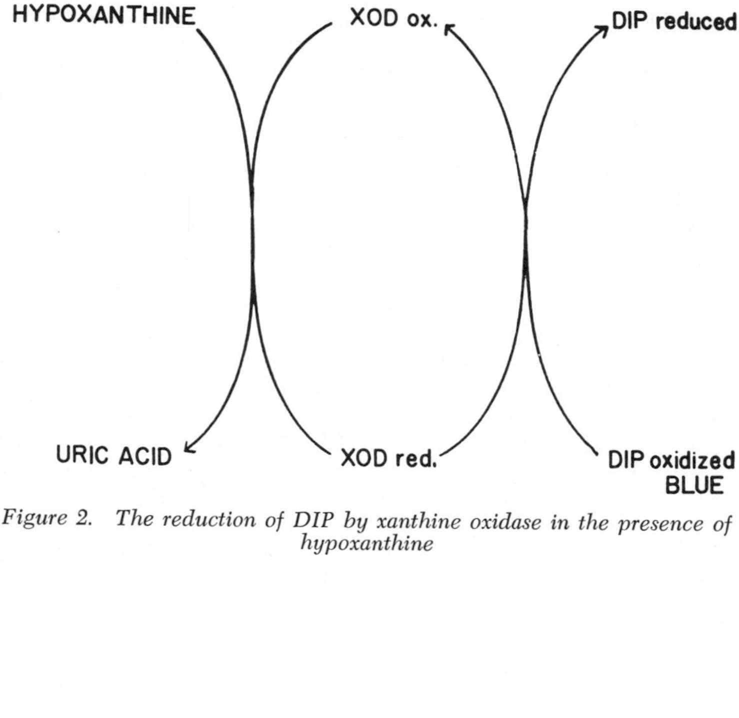

Figure 2. The reduction of DIP by xanthine oxidase in the presence of hypoxanthine

fish were no longer fresh.

Semi-automated hypoxanthine analysis

The manual xanthine oxidase colorimetric method has been further modified for quick routine analysis in the laboratory by employing an automatic pipet and a Spectronic 20 colorimeter, outfitted with a flow-thru cuvette. Briefly, the method is as follows:

1. A 10 g sample of fish is blended for one minute with 90 ml of cold 0.6N perchloric acid.
2. The homogenate is then filtered through Whatman #2 paper to remove protein and tissue debris.
3. A 10 ml aliquot of the filtrate is neutralized to pH 7.6 with 20% KOH and buffered with 0.1M Phosphate buffer, pH 7.6, making the volume up to 100 ml.
4. Approximately 12 ml is added to a centrifuge tube. This is chilled in a freezer until a layer of ice begins to form at the top of the tube.
5. The sample is then centrifuged in a clinical centrifuge for approximately 3 minutes to remove $KClO_3$.
6. The centrifugate is poured off and allowed to temper to 25°C. One ml is used in the analysis for hypoxanthine, and contains 10 mg of fish.
7. A 4 ml portion of buffered xanthine oxidase-DIP solution at 25°C is injected into the 1 ml sample of fish extract with an automatic pipet. The enzyme reagent is comprised of 60 ml of 0.15M phosphate buffer (pH 7.6) containing DIP @ 23 μg/ml, and 20 ml of xanthine oxidase solution @ 0.02-0.025 units per ml.
8. The absorbance at 600 nm is read in the flow-thru cuvette of the Spectronic 20 at exactly 2.5 min. The absorbance change is compared against a standard plot obtained by analyzing known hypoxanthine concentrations.

Employing this method we were able to assess the hypoxanthine values found in Whiting (Merluccium bilinearis) during iced storage, as shown in Figure 3. Hypoxanthine was monitored over a 32 day period, in order to evaluate concentration well past the fresh and acceptable periods. Levels of hypoxanthine increased to a maximum of 6 μM/g between 17 and 21 days. It then gradually declined by 32 days to slightly above 2 μM/g. Figure 4 depicts the results obtained from two separate trials using Tautog (Tautoga onites). Both trials showed very similar results exhibiting maximum hypoxanthine values near 3 μM/g at from 12-14 days in iced storage. Trial one was caught in November and Trail two was taken during December. The values again diminish after peak hypoxanthine formation.

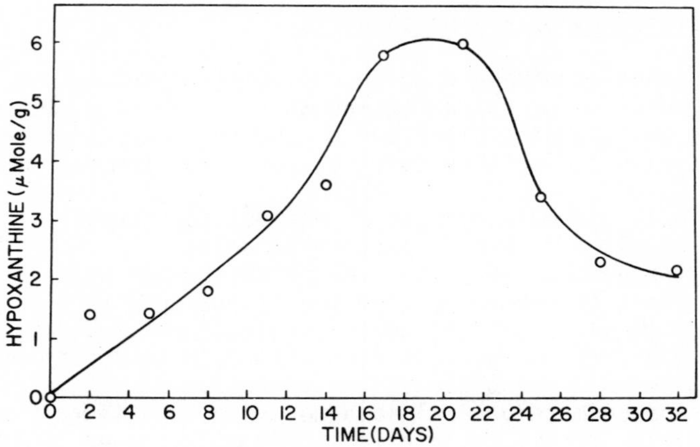

Figure 3. Hypoxanthine concentration in whiting during iced (0°C) storage. Each point represents the mean value for two fish.

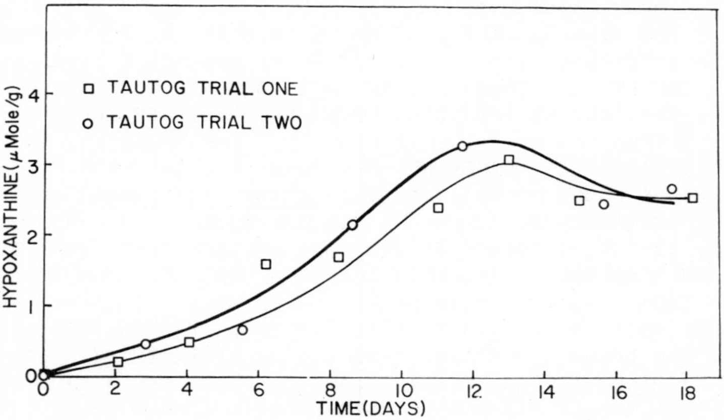

Figure 4. Hypoxanthine concentrations in tautog during two iced (0°C) storage trials. Each point in trial one represents one fish, while trial two shows the mean value for two fish.

Immobilized xanthine oxidase as an analytical tool

Another laboratory method for hypoxanthine analysis, which is still in the preliminary stages of development, is the use of an immobilized xanthine oxidase column. Very few reports are available on the immobilization of xanthine oxidase. Coughlan and Johnson (18) first reported on methods of preparation describing attachment to cellulose, Sepharose and glass. However, they utilized these preparations only to study the flavin chromophore of the enzyme. Petrilli (19) and Singer and Warchut (20), working in our laboratory, studied xanthine oxidase immobilization in acrylamide gel and on glass and Duolite* beads as a method for measuring hypoxanthine concentration. The acrylamide gel entrapment demonstrated the potential of immobilized xanthine oxidase for hypoxanthine analysis, but the enzyme was unstable during continuous analytical use. In a comparison of xanthine oxidase immobilization on Corning zirconia-clad silica glass and Duolite beads, maximum enzyme attachment occurred in a shorter time with the glass than the Duolite. However, the Duolite appeared to retain higher enzyme activity of the two supports, and thus was chosen to prepare an experimental plug flow reactor for hypoxanthine analysis. The immobilized enzyme column was connected to a pump and a continuous flow UV analyzer equipped with a recorder as presented in Figure 5. The buffered substrate, which could be either a sample of fish extract, or a known hypoxanthine solution for calibration, was pumped into the column. The equilibrium absorbance value for the product of uric acid can be directly related to hypoxanthine concentration, as illustrated, over a range of 0-30 μg/ml. This enzyme derivative exhibited remarkable stability when utilized repeatedly over a period of weeks, and retained high activity during months of refrigerated storage.

This analytical tool has great potential in routine monitoring of fish and marine food quality. The equipment and reagent requirements are minimal in comparison to similar approaches (4)(16). The automatic processing and recording of results allows the investigator time to prepare samples and eliminates constant manual operation which is generally involved in the standard colorimetric procedure.

Diamine oxidase as an indicator of marine food spoilage

An enzymatic spoilage test that would help to clarify the hypoxanthine freshness analysis for a given sample of fish was investigated. Spinelli *et al.* (10) compared the pattern of hypoxanthine formation in fish with some other tests, including Total Volatile Bases (TVB). Their results indicated the TVB

*Duolite is a phenolic-formaldehyde resin product of Diamond Shamrock Co.

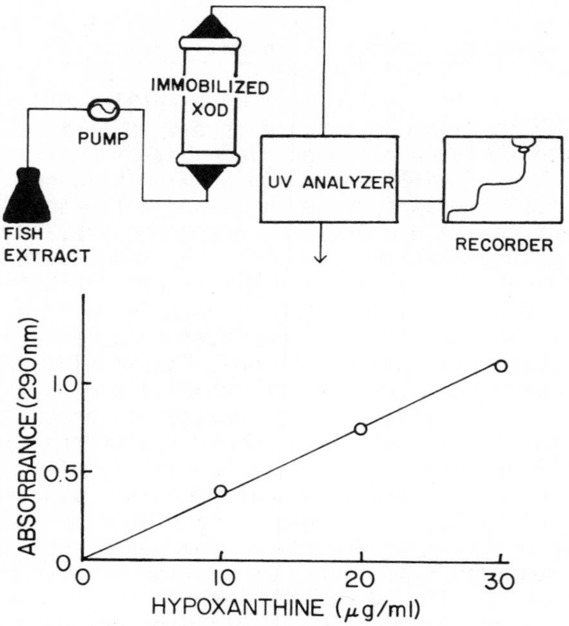

Figure 5. Immobilized xanthine oxidase for the analysis of hypoxanthine

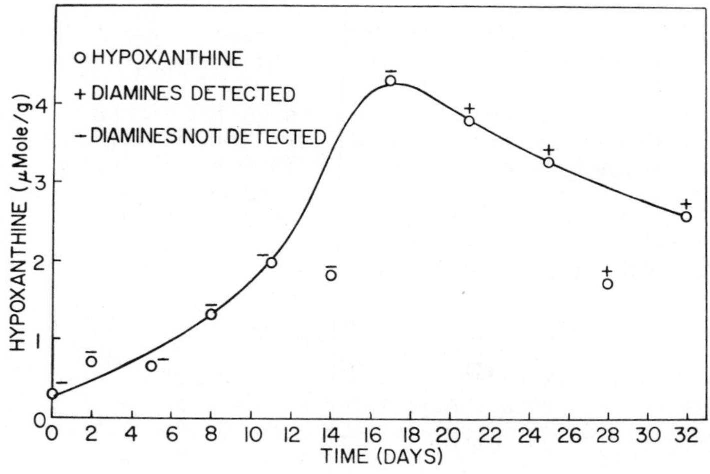

Figure 6. Hypoxanthine concentration and diamine presence in red hake during iced (0°C) storage. Each point represents the mean value for two fish.

values did not begin to measurably increase until the hypoxanthine development had peaked. The composition of the TVB material undoubtedly includes diamine compounds, which are produced by microbial spoilage (21). A recent report by Mietz and Karmas (22) successfully fractionated and quantitated the diamines from tuna by high-pressure liquid chromatography. Their results confirmed that putrescine, cadaverine, and histamine all increased in concentration as the fish reached unacceptable levels. Thus, a non-specific enzyme test for diamines could indicate if these compounds were present and clarify whether the hypoxanthine analysis was indicating fresh or stale fish.

A method for determining diamines with the enzyme diamine oxidase (EC 1.4.3.6) coupled to the o-tolidine-peroxidase colorimetric system was developed. The procedure established is as follows:

1. A 25 g sample of fish is blended for one minute with 75 ml of cold 0.6N Perchloric Acid.
2. This is filtered through a Whatman #2 paper.
3. A 25 ml portion of the filtrate is adjusted to pH 8.5 with 20% KOH and made up to 50 ml with 0.1M borate buffer, pH 8.5.
4. About 12 ml is placed in a centrifuge tube and allowed to chill in a freezer until a layer of ice just begins to form on top.
5. The chilled sample is centrifuged for approximately 3 min in a clinical centrifuge.
6. The centrifugate was poured off and allowed to temper to ambient temperature. Each 1 ml contains 0.125 g of fish.
7. To 1.0 ml of the centrifugate, 2.0 ml of diamine oxidase reagent is added with an automatic pipet, and the mixture is incubated at 37°C for 30 min. The diamine oxidase reagent consists of--10 ml of diamine oxidase solution @ 4 units/ml of 0.1M borate buffer (pH 8.5), 5 ml of o-tolidine dihydrochloride solution @ 500 μg/ml of water, and 5 ml of peroxidase solution @ 226 units/ml of pH 8.5 borate buffer (0.1M).
8. The color development is determined on a Spectronic 20 Colorimeter equipped with a flow-thru cuvette. The results are evaluated with a standard curve prepared from the analysis of known putrescine concentrations.

Since the purpose of the diamine test was only to support hypoxanthine analysis, the results have been presented only in terms of whether diamines were present or not. Figure 6 presents a graph comparing the pattern of hypoxanthine concentration and diamine values in iced Red Hake (*Urophycis chuss*). In Red Hake the hypoxanthine values increased to a peak of approximately 4 μM/g at 17-18 days followed by the usual downward trend. Positive diamine values were found only in samples more than 18 days old, or after the hypoxanthine peak. The hypoxanthine and diamine analysis procedures were also compared on iced Winter Flounder, as shown in Figure 7. The characteristic rise and drop in hypoxanthine concentration with storage time is again evident. The peak value of 7.5 μM/g occurred at approximately 11 days postmortem. In a pattern almost identical to Red Hake, diamines in winter flounder were detected after the storage time for peak hypoxanthine development. Thus, the diamine analysis does show value as a spoilage indicator with the fish tested. Combining the two enzyme procedures, xanthine oxidase for hypoxanthine concentration, and diamine oxidase for diamine detection, can clarify the interpretation of fish quality. Figure 8 illustrates how these two tests generally interact.

Rapid visual enzyme method

The methods that have just been discussed were designed for quick quantitative analysis in the laboratory. However, the greatest need exists for a test of marine food quality that will be simple, inexpensive, and lend itself to usage outside of the laboratory, either on board ship, dockside, or in the food processing plant. Jahns et al. (17) proposed a method for estimating fish freshness which makes use of a strip of absorbent paper impregnated with resazurin, xanthine oxidase, and buffer combination. When the enzyme test strip was moistened with flounder extracts and allowed to react for 5 min at room temperature, the color change from blue to pink correlated with the actual laboratory colorimetric hypoxanthine analysis of the same extracts. This is indicated in Figure 8 by the blue-pink-blue progression of color.

As already indicated the hypoxanthine method is basically a freshness test. If the test strip was set to turn pink at a hypoxanthine concentration of 5 μM/g or 70 mg%, which would be an average value for the five fish analyzed in this study, then a simple color change from blue to a shade of pink could indicate when fish can no longer be considered fresh. However, this test does not seem to be useful in estimating spoilage in seafood. In the results obtained, the hypoxanthine concentration in all fish studied always begins to diminish after the freshness storage period. This makes it difficult, if not impossible, to equate hypoxanthine concentration with storage time beyond 10-15 days in iced conditions. In fact, the color changes from pink back toward blue, as the fish ages, could be misinterpreted. A second visual test to indicate spoilage would compliment the hypoxanthine enzyme test strip.

Since diamines appear to develop after the hypoxanthine peak in a significant quantity, another visual test could be employed on the same strip as indicated in Figure 8. Essentially the hypoxanthine-diamine interaction can be divided into three categories, as illustrated in Figure 9. Marine foods could

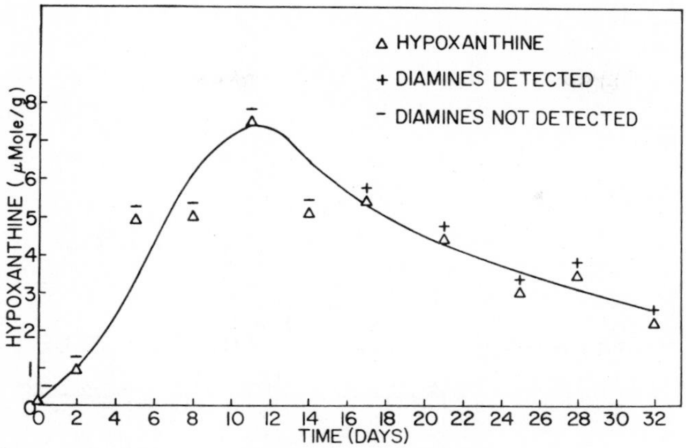

Figure 7. Hypoxanthine concentration and diamine presence in winter flounder during iced (0°C) storage. Each point represents the results for at least two fish.

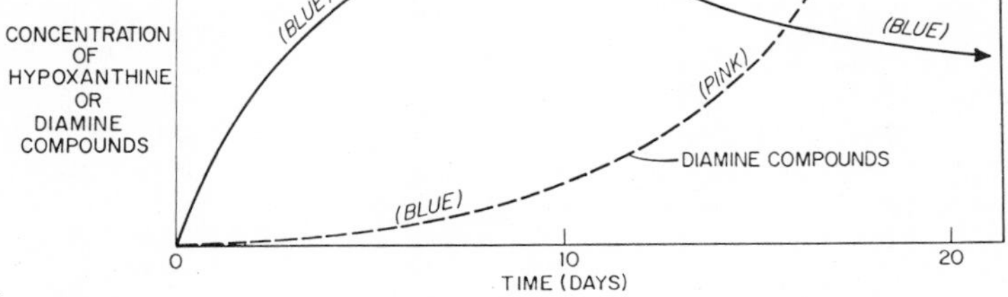

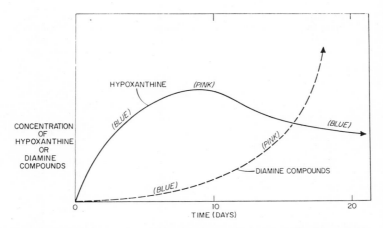

Figure 8. The pattern of hypoxanthine and diamine development in fish

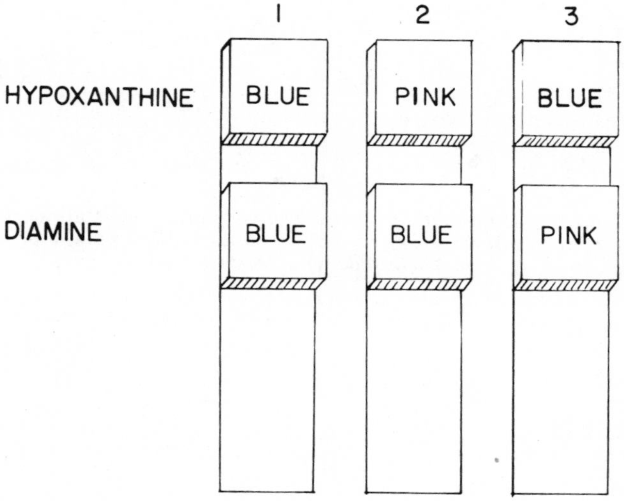

Figure 9. The three phases of rapid visual enzyme analysis of marine foods for freshness (hypoxanthine) and spoilage (diamines)

be simply and quickly evaluated, based on the reaction colors. Class 1 would be the fresh food of highest quality, where hypoxanthine and diamine levels are low, indicated as blue in the test zones. Class 2 would be intermediate quality in which food is acceptable but not fresh. Hypoxanthine levels are high, indicated by the pink test zone and diamines are still low, with a blue test zone. Class 3 occurs when hypoxanthine levels are now lower and a blue test zone, while diamine levels have increased, as indicated by the pink test zone. Marine foods in the 3rd class would be classified as marginal or not acceptable. It may even be possible to sub-divide the 3 test categories further by shades of blue or pink to increase the sensitivity of grading. Perhaps even a third indicator to produce a multi-phase strip might be employed to visually indicate the presence of trimethylamine. This might be accomplished by employing the enzyme trimethylamine dehydrogenase, which Large and McDougall (23) suggested could be used for the microestimation of trimethylamine. When these enzyme test strips are finally employed it should be possible to use these directly on the food, employing the tissue fluid to wet the test zones. This would eliminate the need to sample and prepare a protein-free extract.

The application of enzyme methods to marine food analysis can provide objective, inexpensive and rapid approaches to the assessment of seafood quality. Adaptation of these methods to a visual basis permits extension of enzyme analysis to non-laboratory situations, such as dockside or on board ships. Thus, enzymatic analysis of marine foods can be utilized for quality assessment at critical points, where this type of determination can have maximum impact for grading and purchase.

Acknowledgements

The authors would like to acknowledge the contributions of James A. Hourigan, Richard J. Coduri, Jr. and Jeffrey L. Howe who provided encouragement and technical advise. This research was supported in part by the College of Resource Development, University of Rhode Island; by the Cooperative State Research Service, U.S. Department of Agriculture; and by the NOAA Office of Sea Grant, U.S. Department of Commerce under grants 04-6-158-44002 and 04-6-158-44085. Contribution No. 1709 of the Rhode Island Agricultural Experiment Station.

Literature cited

1. Farber, L. G. in "Fish as Food" Vol. 4, pg. 65, G. Borgstrom, ed. Academic Press, NY, 1965.
2. Hillig, F., Shelton, L. R., Jr., and Loughrey, J. H., J.A.O.A.C. (1962) 45, 724.

3. Dugal, L. C., J. Fish. Res. Bd. Can. (1967) 24, 2229.
4. Ruiter, A., and Weiseman, J. M., J. Fd. Technol. (1976) 11, 59.
5. Burt, J. R., Gibson, D. M., Jason, A. C., and Sanders, H. R., J. Fd. Technol. (1976) 11, 73.
6. Burt, J. R., Gibson, D. M., Jason, A. C., and Sanders, H. R., J. Fd. Technol. (1976) 11, 117.
7. Connell, J. J., Howgate, P. F., Mackie, I. M., Sanders, H. R., Smith, G. L., J. Fd. Technol. (1976) 11, 297.
8. Cheyne, A., Fishing News International, December (1975).
9. Shewan, J. M. and Jones, N. R., J. Sci. Food Agric. (1957) 8, 491.
10. Spinelli, J., Eklund, M. and Miyauchi, D., J. Food Sci. (1964) 29, 710.
11. Jones, N. R., Murray, J., and (in part) Livingston, E. I. and Murray, C. K., J. Sci. Food Agric. (1964) 15, 763.
12. Kalckar, H. M., J. Biol. Chem. (1947) 167, 429.
13. Kassamsarn, B., Sanz Perez, B., Murray, J. and Jones, N. R., J. Food Sci. (1963) 28, 28.
14. Spinelli, J., J. Food Sci. (1967) 32, 38.
15. Beuchat, L. R., J. Agr. Food Chem. (1973) 21, 453.
16. Burt, J. R., Murray, J. and Stroud, G. D., J. Food Technol. (1968) 3, 165.
17. Jahns, F. D., Howe, J. L., Coduri, R. J., Jr. and Rand, A. G. Jr., Food Tech. (1976) 30, (7), 27.
18. Coughlan, M. P. and Johnson, D. B., Biochim. Biophys. Acta (1973) 302, 200.
19. Petrilli, R. "The use of immobilized xanthine oxidase as a method of measuring hypoxanthine concentrations." M.S. Research Project Report, University of Rhode Island, 1975.
20. Singer, A. and Warchut, A. "Immobilization of xanthine oxidase for applications as an analytical tool." M.S. Research Project Report, University of Rhode Island, 1975.
21. Eskin, N. A. M., Henderson, H. M. and Townsend, R. J., "Biochemistry of Foods," pg. 191-201, Academic Press, NY, 1971.
22. Mietz, J. J. and Karmas, E., Inst. Food Tech. Abstracts, Anaheim (1976).
23. Large, P. J. and McDougall, H., Anal. Biochem. (1975) 64, 304.

16

Chemical and Microbiological Changes during Sausage Fermentation and Ripening

S. A. PALUMBO and J. L. SMITH

Eastern Regional Research Center, Agricultural Research Service, U.S. Department of Agriculture, Philadelphia, Pa. 19118

Introduction

The production of sausage began as one of man's earliest attempts at food preservation, possibly as far back as 1500 B.C., when people learned that meat would not spoil if it were finely chopped, mixed with salt and spices, and allowed to dry in rolls. Salami is thought to be named after the city of Salamis on the east coast of Cyprus (1), an early producer of such products. With the passage of time, each individual region developed its characteristic sausage as evidenced in Italy by the well-known Genoa and the lesser known Milano, Sorrento, Lombardi (2), and Siciliano salamis (1). Sausage production involves three of the oldest forms of food preservation: salting, drying, and smoking.

The red or pink color typical of cured meat products, was first noted in Roman times (3). This color was subsequently found to be due to nitrate, occurring as an impurity in the salt, which is reduced by bacteria to nitrite; the nitrite then reacts with the meat pigment myoglobin to give the pink color, nitrosylmyoglobin. The nitrosylmyoglobin is subsequently denatured to give the stable form, nitrosohemochrome.

It was only in the early part of the 20th century that various bacteria (with their enzymes) were discovered to be the agents responsible for two changes that occur during production of dry fermented sausage: lactic acid production and nitrate reduction. Traditionally, the addition of the two bacterial types, *i.e.*, lactic acid bacteria and micrococci (nitrate reducing), was left to chance. Much of the natural flora was contributed by the processor when leftover material from a previous batch was added ("back-slopping") or by reuse of equipment after limited cleaning. The meat was handled in some

---

[1]Agricultural Research Service, U.S. Department of Agriculture.

fashion to permit the development of its own starter culture. This often involved low temperature holding of meat, plus salt and cure in the form of nitrate, in shallow pans to permit the growth of nitrate reducing micrococci with subsequent formation of nitrite to effect curing. After the pan cure, sugars and spices were mixed in and the meat mixture was ground and stuffed into casings. The stuffed sausages were then held in the "green room" where lactic acid bacteria fermented the sugars and some drying occurred. The sausages were then moved into the dry room where further dehydration took place. Much of the chance has been eliminated in modern sausage manufacture by two processing advances: the use of nitrite alone in cures, and the addition of lactic acid starter cultures containing lactobacilli and/or pediococci.

The original development of starter cultures for use in dry fermented sausages followed separate routes in the U.S. and Europe (4). American sausage makers utilized nitrite cures and thus needed to add only lactic acid bacteria. Originally the culture, *Pediococcus cerevisiae*, was lyophilized (5); now it is available as a frozen concentrate (6). European workers favored cultures that would reduce nitrate such as a strain of *Micrococcus*, M 53, available as "Baktofermente" (Rudolf Müller Co.[2], Giessen, Germany). The Europeans changed their views when they observed that the addition of lactobacilli along with the micrococci gave better color development and a somewhat more acid product (7). This mixed culture is available as "Duplofermente," also from R. Müller.

The processing of dry fermented sausages occurs in three principal steps: formulation, fermentation, and drying or ripening. Palumbo *et al.* (8) described a process for the preparation of pepperoni as a fully dry sausage. During formulation of pepperoni, salt, glucose, spices, and cure (nitrate or nitrite) are mixed with the meats and stuffed into casings. An aging period (before addition of other ingredients) of salted meat is included to encourage the development of micrococci and lactobacilli (8, 9). The fermentation period at 35°C and 90% RH is one to three days, depending on the desired pH decrease. After fermentation, the sausages are dried at 12°C and 55-60% RH for six weeks. Major microbiological activity occurs during the second and third steps. In the second step, micrococci reduce nitrate to nitrite within the first 24 hr. of fermentation (9, 10,) with concomitant formation of nitrosylmyoglobin; lactobacilli convert glucose to lactic acid somewhat more slowly than nitrate is reduced, often requiring three days to lower the pH to 4.6-4.7 from a starting value of 5.6 (8).

---

[2]Reference to brand or firm name does not constitute endorsement by the U.S. Department of Agriculture over others of a similar nature not mentioned.

During the third step, drying (or ripening as it is known in the European literature), the lipids and proteins of the sausage are attacked by various bacteria and their enzymes. The sausages lose 20-40% of their starting (green) weight (11). In the older production of sausage, fermentation usually occurred in the green room, generally held at a somewhat higher temperature than the dry room, and some dehydration occurred there. The sausage was then moved to the dry room where further drying occurred, along with desired flavor and textural changes. In the United States, drying room times and temperatures for sausages of a given diameter are specified by the requirements for trichinae inactivation (12).

Although there are almost as many sausage types as there are sausage makers and geographic areas, some classification of their types and characteristics is possible. Fermented sausages are usually described either as semi-dry (loss of 8-15% of starting weight) and fully dry, or simply dry, (loss of 25 to 40% of starting weight). Examples of semi-dry sausages include thuringer, cervelat, Lebanon bologna, and pork roll; examples of dry sausage include summer sausage, pepperoni, and various salamis (11). Some varieties, however, such as pepperoni, can be prepared as both dry and semi-dry forms. As a group, the Italian sausages are quite spicy, not smoked, and have moisture contents of 35-40%, while the northern European varieties are smoked and have somewhat higher moisture contents (13). Dry sausages can also be described by the maximum moisture/protein ratio (M/P) expected for individual varieties (14): salami, 1.9/1; pepperoni, 1.6/1; and Genoa, 2.3/1. (A $\overline{M/P} < 1.6/1$ generally indicates a shelf stable product.) Today, the identity of many dry fermented sausage varieties is blurred and processors are marketing products with two or three varietal names. A sausage labelled "thuringer-cervalat-summer sausage" was observed recently in a local market.

Most of the enzymes that produce the changes during fermentation and ripening of sausages are considered to be associated with bacterial cells, located either within the cells, as in the case of glycolytic enzymes, or outside the cells, as in the case of lipases and proteases. Because of the nature of meat, it is not possible to "pasteurize" or "sterilize" it before sausage manufacture. Thus, it is not possible to do for sausage ripening what Reiter *et al.* (15) did for cheddar cheese ripening; they established that, because the milk was pasteurized, the added bacterial starter culture, with its enzymes, was responsible for the observed changes such as proteolysis, lipolysis, and flavor development that occur during ripening of the cheese.

Alteration of Sausage Components by Bacteria and Their Enzymes

Fermentation. There are two major changes produced in sausages by fermentation: sugars are converted into lactic acid and nitrate is reduced to nitrite.

Lactic acid production. The major activity of the lactic acid bacteria is the conversion of sugars, usually glucose, to lactic acid by the homolactic Embdem-Myerhoff pathway. In their investigation of the stoichiometry of carbohydrate fermentation in Belgian salami, DeKetelaere et al. (16) found that lactic acid was the major acid formed, with minor amounts of acetic, generally in the molar ratio of 10:1 lactic to acetic. They also detected trace amounts of butyric and propionic. Our data for pH decrease and acid production during fermentation of pork-beef pepperoni are shown in Fig. 1. (There is only a very slight pH change in pepperoni during the drying step (8).) Similar data were obtained for Lebanon bologna during fermentation.

The extent to which producers of commercial dry fermented sausage achieve a low pH (acid) product is illustrated in Table I. Some varieties such as Lebanon bologna and summer sausage have low pH values, while others such as pepperoni and Isterband have both higher pH values and a wide range of pH values. Often European dry sausages have pH values indicating virtually no acid production. Their formulations usually have little or no glucose added. For example, Niinivaara et al. (4) added only three grams glucose to 100 kg meat and obtained a pH of 5.3, while Mihalyi and Kormendy (17) added no glucose to their formulation of Hungarian dry sausage and observed a pH of their finished sausage of 6.28.

Table I

pH Values of Some Commercial Fermented/Dry Sausages

Sausage/Type	pH Range	Processors	Reference
Belgian salami	4.48 - 5.10	10	16
Pepperoni	4.7 - 6.1	9	8
Isterband	4.6 - 5.7	10	32
Lebanon bologna	4.6 - 4.9	6	9
Summer sausage	4.6 - 5.0	12	33
Salami	5.1 - 5.5	8	10
Thuringer	4.9 - 5.1	4	10

Nitrate reduction. Many studies have been performed with micrococci, the bacteria that carry out the reduction. Micrococci developed during cold (5°C) aging of salted pork and beef for pepperoni (8) (Fig. 2) and beef for Lebanon bologna (18) and declined during pepperoni fermentation (Fig. 3). A further decline of micrococci occurred during the drying of pepperoni. About midway through the drying period, when the titratable acidity reached 1%, viable micrococci were no longer detected (8). The acid content of the pepperoni was concentrated as the sausages dehydrated in the dry room, thus destroying the acid sensitive micrococci. However, micrococci are usually isolated from less acid pepperoni (8). DeKetelaere et al. (16) stated that micrococci were present through the ripening period of Belgian salami, even though its acid content was above 1%. They utilized a salt-containing medium (S 110 agar), as we did in our study of pepperoni during drying. However, we also employed the gram stain to examine representative colonies on the medium and thus observed that while the count on the medium remained constant, micrococci colonies were replaced by bacilli colonies. It is important for investigators to verify the colony types on the selective agars and not rely solely on the supposed specificity of a given medium.

The reduction of nitrate occurs within the first 24 hr of the fermentation period, usually during the first 2 to 16 hr (10). Zaika et al. (19) observed that formation of most of the cured meat pigment (nitrosylmyoglobin) occurred within the first 24 hr also. They indicated that the micrococci of the natural flora are capable of reducing various levels of nitrate with satisfactory nitrosylmyoglobin formation (Fig. 4). With the lowest level of nitrate tested, 100 ppm, pigment conversion was somewhat slow; levels of 200 up to 1600 ppm nitrate gave equally satisfactory pigment conversion.

The need for nitrate reducing flora in fermented sausage production has been superseded by the use of nitrite. Cures containing only nitrite are adequate for the normal processing of most sausages; however, many sausage makers feel that dry fermented sausages must be cured with nitrate. The occasional lack of natural flora of micrococci that possess desired levels of nitrate reductase was overcome by Niinivaara and coworkers (4) who isolated strains of micrococci having intense nitrate reductase activity, along with other traits desirable for the production of dry fermented sausages.

Bacteria or, more specifically, the end products of their metabolism, can produce two color defects in fermented sausages: "nitrite burn" and "greening." Reliance on microbial reduction of nitrate to yield the nitrite necessary for the curing reactions can lead to instances in which various micrococci and non-pathogenic staphylococci reduce excessive amounts of nitrate (20). The resulting excess nitrite reacts with the meat pigments to give a greenish discoloration. This defect usually occurs on

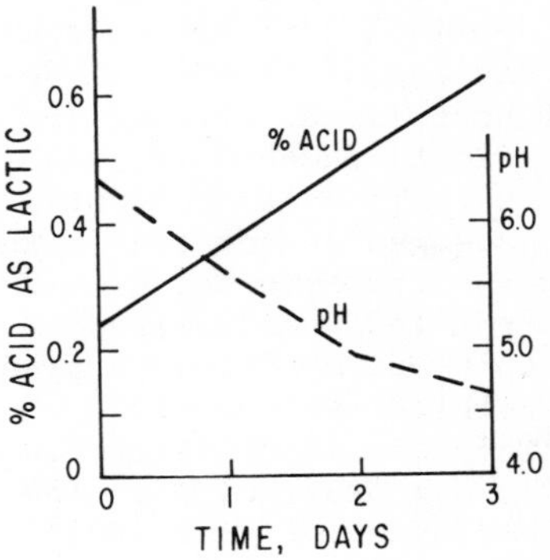

Journal of Food Science

Figure 1. Changes in pH and % acid (as lactic) during the fermentation of a pork-beef pepperoni at 35°C (8)

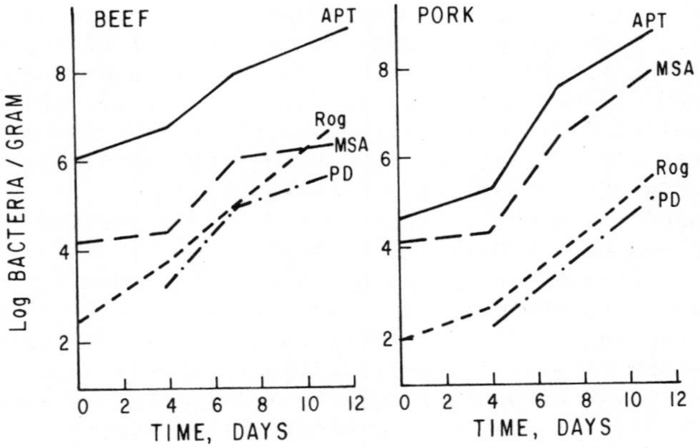

Journal of Food Science

Figure 2. Changes in microflora plated on various media during aging of salted (3% NaCl) beef and pork at 5°C. Media designations: APT *(for total count);* Rog *(Rogosa SL agar for lactic acid bacteria);* MSA *(Mannitol salt agar for micrococci);* PD *(Potato Dextrose agar for yeast) (8)*

the product surface where the aerobic bacteria proliferate. Another surface color defect which also yields a green pigment is caused by Lactobacillus viridescens (21) growing on the surface of cured meat products, especially emulsion and low acid sausages. The organism produces excessive amounts of H_2O_2 (22) which reacts with nitrosylhemochrome to give a green pigment.

Drying (Ripening). Deibel (13) stated that "Essentially, the microbiology associated with the (fermented sausage) product is restricted to the green room" (the fermentation). Palumbo et al. (8) reported that the total count and the number of lactobacilli remained essentially constant during the drying period of pepperoni, while micrococci declined to undetectable levels. DeKetelaere et al. (16) reported that the number of lactobacilli and micrococci remained constant during ripening of Belgian salami. Palumbo et al. (8), Smith and Palumbo (18), and Deibel et al. (10) reported that various commercial dry fermented sausages contained large numbers of viable bacteria. Thus, dry fermented sausage do contain, after processing, large numbers of viable bacteria. Because of the acidity and low water activity of these products, the bacteria cannot multiply.

However, microbial caused changes do occur during the drying or ripening stage. Recent studies on dry fermented sausages have indicated that bacteria and their enzymes can attack various sausage components during the ripening period, so that while the different bacterial populations appear to be stable, important changes are occurring in the lipid and protein components of the sausage. The end products of microbial action on lipids and proteins include compounds that give each individual sausage type its unique flavor and textural characteristics. Recent studies, especially by various European researchers, have yielded some interesting findings on the fate of the lipid and protein components of dry fermented sausages.

Lipolysis. Of the two components present in dry sausages, fat is higher in concentration than protein. In products such as pepperoni fat can comprise more than 50% of the final product weight (8). The fat in many sausage formulations is pork fat.

As indicated above, micrococci are important in reducing nitrate in dry fermented sausages. Work at our laboratory (23) and Cantoni's (24) indicated that micrococci can degrade fresh lard. Smith and Alford (23) found that, of the various microbial cultures surveyed, Micrococcus freudenreichii was the most active in attacking components of fresh lard. This organism substantially increased the peroxide value, and the 2-enals and the 2,4-dienals of fresh lard; further, though not as active as other cultures tested, M. freudenrichii also increased the alkanal content of fresh lard. Thus, M. freudenrichii markedly increased the carbonyl content of the fresh lard, probably by cleaving the unsaturated fatty acids at the site of the double bonds.

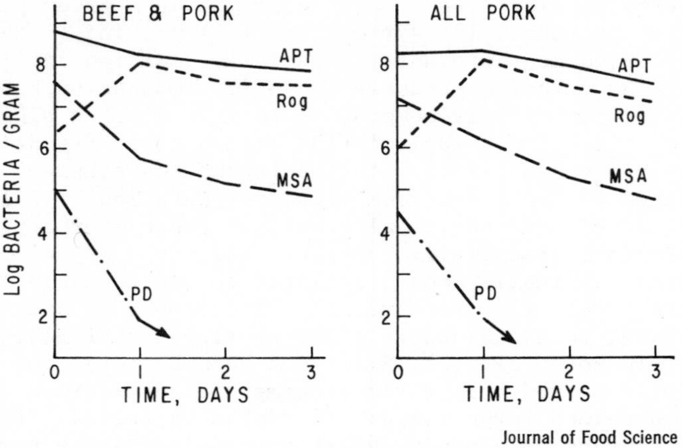

Figure 3. Changes in microflora plated on various media during the fermentation of a pork-beef and an all-pork pepperoni at 35°C (see Figure 2 for media designations) (8)

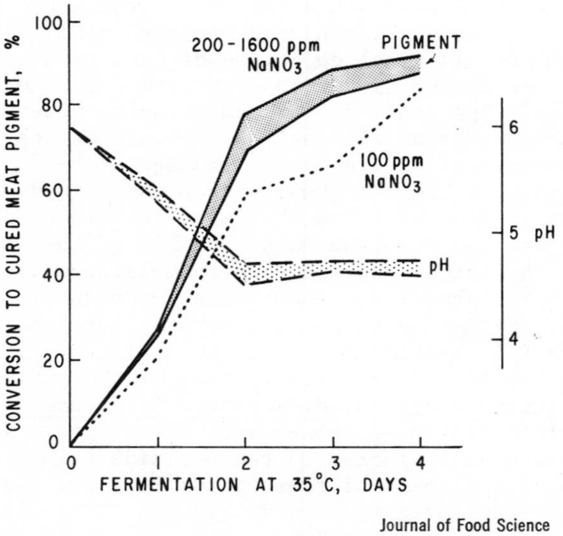

Figure 4. Formation of cured-meat pigment and changes in pH in Lebanon bolognas prepared with 100, 200, 400, 800, and 1600 ppm $NaNO_3$ during fermentation with natural flora (19)

Cantoni et al. (24), studying the action of micrococci on lard (Table II), indicated that micrococci are very active in degrading lard. The organisms produced lipases which cleave both long and short chain fatty acids from the triglycerides. Micrococci can degrade long chain unsaturated fatty acids as indicated by the formation of carbonyls.

The presence of micrococci in finished sausages such as pepperoni (8), as well as others (16, 18), and the work of Smith and Alford (23) and Cantoni et al. (24) on the action of pure cultures of micrococci on lard suggest that the lipids of various sausages should be actively attacked during ripening. Data on various European sausage type support this contention. The most complete study appears to be the work of Demeyer et al. (25) on Belgium salami; other workers have also studied this attack (4, 17, 26).

Table II

Action of Micrococci on Pork Fat (from Reference 24)

	Culture	
	D10	C13
$C_4 - C_{20}$, g FA/100 g fat[a]	23.1	45.2
(FA most easily released: oleic, myristic, palmitoleic, linoleic)		
Volatile FA, mg/100 g fat[b]	87	50
(Principal VFA: propionic, acetic)		
Carbonyls, μM/1 of culture[c]	660	667
(Prinicpal carbonyls: propionaldehyde, isovaleraldehyde)		

[a]After 28 days of culture; FA - fatty acid.

[b]After 28 days of culture; VFA - volatile fatty acid.

[c]After 24 days of culture.

Demeyer *et al*. (25) observed increases in the free fatty acid (FFA) and carbonyls of Belgian salami during ripening. Triglycerides were partially degraded to FFA (Fig. 5) and the unsaturated FFA's were degraded into carbonyls. The triglyceride content decreased with a corresponding increase of FFA's and diglycerides (DG); though based on limited hydrolysis, these data suggest that the enzymes of the microflora removed only one fatty acid from the triglyceride molecule. In a rate study of limited duration, they quantitated and identified the FFA's released (Fig. 6). Though the actual changes were relatively small, the rate of lipolysis appeared to decease in the order: linoleic (18:2) > oleic (18:1) > stearic (18:0) > palmitic (16:0). The FFA distribution (Fig. 6) suggests that the micrococcal lipase has specificity for triglycerides containing linoleic acid.

Demeyer *et al*. (25) then went on to quantitate the total carbonyl content of fat of the sausage during ripening. Using two methods (benzidine derivatives and D.N.P. hydrozones), they observed a several-fold increase in total carbonyls during ripening. They did not record changes in individual carbonyls, as did Halvorson (26) in his study of Isterband, a Swedish fermented sausage. Halvorson observed significant changes in n-hexanals, n-octanals, and some of the 2-alkenals during ripening; these compounds can contribute to the aroma of the finished sausages. He also detected increases in the level of several short chain branched carbonyls which could arise from amino acids by the Strecker degradation.

Though most lipase activity is associated with micrococci, it has also been reported in lactobacilli (Fryer *et al*. (27); Oterholm *et* al. (28)). These investigators used strains of lactobacilli of dairy origin and observed very active degradation of tributyrin, with relatively little activity against other triglycerides. However, the activity of meat lactobacilli against pork fat (lard) and other fats present in dry sausages needs further study.

Proteolysis. After the lipids, the most plentiful component of dry fermented sausages is protein. Less work has been done on the changes in the nitrogenous compounds released during sausage ripening, although progress has been made. Mihalyi and Kormendy (17) reported a 7% decrease in total protein and a 36% increase in non-protein nitrogen (NPN) occurring from the 10th to the 100th day of ripening of Hungarian dry sausage. It was not possible to determine whether the proteolytic enzymes were of meat or microbial origin. They also reported changes in the sacroplasmic and myofibrillar fractions, but since the methods utilized were developed for fresh (not cured or dried) meat, the significance of these changes cannot be properly accessed.

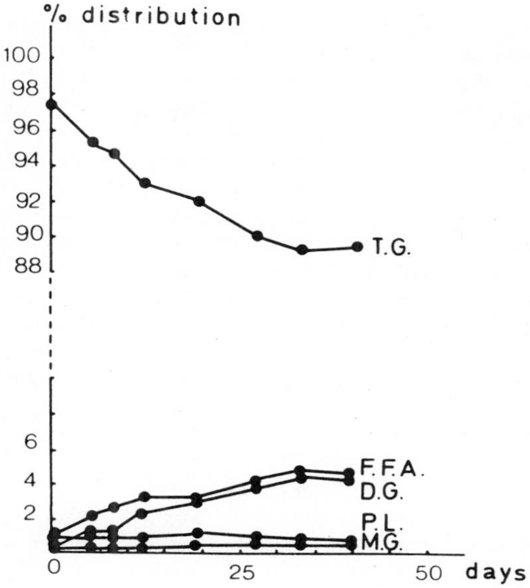

Journal of Food Science

Figure 5. Changes in fatty acid distribution over lipid classes (%) during ripening of Belgian salami: M.G. *(monoglycerides),* D.G. *(diglycerides),* T.G. *(triglycerides),* F.F.A. *(free fatty acids), and* P.L. *(polar lipids) (25)*

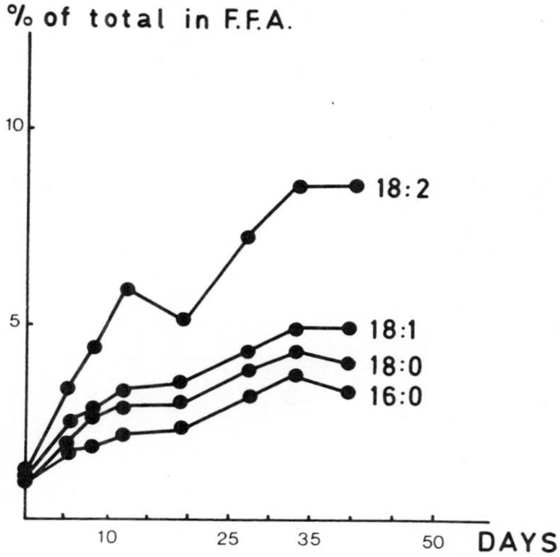

Journal of Food Science

Figure 6. Percent of total palmitic (16:0), stearic (18:0), oleic (18:1), and linoleic (18:2) acid present in F.F.A. *(25)*

Dierick et al. (29) studied the changes in the NPN fractions during the ripening of Belgian salami (Table III) and found an increase of free ammonia from deamination of amino acids and an increase of free α-amino nitrogen (amino acids) from the breakdown of proteins. Peptide-bound nitrogen decreased slightly, with a decrease in nucleotide nitrogen and an increase in nucleoside nitrogen.

Relative changes in the concentration of free amino acids (Table IV), for the sake of convenience, are presented in three groups: those amino acids showing a large increase during ripening, those acids showing a small increase, and those showing a decrease. Of unusual interest is the decrease of glutamic acid that was added to the sausage mix in the form of MSG (monosodium **glutamate**). Even though it was added for flavor, a considerable part of the glutamate was degraded by the time the sausage was fully ripened and ready for consumption.

Table III

Concentration of Nonprotein Nitrogen Compounds during Ripening of Belgian Salami (from Reference 29)

	Days of Ripening		
	0	15	36
NH_3	24*	58	76
Free α-NH_2-N	141	234	255
Peptide bound α-NH_2-N	161	152	145
Nucleot.-N	34	13	12
Nucleos.-N	33	78	83

*mg N/100 g of dry matter.

Table IV

Changes in Free Amino Acids During 36 Days of

Ripening of Belgian Salami (from Reference 29)

Large Increase, >12X	Small Increase, <12X	Decrease
Thre	Asp, Ala, Ileu	Glu
Pro	Ser, Val, Phen	His
Leu	Gly, Met, Lys	Tyr
γ-NH_2-BA		Orn

Dierick et al. (29) observed a decrease in histidine (his), tyrosine (tyr), and ornithine (orn) concentrations and corresponding increases in the concentrations of histamine, tyramine, and putrescine, the decarboxylation products of his, tyr, and orn. Rice and Koehler (30) attempted to identify the organism(s) that possess tyr and his decarboxylase activities. They investigated the lactic acid starter culture organisms, including strains of lactobacilli, streptococci, and Pediococcus cerevisiae. Only the streptococci showed tyr decarboxylase activity; however, streptococci are not usually starter cultures for fermented sausages and the significance of this finding cannot be adequately evaluated. They did not test other organisms such as micrococci that are present in many fermented sausages in large numbers and could conceivably carry out this metabolic activity.

Despite their inability to locate the source of decarboxylase activity, Rice et al. (31) surveyed various dry fermented sausages for their histamine and tyramine content (Table V). The histamine content of most sausages was low. Emulsion-type products such as bologna weiners were low, while braunschweiger, because of its liver content, was slightly higher than ground beef. In contrast, tyramine was found in larger quantities in a variety of dry fermented sausages. Those sausages containing the higher concentrations could provide sufficient tyramine in moderate servings to produce pressor responses in tyramine susceptible individuals.

Table V

Amine Content of Various Sausages

(from Reference 31)

Histamine	Avg.	Tyramine	Avg.
Dry sausage	2.87*	Hard salami	210*
Semidry sausage	3.59	Pepperoni	39
Bologna	1.89	Summer sausage	184
Weiners	1.75	Farmers sausage	314
Braunschweiger	3.60	Genoa salami	534
Ground beef	2.70	Smoked land-Jaeger	396
		Lebanon bologna	224

*μg/g of sausage.

Virtually all the changes during sausage fermentation and ripening are enzyme-catalyzed and most of the enzymes are of bacterial origin. The various types of changes reported here have not all been studied in a single sausage type; however, where overlapping studies exist, their results are similar. In summary, bacteria and their enzymes: ferment sugars to lactic acid; reduce nitrate to nitrite; degrade triglycerides to fatty acids and cleave unsaturated fatty acids and amino acids to form various carbonyls which can contribute to flavor. Proteins are degraded to various simpler nitrogen-containing compounds such as amino acids, ammonia, and amines by both meat and microbial enzymes.

Literature Cited

1. Pederson, C. S., "Microbiology of Food Fermentations," 153-172, The AVI Publishing Company, Inc., Westport, Connecticut, 1971.
2. "Foreign Style Sausage," Sausage and Meat Speciatlities, The Packer's Encyclopedia, part 3, 168-172, The National Provisioner: Chicago, Illinois, 1938.
3. Federal Register (1975) 40(218), 52614-52616.

4. Niinivaara, F. P., Pohja, M.S., and Komulainen, Saima E., Food Tech. (1964) 18(2), 25-31.
5. Deibel R. H., Wilson, G. D., and Niven C. F., Jr. Appl. Microbiol. (1961) 9, 239-243.
6. Everson, C. W., Danner, W. E., and Hammes, P. A., Food Technol. (1970) 24, 42, 44.
7. Nurmi, E., Acta Agralia Fennica, (1966) 108, 1-77.
8. Palumbo, S. A., Zaika, L. L., Kissinger, J. C., and Smith J. L., J. Food Sci. (1976) 41, 12-17.
9. Palumbo, S. A., Smith, J. L., and Ackerman, S. A., J. Milk & Food Tech. (1973) 36(10), 497-503.
10. Deibel, R. H., Niven, C. F., Jr., and Wilson, G. D., Appl. Microbiol. (1961) 9, 156-161.
11. MacKenzie, D. S., "Prepared Meat Product Manufacturing," 62-64, American Meat Institute Center for Continuing Education, 1966.
12. Anonymous, Meat and Poultry Inspection Regulations, Meat and Poultry Inspection Program, Animal and Plant Health Inspection Service, U.S. Department of Agriculture, Washington, D.C., 1973.
13. Deibel, R. H., "Technology of Fermented, Semi-Dried and Dried Sausages," Proceedings of the Meat Industry Research Conference, Chicago, Illinois, March 21-22, 1974; 57-60, American Meat Institute Foundation, Arlington, Virginia, 1974.
14. Anonymous, Chemistry Laboratory Guidebook, Consumer and Marketing Service, Laboratory Services Division of U.S. Department of Agriculture, Washington, D.C., 1971.
15. Reiter, B., Fryer, T. F., and Sharpe, M. E., J. Appl. Bact. (1966) 29(2), 231-243.
16. DeKetelaere, A., Demeyer, D., Vandekerckhove, P., and Vervaeke, I., J. Food Sci. (1974) 39, 297-300.
17. Mihalyi, V. and Kormendy, L., Food Tech. (1967) 21, 1398-1402.
18. Smith, J. L. and Palumbo, S. A., Appl. Microbiol. (1973) 26(4), 489-496.
19. Zaika, L. L., Zell, T. E., Smith, J. L., Palumbo, S. H., and Kissinger, J. C., J. Food Sci. (1976) 41, 1457-1460.
20. Bacus, J. N. and Deibel, R. H., Appl. Microbiol. (1972) 24(3), 405-408.
21. Niven, C. F., Jr. and Evans, J. B., J. Bact. (1957) 73, 758-759.
22. Niven, C. F., Jr., Castellani, A. G., and Allanson, V., J. Bact. (1949) 58(5), 633-641.
23. Smith, J. L. and Alford, J. A., J. Food Sci. (1969) 34(1), 75-78.
24. Cantoni, C., Molnar, M. R., Renon, P., and Giolitti, G., J. Appl. Bact. (1967) 30(1), 190-196.
25. Demeyer, D., Hoozee, J., and Mesdom, J., J. Food Sci. (1974) 39, 293-296.

26. Halvarson, H., J. Food Sci. (1973) 38, 310-312.
27. Fryer, T. F., Reiter, B., and Lawrence, R. C., J. Dairy Sci. (1967) 50, 388-389.
28. Oterholm, A., Ordal, Z. J., and Witter, L. D., Appl. Microbiol. (1968) 16, 524-527.
29. Dierick, N., Vandekerckhove, P., and Demeyer, D., J. Food Sci. (1974) 39, 301-304.
30. Rice, S. L. and Koehler, P. E., J. Milk Food Tech. (1976) 39(3), 166-169.
31. Rice, S., Eitenmiller, R. R., and Koehler, P. E., J. Milk Food Tech. (1975) 38(4), 256-258.
32. Ostlund, K. and Regner, B., Nord. Vet.-Med. (1968) 20(10), 527-542.
33. Goodfellow, S. J. and Brown, W. L., Food Product Develop. (1975) 9(6), 80 and 82.

17

Enzymic Hydrolysis of Ox-Blood Hemoglobin

K. J. STACHOWICZ

Institute of Fermentation Industry, Rakowiecka 36, Warsaw, Poland

C. E. ERIKSSON and S. TJELLE

SIK, The Swedish Food Institute, Fack, S-40021 Göteborg, Sweden

In the search for novel, cheap proteins to be used in food either as substitutes for meat, for fortification purposes or as functional additives, the blood of slaughtered animals has been surprisingly little considered. Even though some whole blood is used as an ingredient in a few types of foods in a few areas in the world, wider use of blood is hampered primarily for religious and ethical reasons and by its characteristic flavor and color. In addition to this, attempts to use blood in certain foods as a combined protein and iron fortifier and coloring agent have failed due to the enhancing of lipid oxidiation and off-flavor development catalyzed by the iron-porphyrin group present in the hemoglobin (1). This kind of catalytic activity may also be increased by denaturation of hemoproteins e.g. by heat (2). The total amount of blood represents a great deal of protein, which to day is used to only a limited extent as an animal feed. Often, however, it gives rise to serious pollution of water recipients and should therefore be subject to intense studies concerning its processing into various protein products for food use.

In Sweden most blood is collected under strict hygienic conditions in immediate connection with the slaughter of cattle and pigs. This blood is then stabilized by sodium citrate, transported cold to central plants where the plasma is separated from the red cells. Plasma, either liquid, concentrated, frozen, or dried is used in the manufacture of meat products, e.g. sausage mainly due to its good emulsification and binding properties. The red cell fraction is dried and mostly used in animal feeding, a small part in food. In Sweden about 2000 tons of blood protein mainly hemoglobin, is available annually in the red cell fraction. This quantity represents, however, less than 1% of the Swedish protein consumption to day. Since protein is overconsumed in our country, production of blood protein for human nutrition is not relevant. Hence, utilization of blood proteins should be primarily based more on their possible technical functions in food like their binding properties, their use as flavor precursors, consistency regulators, meat extenders, etc. It is thus believed that access to a wide range of blood protein products with different properties should be advantageous. The oldest and still most widely employed approach is the preparation of a decolored protein product from the red cell fraction by removal of the heme group from hemoglobin through the acid-acetone method (3, 4) or modifications of it. In this process apohemoglobin can be precipitated while the hematin remains in the mixed acetone-acid-water phase.

$$\text{Hemoglobin} \xrightarrow{\text{Acid acetone}} \text{Apohemoglobin(s)} + \text{hematin}$$

A great deal of work has been carried out to optimize this process according to protein yield, product color, functional properties, solvent recycling, explosion safety, etc. In Australia the cost of a full scale plant for production of a decolored protein product from whole blood has been estimated, on the basis of extensive pilot plant investigations (5).

In our laboratory another method has been applied for obtaining protein products from the red cell fraction, including a decolored product. Preliminary work revealed that enzymic hydrolysis of hemoglobin followed by gel filtration or ultrafiltration yielded hydrolysate fractions which contained only small amounts of heme. These investigations will be publised later. However, some large-scale work was also carried out in order to produce single, one-kilogram batches of ox blood protein hydrolysate for experimental food use. This work will be presented in what follows.

Pretreatment of red cells.

Most proteolytic enzymes attack denatured proteins better than native ones due to the inherent steric unavailibility of both the enzyme and the native substrate. The red cell fraction obtained after plasma separation forms a viscous liquid containing 35-40% of dry matter mainly protein and some inorganic constituents. The protein within the cells is released by hemolyzing them, simply by diluting them with water. Denaturation of the dissolved protein was accomplished by alkali treatment, after a pre-investigation performed in order to compare denaturation by alkali treatment and by previously reported denaturation methods involving combined ethyl alcohol and heat treatment applied to soy protein (6, 7). The denaturation effect was estimated by measurement of the hydrolysis rate of the treated protein in the presence of a proteolytic enzyme.

Alkali treatment was performed by adding 5 N NaOH to a stirred sample of hemolyzed red cell fraction fram ox blood until pH 11 was reached. This pH was retained up to 24 h at 20°C and samples were withdrawn after 0.5, 1, 2, and 24 h for kinetic measurements. The samples were first adjusted to pH 9 by the addition of 1 N HCl and then centrifuged. The supernatant was further diluted with water to a hemoglobin content of about 4%. The dilution factor was always 9.

The enzyme activity measurements were performed in a pH-stat (Mettler system: DK 10 + DK 11 + DV 11, Switzerland). Twenty ml of substrate containing about 4% of hemoglobin, was first adjusted to pH 9 in the titration vessel of the pH-stat. Then 5 mg of alcalase ® (NOVO, Copenhagen, Denmark) in 1 ml of water (pre-adjusted to pH 9) was added, the μmol OH^- consumed being continuously recorded during the first 10 min. The titration was carried out with 0.01 N NaOH at 50°C and pH 9. By this method no difference was found between alkali treatment at pH 11 for one to two h and the combined ethyl alcohol-heat treatment (Table I). Hence, owing to its superior simplicity and reproducibility, alkali treatment was preferred in the subsequent large scale operations.

Large-scale operations.

Alkali denatured, filtrated ox blood protein (45-46.5 l, 3.3-3.9% in water) from hemolyzed red cells was hydrolyzed at pH 9 and 50°C by means of the food grade, proteolytic enzyme alcalase ® (NOVO, Copenhagen, Denmark) in a membrane reactor principally designed as shown in Figure 1. The enzymic reaction was carried out both during a certain time in the temperature-controlled feed tank before starting ultrafiltration and during the volume reduction period that

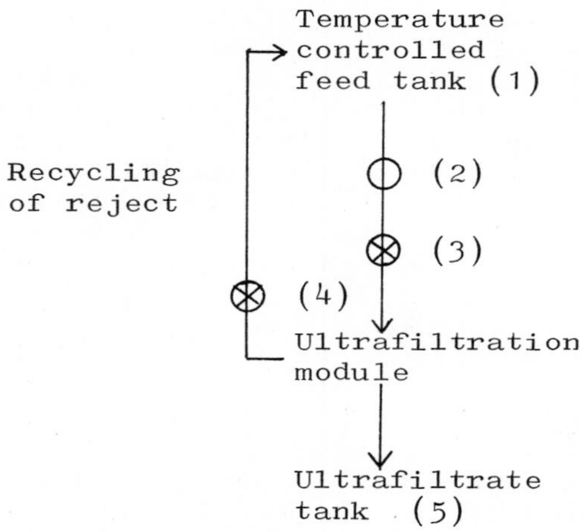

Figure 1. Principal procedure for the enzymic hydrolysis and ultrafiltration of protein from ox-blood red cells: (1) enzyme, substrate (2) circulating pump; (3) inlet pressure regulator and gauge; (4) outlet pressure regulator and gauge; and (5) protein hydrolysate

occurred when the ultrafiltration was going on. The pH drop in the reaction medium was compensated for by frequent addition of measured volumes of 5 N NaOH.

Table I. Effect of pre-treatment on the rate of enzymic hydrolysis of blood protein.

Pre-treatment	Hydrolysis rate µmol OH^-/10 min
Control, untreated	22
pH 11 for 0.5 h	43
1 h	48
2 h	48
24 h	26
Ethyl alcohol 1 h at 70°C	
Freeze-dried	46
Air-dried	43
pH 11, 44 h, freeze-dried	40

Enzyme: Alcalase 5 mg per 20 ml of 4% protein in water
Temperature: 50°C
pH: 9.0

The ultrafiltration device was of the hollow fiber type (Romicon Inc., USA). Three different ultrafiltration membranes, XM 50, PM 10, and PM 5 with nominal cut-off values of 50.000, 10.000, and 7.000, respectively, were used. It was expected that the different characteristics of the membranes and the time of reaction would affect the properties of the protein hydrolysates. The circulation flow and ultrafiltrate flow were maintained by a pump. The inlet and outlet pressure of the ultrafiltration unit were kept constant and the circulation flow rate adjusted by means of two pressure regulators. (See Figure 1.)

The ultrafiltrate was collected in 5 l batches from which samples were taken for further analysis. The rejected material was continuously recycled from the ultrafiltration unit back to the feed tank. The total hydrolysis time ranged from 52 min to 2 h, depending on the flow capacity of the membrane unit used and the pre-hydrolysis time. The number of split peptide bonds, as measured by the total consumption of NaOH, was about the same in all three experiments. The starting material and the different fractions of the ultrafiltrate, collected at regular intervals during the hydrolysis, were subjected to dry weight and hemin determination. Hemin determination was based on the formation of a pyridine hemochromogen complex in strong alkali under reducing conditions (8). An average ultrafiltrate was produced by mixing equal volumes of the ultrafiltrate fractions. Both the starting material and the average ultrafiltrate were also analyzed on their ash content. The principal ultrafiltrate contents and running conditions are shown in Tables II - IV.

In all three runs a fairly large, final yield of hydrolysate dry matter was obtained; 83%, 73, and 72 in the XM 50, PM 10, and PM 5 runs respectively. Looking at the intermediate yields, one can see that both the dry matter and the hemin concentration increased with time. The XM 50 run showed only a small increase while in both the PM 10 and PM 5 runs the concentration in-

Table II. Hydrolysis and ultrafiltration of ox blood proteins with alcalase in an XM 50 membrane reactor. Composition of starting material and ultrafiltrate. Running conditions.

	Volume (l)	Total dry matter (g)	Total hemin (g)	Dry matter hemin (%)
Protein start	45	1661	48	2.9
Ultrafiltrate				
fraction No. 1	5	170	1.1	0.6
2	5	174	1.2	0.7
3	5	175	0.9	0.5
4	5	184	1.4	0.7
5	5	176	1.5	0.9
6	5	177	1.5	0.9
7	5	180	1.5	0.9
8	5	188	2.9	1.5
Sum, fractions	40	1424	12.0	0.8
Average filtrate	40	1416	12.2	0.9
Recovery (%)	89	85	25	

Membrane: Romicon XM 50. Cut off: 50.000. Area: 2.46 m^2
Pressure, inlet: 103 kPa (1.1 kp/cm^2, 15 lb/in^2).
Pressure, outlet: 69 kPz (0.7 kp/cm^2, 10 lb/in^2).
Flux: 89 l/m^2 x h. Enzyme: Alcalase, 10.1 g.
Temperature: 48.5 - 51 °C. Time: Pre-hydrolysis 107 min. ultrafiltration 13 min. Total NaOH consumption: 3.0 mol.

Table III. Hydrolysis and ultrafiltration of ox blood proteins with alcalase in a PM 10 membrane reactor. Composition of starting material and ultrafiltrate. Running conditions.

	Volume (l)	Total dry matter (g)	Total hemin (g)	Dry matter hemin (%)
Protein start	46,5	1548	38.8	2.5
Ultrafiltrate fraction No. 1	10.0	109	0.007	0.006
2	10.0	259	0.007	0.003
3	10.0	325	0.014	0.004
4	11.0	428	0.033	0.007
Sum, fractions	41.0	1121	0.061	0.005
Average filtrate	41.0	1121	0.117	0.010
Recovery (%)	88	72	0.3	

Membrane: Romicon PM 10. Cut off: 10.000. Area: 2.46 m^2 (26.5 ft^2). Pressure, inlet: 138 kPz (1.1 kp/cm^2, 20 lb/in^2). Pressure, outlet: 103 kPa (1.1 kp/cm^2, 15 lb/in^2). Flux: 12,7 l/m^2 x h. Enzyme: Alcalase, 22.5 g. Temperature: 49.5 - 50.5°C. Time: Pre-hydrolysis 21 min, ultrafiltration 64 min. Total NaOH consumption: 3.0 mol.

crement was much larger from the first to the last fraction. The hemin content of the ultrafiltrate dry matter was always significantly lower than that of the starting material. The average ultrafiltrate dry matter contained 0.9, 0.01, and 0.02% of hemin, thus a 3-, 250-, and 150-fold heme reduction in the XM 50, PM 10, and PM 5 runs respektively. It should be noted that the figures of heme content, particularly in the PM 10 and PM 5 experiments, are somewhat uncertain since it is just on the verge of detectability with the method used (8).

The three average ultrafiltrates were freeze-dried after having been neutralized by the addition of HCl, citric or lactic acid. The moisture and ash content were determined in the freeze-dried products. The results in Table V show that the dry matter contained 13-16% of ash. The added NaOH for pH-adjustment during the reaction and the acid used for neutralization before freeze-drying account for about two thirds of the

Table IV. Hydrolysis and ultrafiltration of ox blood proteins with alcalase in a PM 5 membrane reactor. Composition of starting material and ultrafiltrate. Running conditions.

	Volume (l)	Total dry matter (g)	Total hemin (g)	Dry matter hemin (%)
Protein start	45	1767	52.5	3.0
Ultrafiltrate				
fraction No. 1	5	73	0.016	0.02
2	5	97	0.011	0.01
3	5	120	0.012	0.01
4	5	143	0.018	0.01
5	5	168	0.025	0.01
6	5	195	0.037	0.02
7	5	226	0.052	0.02
8	5	263	0.081	0.03
Sum, fractions	40	1285	0.252	0.02
Average filtrate	40	1307	0.292	0.02
Recovery (%)	89	73	0.6	

Membrane: Romicon PM 5. Cut off: 7.000. Area: 2.18 m^2 (23.5 ft^2). Pressure, inlet: 138 kPa (1.4 kp/cm^2, 20 lb/in^2). Pressure, outlet: 103 kPa (1.1 kp/cm^2, 15 lb/in^2). Flux: 28 l/m^2 x h. Enzyme: Alcalase 22.5 g. Temperature: 50 - 51°C. Time: Pre-hydrolysis 7.5 min, ultrafiltration 44.5 min. Total NaOH consumption: 3.2 mol.

Table V. Some properties of hydrolyzed, ultrafiltrated, neutralized and freeze-dried protein hydrolysates from ox blood red cells. Neutralization was made either with hydrochloric acid or citric acid.

Average ultrafiltrate	Hemin (%)	Moisture (%)	Ash (%)	Color: Chloride	Color: Citrate
XM 50	0.9	5.8	12.9	dark brown	light brown
PM 10	0.01	4.9	16.3	beige	cream
PM 5	0.02	6.1	14.0	beige	cream

ash content, while the rest originates from the mineral content of the red cell fraction. The color of the dry product depended on the acid used for neutralization. Hydrochloric acid (and lactic acid) gave brown products, whilst citric acid gave cream-colored ones. Particularly the product treated with lactic acid was very hygroscopic. Products from the PM 10 and PM 5 runs were somewhat bitter, no significant difference in bitterness being found when early and late filtrate fractions were compared. Further work on the use of blood protein hydrolysates will include investigations on the possible control of the amount of inorganic constituents and bitterness as well as investigations on the functional properties of these products.

Abstract

This paper presents evidence that enzymic hydrolysis under slightly alcaline conditions can be used to prepare protein hydrolysates with a very low heme content from the protein mainly hemoglobin, in the red cells of cattle blood. The reactions were carried out in a hollow fiber membrane reactor, both the reaction and the properties of the product being affected by the choice of membrane. The appearance of the products could be further affected by the choice of acid for neutralization of the hydrolysate. The products obtained had a fairly high content of ash and were somewhat bitter.

Literature cited

(1) Tappel, A.L., in "Autioxidation and antioxidants" Lundberg, W.O., ed., Vol. I, p. 325, Interscience Publ., New York 1961.

(2) Eriksson, C.E., Olsson, P., and Svensson, S., J. Am. Oil Chem. Soc. (1971) 48, 442.

(3) Lewis, U.J., J. Biol. Chem. (1954) 206, 109.

(4) Anson, M.L., and Mirsky, A.E., J. Gen. Physiol. (1930) 13, 469.

(5) Wilson, B.W. in Process Industries of Australia - Impact and growth, National Chemical Engineering Conference, p. 431, Surfers Paradise, Queensland, 1974.

(6) Fukushima, D., Cereal Chem. (1969) 46, 156.

(7) Fukushima, D., Cereal Chem. (1969) 46, 405.

(8) Paul, K.G., Theorell, H., and Åkesson, Å., Acta Chem. Scand. (1953) 7, 1284.

18

Utility of Enzymes in Solubilization of Seed and Leaf Proteins

E. A. CHILDS,[a] J. L. FORTE,[b] and Y. KU

Department of Food Technology and Science, University of Tennessee, Knoxville, Tenn. 37901

The use of enzymes for solubilization of seed and leaf proteins has been studied as a means of overcoming difficulties presented by the varying condition of seed and leaf material available for processing. The variations are the result of differences in the original condition of the material and its post-harvest treatment.

Extraction of Protein from Oilseeds

Oilseed aleurin proteins are found in cytoplasmic vacuoles. Examples include the 7S and 11S globulins of soybeans (1), arachin from peanuts (2,3), edistin from hemp seed (4), vicilin and legumin from pea seeds (5), and vicilin and legumin from Vicia fava (6). Based on the similarity of amino acid sequences as measured by the Difference Index suggested by Metzger (7), Dieckert and Dieckert (8) have suggested the reserve proteins of leguminous seeds are quite homologous and are usually hexamers or tetramers of a disulfide bridged subunit with an A-S-S-B structure. These similarities in (a) cytoplasmic location and compartmentalization and (b) amino acid sequence of a variety of oilseed proteins offers a unifying view of the conditions faced in extracting proteins from oilseeds for production of protein concentrates or isolates.

Several techniques are available for production of seed protein concentrates and isolates. Protein isolates can be prepared from defatted seeds by extraction in dilute alkali, followed by isoelectric precipitation of

[a]Current Address: Dederich Corporation, P.O. Box 7, Hubertus, WI 53033

[b]Current Address: Biology Division, Oak Ridge National Laboratories, Oak Ridge, TN 37830

the protein at an acidic pH, and drying the precipitated material after washing. Protein concentrates may be produced from seeds by extraction of low molecular weight molecules at pH 4.0 with the protein remaining in the insoluble material for drying. Sugarman (9) developed methodology for producing protein concentrates from full fat seeds. An aqueous extraction is performed and the resulting emulsion broken by shear forces at an appropriate temperature and pH.

These methods are efficient if the seed has not been heated to an extent that will cause denaturation of the protein. If denaturation has occurred, extractable protein yields will drop to less than 20% in soybeans or cottonseed (10,11).

Because most oilseeds are also an oil source, they may undergo a hexane extraction and then be subjected to heat and moisture to aid in desolventization (13,14). In specific commodities such as soybeans, heat may also be utilized to maximize the nutritional quality of the protein (15). Thus the probability of utilizing seeds containing denatured proteins for production of protein isolates and concentrates seems likely.

In cottonseed, Martinez et al (12) have suggested the lowered solubility of protein from heated seeds is the result of cytoplasmic proteins "gluing" the cytoplasmic vacuoles containing the storage proteins together so they cannot disperse when the cell is ruptured during grinding. In order to increase the extraction of protein from heat denatured substrates, techniques which chemically or mechanically break the cytoplasmic proteins holding the aleurin vacuoles together must be developed. Should the insolubilization be the result of the binding of individual proteins or storage vacuoles to other cellular constituents (e.g. fiber polymers), these bonds would also have to be broken.

Extraction of Protein from Leaves

Leaf proteins occur throughout the cytoplasm of individual cells with 30-40% of the protein localized in the chloroplast (16). Extraction of these proteins has been accomplished by pulping the leaves followed by pressing to separate the proteinaceous juice from the leaf fiber. The juice is then heated to coagulate the protein and the coagulated protein and juice are separated by appropriate means (17). When these chloroplast proteins are extracted, large amounts of pigments are also freed from the cellular matrix to yield a green product.

The presence of the pigments has lowered leaf protein concentrate acceptability as a human food. This has led to development of processes which produce a fiber free protein-carotenoid complex for use as an animal feed (18,19,20,21,22). Recently, a number of solvent extraction and differential heating techniques have been developed to produce colorless or low-colored protein concentrates (23,24,25,26,27,28,29,30,31,32,33).

Production of any of the above products requires the availability of non-heated leaf products. Approximately 45% of the leaf protein can be extracted from fresh alfalfa leaves (33) while only 5-15% of available protein can be extracted from alfalfa dried at 75-150°C (34).

Approaches to Extraction of Heat Denatured Protein From Seeds and Leaves

Two experimental approaches have been evaluated in experiments to increase the yield of extracted protein from heat-denatured seeds and leafs: enzymatic digestion and ultrasonic dispersion.

Cellulases and proteases have been used to hydrolyze portions of the plant matrix to allow more efficient dispersion and extraction of proteins. Hang *et al* (35) showed pea bean protein extractability was increased by treatment with *Aspergillus niger* cellulase and Abdo and King (36) and Sreekantiah *et al*, (37) reported increased nitrogen extraction from soybeans, chickpeas and sesame following cellulase treatment. Conversely Lu and Kinsella (32) found a non-significant increase in alfalfa meal extractability following cellulase treatment. In our laboratories, cellulase has not increased protein extraction yields from alfalfa meal, screw expressed cottonseed meal, or hexane defatted soybean meal.

Hydrolysis of seed proteins has been intensively investigated. Arzu *et al* (38) described the hydrolysis of cottonseed cake proteins by 10 different proteolytic enzyme preparations of bacterial, fungal, plant, and animal origin. All enzymes hydrolyzed cottonseed protein. Proteolytic enzymes have also been shown to hydrolyze soybean (39) and sesame meal (37) proteins. In none of these studies was the efficiency of protein extraction from a food substrate measured.

The use of ultrasonic energy to increase solubilization of protein from heat denatured sources is a new development. Wang (10) increased the efficiency of protein extraction from autoclaved soybean flakes from 16% to 58% by application of ultrasonic waves. In addition, the ultracentrifuge patterns of the extracted proteins were the same of those of proteins extracted from non-autoclaved samples by conventional techniques.

Molina and La Chance (40) increased the extractability of protein from heat treated coconut meal by treatment with a variety of proteolytic enzymes. In those experiments, the meal was incubated with the protease and the residue extracted with dilute alkali. Yields of up to 80% were obtained.

In experiments with cottonseed meal and alfalfa meal (11,41), Molina and La Chanes's techniques were modified (Figure 1). Instead of shaking, the samples were stirred and separation of the solid substrate from the liquid was accomplished by gravity filtration through cotton organdy rather than by centrifugation.

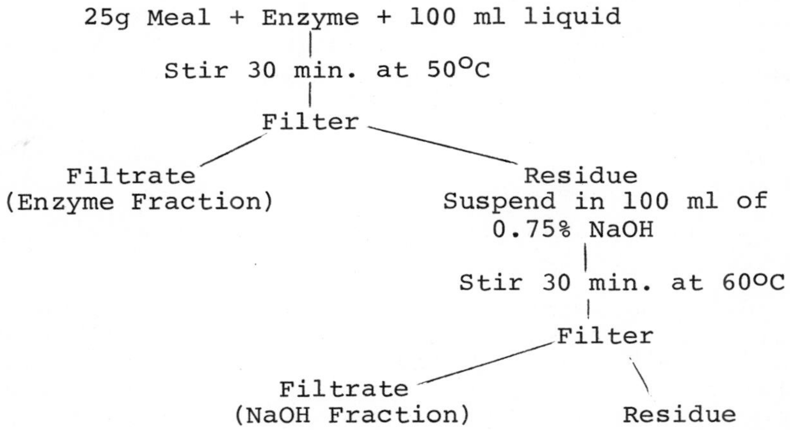

Figure 1. Flow sheet for proteolytic enzyme-chemical extraction of protein from heat-denatured substrates

Proteolytic enzyme pretreatment increased extractability of protein from heat denatured cottonseed meal (Table I). Without enzyme treatment, only 15% of the cottonseed protein was extracted by a two-step chemical extraction (11). Ficin treatment increased the amount of protein extraction approximately 2.5 times. Trypsin treatment was most effective, causing an approximate fivefold increase in extraction with greater than 65% of protein being extracted.

Table I. Efficiency of a two-step proteolytic enzyme-chemical technique for extraction of protein from cottonseed and alfalfa meal proteins. All data are the mean of three replicates ± standard deviation (From 11).

Enzyme	Total % Kjeldahl Protein Extracted	
	Cottonseed Meal	Alfalfa Meal
None	15.1 ± 2.3	21.9 ± 4.6
Papain	16.0 ± 2.4	23.6 ± 2.8
Ficin	42.7 ± 1.8	34.1 ± 11.3
Trypsin	69.4 ± 5.7	60.2 ± 5.7

Without enzyme treatment, only 21.9% of the alfalfa protein was extracted (Table I). As with cottonseed meal, treatment with papain did not markedly improve the efficiency of extraction (23.6%). Ficin treatment increased the amount of protein extracted almost twofold (34.1). Trypsin treatment was the most effective, causing an approximate threefold increase in extraction with greater than 60% of available protein being extracted.

Parameters Affecting the Enzymatic-Chemical Extraction Process

pH. Experiments have been performed in which the amount of protein extracted during the enzymatic hydrolysis step and the dilute alkali extraction steps were quantified separately as a function of the pH of the enzyme incubation. Trypsin treatment was more effective than ficin or bromelain at pH's from 4.0 - 9.5 (Table II). There was finite but little variation amongst the enzyme fractions. Trypsin treatment, however, resulted in a two to threefold increase in the protein content of the NaOH fraction relative to ficin or bromelain treatment. This would suggest that trypsin acts on the substrate in a manner which allows the protein to become more dispersable in dilute alkali.

At less than pH 6.5, bromelain was the most efficient in the enzyme extraction step in alfalfa meal (Table III). At pH's greater than 6.5, trypsin and bromelain were more effective than ficin. As with cottonseed, the majority of the protein was extracted in the dilute alkali step suggesting the enzymes are acting in a manner which allows the protein to become more dispersable in the alkali.

Table II

Grams of protein extracted from 25 g of cottonseed meal in a two step proteolytic enzyme-chemical extraction technique utilizing trypsin, bromelain, or ficin. Extractability is presented as a function of the pH of the enzyme reaction media.

	Trypsin		Ficin		Bromelain	
pH	Enz. Frac.	NaOH Frac.	Enz. Frac.	NaOH Frac.	Enz. Frac.	NaOH Frac.
3.0	1.20±0.2	0.50±0.2	0.75±0.3	0.50±0.2	0.80±0.2	1.40±0.3
4.0	1.80±0.4	2.05±0.1	1.20±0.8	0.70±0.2	1.30±0.6	0.70±0.2
5.0	2.45±0.5	2.30±0.4	1.10±0.6	0.80±0.2	1.30±0.2	1.10±0.3
6.0	2.40±0.4	2.60±0.6	1.15±0.7	0.40±0.1	1.60±1.2	1.30±0.2
7.0	2.50±0.6	2.55±0.4	1.30±0.6	0.60±0.1	1.80±0.9	1.20±0.2
8.0	2.48±0.9	3.40±0.4	1.30±0.7	0.60±0.2	1.80±0.9	1.10±0.1
9.0	2.51±0.8	3.40±0.2	1.40±0.6	0.70±0.2	2.10±0.7	1.30±0.3

Table III

Grams of protein extracted from 25 g of alfalfa meal in a two step proteolytic enzyme-chemical extraction technique utilizing trypsin, bromelain or ficin. Extractability is presented as a function of the pH of the enzyme reaction media.

pH	Trypsin Enz. Frac.	Trypsin NaOH Frac.	Ficin Enz. Frac.	Ficin NaOH Frac.	Bromelain Enz. Frac.	Bromelain NaOH Frac.
3.0	0.75±0.1	0.52±0.2	0.34±0.1	0.72±0.1	0.72±0.2	1.20±0.3
4.0	0.72±0.1	0.53±0.1	0.38±0.0	0.74±0.1	0.68±0.1	1.00±0.1
5.0	0.78±0.1	0.62±0.1	0.39±0.1	0.80±0.0	0.68±0.1	1.10±0.1
6.0	0.92±0.2	0.95±0.1	0.41±0.0	0.81±0.1	0.52±0.1	0.90±0.2
7.0	1.00±0.2	0.97±0.1	0.43±0.2	0.82±0.1	0.53±0.1	0.92±0.1
8.0	1.05±0.2	1.10±0.1	0.55±0.2	0.83±0.1	0.55±0.2	0.98±0.1
9.0	1.05±0.2	1.22±0.2	0.48±0.1	0.80±0.1	0.49±0.2	0.98±0.1

One possible hypothesis to explain the above data is that the proteins are bound to a polymer ligand after heat treatment and the proteolytic enzyme cleaves the two molecules. A candidate polymer might be dietary fiber. To test this hypothesis, a Neutral Detergent Fiber (NDF) extract of cottonseed meal was prepared (Van Soest, 1973). Cottonseed meal samples were also produced from the extraction residues. It was noted that the N content of the NDF samples from trypsin treated cottonseed meal was much lower than that from other samples (41). Therefore, trypsin was able to cleave protein-fiber bonds more efficiently than other enzymes. This strongly suggests, but not clearly prove, the superior performance of trypsin may be based on its ability to cleave fiber-protein bonds. This action could release non-hydrolyzed protein, rather than water-soluble fragments; and the whole protein might then exhibit its normal non-denatured solubility characteristics in dilute alkali. This hypothesis is given additional credence by recent observations that 60% of Nitrogen solubilized from cottonseed by trypsin treatment has a molecular weight greater than 1000 daltons (42).

Relationship of enzyme concentration to protein extraction. The relationship of enzyme concentration to protein extraction in cottonseed was curvilinear (Table IV). Treatment with a trypsin concentration of 0.03g/25g meal caused greater than 75% of the protein to be extracted. At greater than 0.03g enzyme/25g meal, there was a lowered rate of increase in protein extractability.

Table IV

Effect of Trypsin Concentration on the Efficiency of a Two-Step Trypsin-Chemical Extraction or Protein From Cottonseed and Alfalfa Meal

Amount of Trypsin/25g Substrate	Total % Kjedahl Protein Extracted	
	Cottonseed Meal	Alfalfa Meal
0	16.4 $\pm$ 2.3	21.8 $\pm$ 2.1
0.01	18.2 $\pm$ 1.7	27.3 $\pm$ 1.4
0.03	78.4 $\pm$ 3.1	41.7 $\pm$ 4.8
0.06	82.6 $\pm$ 4.3	57.5 $\pm$ 7.3
0.12	86.5 $\pm$ 6.1	64.3 $\pm$ 2.8
0.24	------	73.4 $\pm$ 4.8

The relationship of trypsin concentration to extractable protein was quite different for alfalfa meal (Table IV). The efficiency of the process increased with increasing amounts of trypsin up to 0.24g trypsin/25g alfalfa meal. This is 8 times the concentration used in cottonseed meal.

Effect of Ultrasonic Energy on the Enzyme-Chemical Technique

Because of the potential advantages associated with use of the ultrasonic technique (e.g., cavitation of sample to create new sites for enzyme activity, increased mixing efficiency, decreased reaction times, and reduced energy expenditures), experimental studies have been performed to determine the effects of ultrasonic energy on the proteolytic enzyme-chemical technique in cottonseed and soybean meal. The combined enzymatic-chemical-ultrasonic technique increased the efficiency of total protein extraction from cottonseed meal over enzyme use alone (Table V). In soybean meal, the increase in extraction efficiency was approximately 10%.

Table V

Solubilization of cottonseed and soybean meal protein by enzymatic-chemical techniques with and without ultrasonic energy

Technique	Total % Kjeldahl Protein Extracted	
	Cottonseed Meal	Soybean Meal
Trypsin, no ultrasonic energy	65.66 $\pm$ 4.98	80.72 $\pm$ 3.85
Trypsin, 200 accoustical watts	72.99 $\pm$ 3.18	91.32 $\pm$ 4.70

The presence of ultrasonic energy also increases the rate of the enzymatic-chemical procedure without ultrasonic energy, approximately 120 minutes was required for the extraction (11,42,43). With the addition of ultrasonic energy, there were no significant differences in the in the percentage protein extracted for sonication times varying from 1 - 16 min of trypsin-cottonseed meal mixtures (Table VI). These data suggest that sonication times of 1 min are adequate. The same was true for soybean meal where 90% of available protein was extracted in one minute. Therefore, ultrasonic energy increases the efficiency of enzyme solubilication procedures.

Table VI

Effect of Sonication Time on Extraction of Protein From Cottonseed and Soybean Meal

Sonication Time (Min)	Total % Kjeldahl Protein Extracted Cottonseed Meal	Soybean Meal
1	70.0 ± 1.2	93.7 ± 4.2
2	67.5 ± 1.8	93.9 ± 2.4
4	67.5 ± 3.4	95.3 ± 3.8
8	73.2 ± 6.1	94.8 ± 2.7
12	61.3 ± 9.0	-------
16	71.6 ± 7.8	-------

Functionality of Extracted Protein

Protein extracted from cottonseed meal by the enzymatic chemical process is quite functional. The solubility of protein concentrates produced by isoelectric precipitation was typical of solubility of cottonseed protein as a function of pH (42). Solubility was minimal at acidic pH's and increased in basic pH's (Crenwelge et al., 1974). There were no marked differences in solbuility in 0.1 M NaCl vs water. This suggests little or no interaction between pH and ionic strength.

There were no significant differences in the water- or oil-holding capacities of the samples (Table VII). However, the emulsifying capacity of the NaOH fraction produced by the combined technique was significantly greater than that of other fraction concentrates. It has often been suggested that emulsifying capacity is a function of molecular radius (45) and these data tend to confirm that observation since the highest emulsification capacity was noted with the concentrate having the lowest amount of low molecular weight nitrogen (42).

Table VII

Functionality of cottonseed meal protein isolates prepared by the protelytic enzyme-chemical technique (PC) and the ultrasonic-enzymatic (UE) technique. (All data are the mean of three replicates ± standard deviation)

Sample	Oil-holding capacity (ml/g)	Water-holding capacity (ml/g)	Emulsifying capacity (ml/g)
Enzyme fraction (PC)	4.03 ± 1.10	2.43 ± 0.86	333 ± 4.0
NaOH fraction (PC)	3.53 ± 0.32	2.80 ± 0.26	443 ± 4.0
NaOH fraction (UE)	3.66 ± 0.15	2.73 ± 0.15	586 ± 2.5

In summary, investigations with extremely heat denatured substrates (NSI 0.15) such as cottonseed meal, alfalfa meal and soybean meal have shown that trypsin treatment will allow efficient protein extraction. The efficiency of this extraction can be markedly increased by interfacing ultrasonic energy with the system. Isolates produced from the extracted proteins are functional.

REFERENCES

1. Koshiyama, I. 1972. Agri. Biol. Chem. 36:62.
2. Altschul, A.M., N.J. Neucere, A.A. Woodham, and J.M. Dechary. 1964. Nat. 203:501.
3. Daussant, J., N.J. Neucere, and L.Y. Yatsu. 1969. Plant Physiol. 44:471.
4. St. Angelo, A.J., L.Y. Yatsu, and A.M. Altschul. 1968. Arch. Biochem. Biophys. 124:199.
5. Varner, J.E. and G. Schidlovsky. 1963. Plant Physiol. 38:139.
6. Graham, T.A. and B.E.S. Gunning. 1970. Nat. 228:81.
7. Metzger, H., M.B. Shapiro, J.E. Mosiman, and J.E. Vinton. Nat. 219:1166.
8. Dieckert, J.W. and M.C. Dieckert. 1976. J Fd Sci. 41:475.
9. Sugarman, N. 1956. U.S. Patent 2,762,820.

10. Wang, L.C. 1975. J Fd Sci. 40:549.
11. Childs, E.A. 1975. J Fd Sci. 40:78.
12. Martinez, W.H., Berardi, L.C. and L.A. Goldblatt. 1970. J Agr Fd Chem. 18:961.
13. Vix, H.L.E. 1968. Oil Mill Gaz. 72:53.
14. Martinez, W.H. 1969. Proc. XVII Cotton Proc Clin. ARS Report 72-69:18.
15. Circle, S.J. and A.K. Smith. 1972. "Soybeans:Chemistry and Technology." 638. AVI Publishing Co.
16. Neurath, H. and K. Bailey. 1973. "The Proteins." 542. Academic Press.
17. Pirie, N.W. 1971. "Leaf Protein: Its Agronomy, Preparation, Quality and Use." IBP Handbook #20. Blackwell Scientific.
18. Knuckles, B.E., R.R. Spencer, M.E. Lazar, E.M. Bickoff. and G.O. Kohler. 1970. J Agr Fd Chem. 18:1086.
19. Kohler, G.O. and E.M. Bickoff. 1970. Third Int'l Congress of Food Science and Technology. Wash, D.C.
20. Lazar,M.E., R.R. Spencer, B.E. Knuckles, and E.M.J. Bickoff. 1971. J Agr Fd Chem. 19:944.
21. Miller, R.E., R.H. Edwards, M.E. Lazar, E.M. Bickoff, and G.O. Kohler. 1972. J Agr Fd Chem. 20:1151.
22. Spencer, R.R., A.C. Mottola, E.M. Bickoff, J.P. Clark, and G.O. Kohler. 1971. J Agr Fd Chem. 19:504.
23. Chayen, I.H., R.H. Smith, G.R. Tristram, D. Thirkell, and T. Webb. 1961. J Sci Fd Agr. 12:502.
24. Hartman, G.H., W.R. Akeson, and M.A. Stohmann. 1967. J Agr Fd Chem. 15:74.
25. Huang, K.T., M.C. Tao, M. Boulet, R.R. Riel, J.P. Julien, and G.J. Grissom. 1971. Can Inst Fd Tech. 4:85.
26. Wilson, R.F., and J.M.A. Lilley. 1965. J Sci Fd Agr. 16:173.
27. Byers, M. 1967. J Sci Fd Agr. 18:34.
28. Cowlishaw, S.J., D.E. Eyles, W.F. Raymond, and J.M.A. Tilley. 1956. J Sci Fd Agr. 7:775.
29. deFremery, D., E.M. Bickoff, and G.O. Kohler. 1973. J Agr Fd Chem. 20:1155.
30. Henry, K.M. and J.E. Ford. 1965. J Sci Fd Agric. 16: 425.
31. Lexander, K., R. Carlson, V. Schalen, A. Simonsson, and T. Lundborg. 1970. Annal. Appl. Biol. 66:193.
32. Subba-Rau, B.H., S. Mahadeviah, and N. Singh. 1969. J Sci Fd Agr. 20:355.
33. Edwards, R.H., R.E. Miller, D. deFremery, B.E. Knuckles, E.M. Bickoff, and G.O. Kohler. 1975. J Agr Fd Chem. 23:620.
34. Lu, P.S. and J.E. Kinsella. 1972. J Fd Sci. 37:94.
35. Hang, Y.D., W.F. Wilkens, A.S. Hill, K.H. Steinkraus, and L.R. Hackler. 1970. J Agr Fd Ch. 18:9.

36. Abdo, K.M. and K.W. King. 1967. J Agr Fd Chem. 15:83.
37. Sreekantiah, K.R., H. Ebine, J. Ohta, and M. Nakans. 1969. Fd Tech. 23:1055.
38. Arzu, A., H. Mayorga, J. Gonzales, and C. Rolz. 1972. J Agr Fd Chem. 20:805.
39. Fujimaki, M. H. Kato, S. Arai, and E. Tamaki. 1968. Fd Tech. 22:889.
40. Molina, M.R. and P.A. LaChance. 1973. J Fd Sci. 38: 607.
41. Childs, E.A. 1976. Unpublished data.
42. Childs, E.A. and J.L. Forte'. 1976. J Fd Sci. 41: 652.
43. Ku, Y. and E.A. Childs. 1976. Tenn Fm Hm Sci. 97:26.
44. Crenwelge, D.D., C.W. Dill, P.T. Tybor, and W.A. Landmann. 1974. J Fd Sci. 39:175.
45. Carpanter, J.L. and R.L. Saffle. 1965. Fd Tech. 1567.

INDEX

F

G

H

I